Objective Agronomy

Anand Kumar Vishwakarma
Senior Scientist (Agronomy)
Brajendra
Senior Scientist (Soil Science)
T. Ramesh
Scientist (Soil Science)
S.V. Ngachan
Director

ICAR Research Complex for NEH Region
Umiam - 793 103, Meghalaya, India

A Paperback Division of

New India Publishing Agency

101, Vikas Surya Plaza, CU Block, LSC Market
Pitam Pura, New Delhi 110 034, India
Phone: + 91 (11)27 34 17 17 Fax: + 91(11) 27 34 16 16
Email: info@nipabooks.com
Web: www.nipabooks.com

Feedback at feedbacks@nipabooks.com

ISBN: 978-93-80235-12-7

Composed and Designed by NIPA

Objective Agronomy

Preface

Agronomy is such science of agriculture as encompassing all the branches of it. The main focus of agriculture is also agronomy. Therefore the soul of agriculture is agronomy and its knowledge is essential at all levels and intensity. There is an abundance of literature in the subject. Several books cover the core syllabus of the undergraduate and postgraduate courses being taught in the universities. Several textbooks have been written which are suitable for undergraduate and postgraduate students. A good grasp of the subject, however requires a drilled mind to be aware of so many aspects. Newer dimensions are adding up fast in the science of agronomy. Integrating the ever widening dimensions of basic and applied aspects of agronomy subject cannot be merely met by a textbook or a manual or a periodical. Since to finetune his agronomy knowledge and deliberations a student is often in a stress to locate his resources. Further for a far flung university student and the colleges which are not agricultural such need is acutely felt. The question bank fulfills such need.

There is a great need of a question bank for all sorts of examinations in agronomy discipline. Since often the subject comprehended has to be moulded into a form of inquisitivity to satisfy the examiner's mind and psyche. For overall success of a student it is quite essential that he has gone through all those finer points which an examiner is looking for. Such a book will go a long way to overcome all those difficulties of a student who is having the adequate resources as well. However there is great void in agronomy of such good question bank. The present deliberation is an attempt to compile, frame such questions which an examiner is looking for. Our main aim is to come up with an intensive as well as exhaustive documentation which integrates all the information's in agronomy subject which a student is looking for. The question bank covers such need as it has seventeen chapters and a model question paper. The book is best suited to those students preparing for competitive exams such as JRF,SRF,IARI entrance exam, civil services, ARS and host of other exams being conducted by the universities in agronomy. **The question bank consists of almost 4500 solved questions.** Since this is an exhaustive coverage of the subject mistakes here and there can occur. So we earnestly request the users of the book to bring in notice of such things either to the authors or to the publisher.We hope that the book will be of immense benefit to all those users aiming to further their career in agronomy.

Authors

Contents

Preface .... *v*

Chapter 1: Introductory Agronomy 1
Multiple Choice Questions (MCQ's)

Chapter 2: Agro Meteorology 18
Multiple Choice Questions (MCQ's)

Chapter 3: Soil and Tillage 48
Multiple Choice Questions (MCQ's)

Chapter 4: Soil and Water Conservation 88
Multiple Choice Questions (MCQ's)

Chapter 5: Mineral Nutrition and Soil Fertility 99
Multiple Choice Questions (MCQ's)

Chapter 6: Manures and Fertilizers 143
Multiple Choice Questions (MCQ's)

Chapter 7: Irrigation Water Management 160
Multiple Choice Questions (MCQ's)

Chapter 8: Dryland Agriculture 185
Multiple Choice Questions (MCQ's)

Chapter 9: Weed Management 196
Multiple Choice Questions (MCQ's)

Chapter 10: Cropping System 232
Multiple Choice Questions (MCQ's)

Chapter 11: Cereal Crop Production 245
Multiple Choice Questions (MCQ's)
Rice 245
Bajra 258
Barley 262
Maize 266
Millets 272
Oat 275
Sorghum 277
Wheat 280

Chapter 12: Oilseed Crop Production 289
Multiple Choice Questions (MCQ's)
Groundnut 289
Soybean 293
Sesamum 296
Mustard 299
Castor 303
Sunflower 305
Safflower 307
Niger 310
Linseed 312

Chapter 13: Pulse Crop Production 316
Multiple Choice Questions (MCQ's)
Blackgram 316
Gram 320
Greengram 323
Lentil 326
Pea 328
Pigeonpea 331

Chapter 14: Fiber Crop Production 336
Multiple Choice Questions (MCQ's)
Cotton 336
Jute 346
Sunnhemp 350

Chapter 15: Tuber Crop Production 354
Multiple Choice Questions (MCQ's)
Potato 354

Chapter 16: Plantation Crop Production 361
Multiple Choice Questions (MCQ's)
Coffee 361
Tea 364
Tobacco 367

Chapter 17: Sugar Crop Production 375
Multiple Choice Questions (MCQ's)
Sugarcane 375
Sugarbeet 385

Chapter 18: Model Question Papers 389
Model Paper-I 389
Model Paper-II 409
Model Paper-III 429
Model Paper-IV 450
Model Paper-V 470

1 Chapter

Introductory Agronomy

Multiple Choice Questions (MCQ's)

1. Agriculture is a
   a) Greek word b) Latin word
   c) German word d) Spanish **Ans : b**

2. The term agronomy is derived from
   a) Greek word b) Latin word
   c) German word d) Spanish **Ans : a**

3. Agriculture is a
   a) Physical process b) Chemical process
   c) Biological process d) All **Ans : d**

4. Which is not a primary resource for agriculture
   a) Forest b) Land
   c) Air d) Water **Ans : a**

5. Which is a primary resource for agriculture
   a) Seed b) Land
   c) Fertilizer d) Pesticide **Ans : b**

6. Which is a primary resource for agriculture
   a) Manures b) Water
   c) Fertilizer d) Herbicide **Ans : b**

7. Which is a primary resource for agriculture
   a) Air b) Tractor
   c) Fertilizer d) Pesticide **Ans : a**

8. Which of the following are primary resources for agriculture
   a) Land, Air and Water
   b) Seed, Fertilizer and Pesticide
   c) Seed, Air and Water
   d) Land, Fertilizer and Water **Ans : a**

9. Which is not a component of aerial environment for agriculture
   a) Wind b) Solar radiation
   c) Relative humidity d) None **Ans : d**

10. Which is a component of aerial environment for agriculture
    a) Rainfall b) Temperature
    c) Relative humidity d) All **Ans : d**

11. The environment of crop constitutes of
    a) Soil b) Aerial
    c) Both d) None **Ans : c**

12. The environment of crop amenable to modification
    a) Soil b) Aerial
    c) Both d) None **Ans : a**

13. Which of the following is a soil physical environment
    a) Soil pH b) Soil temperature
    c) Electrical conductivity d) All **Ans : b**

14. Which of the following is a soil chemical environment
    a) Soil water b) Soil temperature
    c) Electrical conductivity d) All **Ans : c**

15. Which of the following is a soil physical environment
    a) Soil pH b) Soil air
    c) Electrical conductivity d) All **Ans : b**

16. Which of the following is not a soil physical environment
    a) Soil nutrient b) Soil air
    c) Soil temperature d) Soil water **Ans : a**

17. Which of the following is a soil chemical environment
    a) Soil nutrient b) Soil air
    c) Soil temperature d) Soil water **Ans : a**

18. Which of the following is not a soil chemical environment
    a) Soil nutrient b) Soil pH
    c) Soil Electrical conductivity d) Soil water **Ans : d**

19. Which of the following is not amenable to modification
    a) Soil environment b) Aerial environment
    c) Both d) None **Ans : b**

20. Cultivation of wheat and Barley begins in
    a) 8700 BC b) 7500 BC
    c) 7700 BC d) 5700 BC **Ans : b**

21. Cultivation of maize begins in
    a) 8700 BC b) 4400 BC
    c) 7700 BC d) 5700 BC **Ans : b**

22. Cultivation of potato begins in
    a) 8700 BC b) 3500 BC
    c) 7700 BC d) 5700 BC **Ans : b**

23. Cultivation of chickpea and cotton begins in
    a) 8700 BC b) 2300 BC
    c) 7700 BC d) 5700 BC **Ans : b**

24. Cultivation of rice begins in
    a) 8700 BC b) 2200 BC
    c) 7700 BC d) 5700 BC **Ans : b**

25. Cultivation of ragi begins in
    a) 8700 BC b) 1800 BC
    c) 7700 BC d) 5700 BC **Ans : b**

26. Cultivation of sorghum begins in
    a) 8700 BC b) 1725 BC
    c) 7700 BC d) 5700 BC **Ans : b**

27. Wheel was invented in
    a) 8700 BC b) 3400 BC
    c) 7700 BC d) 5700 BC **Ans : b**

28. Experiments pertaining to plant nutrition in systematic way were initiated by
    a) Jethro tull b) Gregor Johann Mendel
    c) Van Helmont d) None **Ans : c**

29. Who concluded that the main 'principle of vegetation' is water
   a) Jethro tull   b) Gregor Johann Mendel
   c) Van Helmont   d) None   **Ans : c**

30. The book 'Horse hoeing husbandry' was published by
   a) Jethro tull   b) Gregor Johann Mendel
   c) Van Helmont   d) None   **Ans : a**

31. The 'Annals of Agriculture' was published by
   a) Jethro tull   b) Aurthur Young
   c) Van Helmont   d) None   **Ans : b**

32. The book 'Elements of Agricultural Chemistry' was published by
   a) Jethro tull   b) Sir Humphry Davy
   c) Van Helmont   d) None   **Ans : b**

33. Experiments on effects of manures on crop were initiated by
   a) Sir John Bannet Lawes   b) Gregor Johann Mendel
   c) Van Helmont   d) None   **Ans : a**

34. Who initiated the synthetic fertilizer industry
   a) Lawes   b) Gregor Johann Mendel
   c) Van Helmont   d) None   **Ans : a**

35. Who patented the process of treating rock phosphate to produce superphosphate
   a) Jethro tull   b) Lawes
   c) Van Helmont   d) None   **Ans : b**

36. The laws of heredity were discovered by
   a) Jethro tull   b) Gregor Johann Mendel
   c) Van Helmont   d) None   **Ans : b**

37. Imperial Agricultural Research Institute was established in the year
   a) 1905   b) 1913
   c) 1923   d) None   **Ans : a**

38. .... (1674- 1741 AD) conducted several experiments on plant nutrition and published a book, "Horse Hoeing Husbandry"
   a) Sir Humphry Davy   b) Aristital
   c) Jethro Tull   d) Van Helmounst   **Ans : c**

39. Which of the following is a primitive form of agriculture
 a) Shifting agriculture b) Jhuming
 c) Hunting and gathering d) All **Ans : d**

40. Cereals occupy nearly............ of cultivated lands
 a) 1/3 b) 1/2
 c) 2/3 d) 3/4 **Ans : c**

41. The term agronomy has been derived from ....word
 a) Urdu b) Persian
 c) Greek d) French **Ans : c**

42. Kharif, Rabi and Ziad words belong to:
 a) Urdu b) Persian
 c) Arabic d) French **Ans : c**

43. Famous Rosthamsted Agricultural Experiment Station was established by:
 a) Liebig b) DeSaussure
 c) Jesthro Tull d) Lawes and Gilbert ....

44. Natural farming was advocated by:
 a) Gilbert b) Fakuoka
 c) Lawes d) Norman E. borlaug **Ans : b**

45. Imperial Agricultural Research Institute was established in
 a) 1905 b) 1915
 c) 1925 d) 1935 **Ans : a**

46. Imperial Agricultural Research Institute was located at
 a) Pusa b) Pondicherry
 c) Patna d) Punjab **Ans : a**

47. Department of Food was created in
 a) 1905 b) 1942
 c) 1925 d) 1935 **Ans : b**

48. AICRP on Maize was started in
 a) 1955 b) 1942
 c) 1965 d) 1957 **Ans : d**

49. AICRP on Dryland Agriculture was started in
 a) 1955 b) 1971
 c) 1965 d) 1957 **Ans : b**

50. NDP was started in
 a) 1955 b) 1942
 c) 1965 d) 1957 **Ans : c**

51. NSC was established in
a) 1963 b) 1942
c) 1965 d) 1957 **Ans : a**

52. The apex agency for coordinating agricultural research and education in India is
a) IARI b) ICAR
c) SAUs d) All **Ans : b**

53. The average land availability per person globally is about.....ha
a) 3.0 b) 3.8
c) 4.5 d) 0.8 **Ans : b**

54. The average land availability per person in India is about...ha
a) 3.0 b) 3.8
c) 4.5 d) 0.8 **Ans : d**

55. The cultivable per capita land availability world wide is about ....ha
a) 0.3 b) 3.8
c) 0.5 d) 0.8 **Ans : a**

56. The cultivable per capita land availability in India is about...ha
a) 0.3 b) 3.8
c) 0.15 d) 0.5 **Ans : c**

57. The total global volume of water is about...............million cubic kilometers
a) 1400 b) 2000
c) 424 d) 328 **Ans : a**

58. The total volume of water in the sea is about.............%
a) 3 b) 77
c) 87 d) 97 **Ans : d**

59. The total volume of water in inland is about.............%
a) 3 b) 77
c) 87 d) 97 **Ans : a**

60. Of the total volume of inland water.............% is in the form of glaciers
a) 1 b) 77
c) 87 d) 22 **Ans : b**

61. Of the total volume of inland water.............% is in the form of ground water
a) 1 b) 77
c) 87 d) 22 **Ans : d**

62. Of the total volume of inland water.............% is in the form of fresh water
a) 1 b) 77
c) 87 d) 22 **Ans : a**

63. Total irrigated area world wide is about................M ha
a) 328 b) 240
c) 200 d) 255 **Ans : d**

64. The total geographical area of India is about.............M ha
a) 1400 b) 2000
c) 424 d) 328 **Ans : d**

65. The total area subjected to soil erosion and degradation in India is about.............M ha
a) 140 b) 188
c) 124 d) 328 **Ans : b**

66. ....% of total area subjected to soil erosion and degradation in India
a) 40 b) 88
c) 24 d) 54 **Ans : d**

67. The total irrigated area of India is about.............M ha
a) 50 b) 20
c) 42 d) 328 **Ans : a**

68. Dryland contribute about........... %of the total cultivated area
of India
a) 72 b) 20
c) 42 d) 28 **Ans : a**

69. Dryland contribute about .... %of the total food production of India
a) 72 b) 50
c) 43 d) 28 **Ans : c**

70. India ranks............. in total irrigated area in the world
a) Tenth b) Second
c) Third d) First **Ans : d**

71. India share is.............% of total irrigated area in the world
a) 20 b) 10
c) 15 d) 30 **Ans : a**

72. India ranks............. in wheat production in the world
a) Tenth b) Second
c) Third d) First **Ans : b**

73. India ranks............. in rice production in the world
a) Tenth b) Second
c) Third d) First **Ans : b**

74. India ranks............. in cotton production in the world
a) Tenth b) Second
c) Third d) First **Ans : c**

75. India ranks............. in sugarcane production in the world
a) Tenth b) Second
c) Third d) First **Ans : b**

76. India ranks............. in pulses production in the world
a) Tenth b) Second
c) Third d) First **Ans : d**

77. India ranks............. in groundnut production in the world
a) Tenth b) Second
c) Third d) First **Ans : b**

78. India ranks............. in tea production in the world
a) Tenth b) Second
c) Third d) First **Ans : d**

79. India ranks............. in milk production in the world
a) Tenth b) Second
c) Third d) First **Ans : d**

80. Fertilizer consumption in India is about ....kg/ha
a) 75 b) 120
c) 142 d) 228 **Ans : a**

81. Maximum fertilizer consumption in the world is in
a) Netherlands b) Japan
c) China d) India **Ans : a**

82. Maximum reduction in yield due to weeds has been observed in
a) Sugarbeet b) Rice
c) Cotton d) Wheat **Ans : a**

83. The total food loss due to weeds is estimated to be .... Mt globally
a) 75 b) 120
c) 142 d) 287 **Ans : d**

84. The weeds accounts for ....% of the total food loss in India
a) 75 b) 30
c) 45 d) 20 **Ans : c**

85. The insects accounts for ....% of the total food loss in India
a) 75 b) 30
c) 45 d) 20 **Ans : b**

86. The diseases accounts for ....% of the total food loss in India
a) 75 b) 30
c) 45 d) 20 **Ans : d**

87. The weeds accounts for ....% reduction in yield of sugarbeet
a) 70 b) 30
c) 45 d) 20 **Ans : a**

88. Central Arid Zone Research Institute is located at
a) Jodhpur, Rajasthan b) Nagpur, Maharashtra
c) Lucknow, U.P. d) Simla, H.P. **Ans : a**

89. Central Institute for Cotton Research is located at
a) Jodhpur, Rajasthan b) Nagpur, Maharashtra
c) Lucknow, U.P. d) Simla, H.P. **Ans : b**

90. Central Institute for Sub-Tropical Horticulture is located at
a) Jodhpur, Rajasthan b) Nagpur, Maharashtra
c) Lucknow, U.P. d) Simla, H.P. **Ans : c**

91. Central Potato Research Institute is located at
a) Jodhpur, Rajasthan b) Nagpur, Maharashtra
c) Lucknow, U.P. d) Simla, H.P. **Ans : d**

92. Central Research Institute for Jute and Allied Fibres is located at
a) Barrackpore, W.B. b) Port Blair
c) Bhopal, M.P. d) Kasargod, Kerela **Ans : a**

93. Central Agricultural Research Institute is located at
a) Barrackpore, W.B. b) Port Blair
c) Bhopal, M.P. d) Kasargod, Kerela **Ans : b**

94. Central Institute of Agricultural Engineering is located at
a) Barrackpore, W.B. b) Port Blair
c) Bhopal, M.P. d) Kasargod, Kerela **Ans : c**

95. Central Plantation Crops Research Institute is located at
a) Barrackpore, W.B. b) Port Blair
c) Bhopal, M.P. d) Kasargod, Kerela **Ans : d**

96. Central Research Institute for Dry lands Agriculture is located at
a) Hyderabad, A.P. b) Cuttack, Orissa
c) Dehradun, U.P. d) Karnal, Haryana. **Ans : a**

97. Central Rice Research Institute is located at
a) Hyderabad, A.P. b) Cuttack, Orissa
c) Dehradun, U.P. d) Karnal, Haryana. **Ans : b**

98. Central Soil and Water Conservation Research and Training Institute is located at
a) Hyderabad, A.P. b) Cuttack, Orissa
c) Dehradun, U.P. d) Karnal, Haryana. **Ans : c**

99. Central Soil Salinity Research Institute is located at
a) Hyderabad, A.P. b) Cuttack, Orissa
c) Dehradun, U.P. d) Karnal, Haryana. **Ans : d**

100. Central Tobacco Research Institute is located at
a) Rajahmundry, A.P. b) Mumbai, Maharashtra
c) Hyderabad, A.P. d) Karnal, Haryana **Ans : a**

101. Central Institute for Research on Cotton Technology is located at
a) Rajahmundry, A.P. b) Mumbai, Maharashtra
c) Hyderabad, A.P. d) Karnal, Haryana **Ans : b**

102. Directorate of Oilseeds Research is located at
a) Rajahmundry, A.P. b) Mumbai, Maharashtra
c) Hyderabad, A.P. d) Karnal, Haryana **Ans : c**

103. Directorate of Rice Research is located at
a) Rajahmundry, A.P. b) Mumbai, Maharashtra
c) Hyderabad, A.P. d) Karnal, Haryana **Ans : c**

104. Directorate of Wheat Research is located at
a) Rajahmundry, A.P. b) Mumbai, Maharashtra
c) Hyderabad, A.P. d) Karnal, Haryana **Ans : d**

105. Directorate of Water Management Research is located at
a) Rahuri, Maharashtra b) Pusa, New Delhi
c) Jhansi, U.P. d) Lucknow, U.P. **Ans : b**

106. Indian Agricultural Research Institute (Deemed University) is located at
a) Rajahmundry, A.P. b) New Delhi
c) Hyderabad, A.P. d) Karnal, Haryana **Ans : b**

107. Indian Grassland and Fodder Research Institute is located at
a) Rajahmundry, A.P. b) Mumbai, Maharashtra
c) Jhansi d) Karnal, Haryana **Ans : c**

108. Indian Institute of Sugarcane Research is located at
a) Rajahmundry, A.P. b) Mumbai, Maharashtra
c) Hyderabad, A.P. d) Lucknow, U.P. **Ans : d**

109. Indian Institute of Soil Science is located at
a) Bhopal, M.P. b) Kanpur, U.P.
c) Meghalaya d) New Delhi **Ans : a**

110. Indian Institute of Pulses Research is located at
a) Bhopal, M.P. b) Kanpur, U.P.
c) Meghalaya d) New Delhi **Ans : b**

111. ICAR Research Complex for NEH Region is located at
a) Bhopal, M.P. b) Kanpur, U.P.
c) Meghalaya d) New Delhi **Ans : c**

112. Indian Agricultural Statistics Institute is located at
a) Bhopal, M.P. b) Kanpur, U.P.
c) Meghalaya d) New Delhi **Ans : d**

113. Indian Institute of Horticulture Research is located at
a) Bangalore, Karnataka b) Ranchi, Bihar
c) Calcutta, W.B. d) Solan, H.P. **Ans : a**

114. Indian Lac Research Institute is located at
a) Bangalore, Karnataka b) Ranchi, Bihar
c) Calcutta, W.B. d) Solan, H.P. **Ans : b**

115. Jute Technological Research Laboratory is located at
a) Bangalore, Karnataka b) Ranchi, Bihar
c) Calcutta, W.B. d) Solan, H.P. **Ans : c**

116. National Centre for Mushroom Research and Training is located at
a) Bangalore, Karnataka b) Ranchi, Bihar
c) Calcutta, W.B. d) Solan, H.P. **Ans : d**

117. National Research Centre Center for Cashew is located at
a) Nagpur, Maharashtra b) Indore, M.P.
c) Junagadh, Gujarat d) Puttar, Karnataka **Ans : d**

118. National Research Centre for Groundnut is located at
a) Nagpur, Maharashtra b) Indore, M.P.
c) Junagadh, Gujarat d) Puttar, Karnataka **Ans : c**

119. National Research Centre on Soybean is located at
a) Nagpur, Maharashtra b) Indore, M.P.
c) Junagadh, Gujarat d) Puttar, Karnataka **Ans : b**

120. National Research Centre on Citrus is located at
a) Nagpur, Maharashtra b) Indore, M.P.
c) Junagadh, Gujarat d) Puttar, Karnataka **Ans : a**

121. National Centre for Integrated Pest Management is located at
a) Nagpur, Maharashtra b) Indore, M.P.
c) Faridabad, Haryana d) Puttar, Karnataka **Ans : c**

122. National Research Centre for Sorghum is located at
a) Nagpur, Maharashtra b) Indore, M.P.
c) Junagadh, Gujarat d) Hyderabad, A.P. **Ans : d**

123. National Centre for Weed Science is located at
a) Nagpur, Maharashtra b) Indore, M.P.
c) Jabalpur, M.P. d) Puttar, Karnataka **Ans : c**

124. National Bureau of Plant Genetic Resources is located at
a) Nagpur, Maharashtra b) Indore, M.P.
c) Junagadh, Gujarat d) New Delhi **Ans : d**

125. National Research Centre for Seed Spices is located at
a) Nagpur, Maharashtra b) Ajmer, Rajasthan
c) Junagadh, Gujarat d) Puttar, Karnataka **Ans : b**

126. Central Institute of Arid Horticulture is located at
a) Bikaner, Rajasthan b) Indore, M.P.
c) Junagadh, Gujarat d) Puttar, Karnataka **Ans : a**

127. National Bureau of Soil Survey and Land Use Planning is located at

a) Nagpur, Maharashtra
b) Indore, M.P.
c) Junagadh, Gujarat
d) Puttar, Karnataka

Ans : a

128. National Research Centre on Rapeseed and Mustard is located at

(a) Bharatpur, Rajasthan
(b) Indore, M.P.
(c) Junagadh, Gujarat
(d) Puttar, Karnataka

Ans : a

129. National Centre for Integrated Pest Management, LBS Centre for Bio and Plant Protection is located at

(a) Nagpur, Maharashtra
(b) Indore, M.P.
(c) Pusa, New Delhi
(d) Puttar, Karnataka

Ans : c

130. National Centre for Agricultural Economics and Policy Research, Library Avenue is located at

a) Nagpur, Maharashtra
b) Indore, M.P.
c) Pusa, New Delhi.
d) Puttar, Karnataka

Ans : c

131. National Biotechnological Centre for Crop Science is located at

a) Nagpur, Maharashtra
b) Indore, M.P.
c) Junagadh, Gujarat
d) Pusa, New Delhi

Ans : d

132. National Research Centre for Oil Palm, Pedavegi is located at

a) Nagpur, Maharashtra
b) West Godavari, A.P.
c) Junagadh, Gujarat
d) Puttar, Karnataka

Ans : b

133. National Research Centre for Onion and Garlic is located at

a) Nagpur, Maharashtra
b) Indore, M.P.
c) Pune
d) Puttar, Karnataka

Ans : c

134. Nuclear Research Laboratory is located at

a) Nagpur, Maharashtra
b) Indore, M.P.
c) Junagadh, Gujarat
d) Pusa, New Delhi

Ans : d

135. Project Directorate for Cropping Systems Research is located at

a) Nagpur, Maharashtra
b) Indore, M.P.
c) Junagadh, Gujarat
d) Meerut, U.P.

Ans : d

136. Plant Quarantine Regional Station, National Bureau of Plant Genetic Resources is located at

a) Nagpur, Maharashtra
b) Hyderabad, A.P.
c) Junagadh, Gujarat
d) Puttar, Karnataka

Ans : b

137. Sugarcane Breeding Institute is located at
a) Nagpur, Maharashtra b) Coimbatore, TN.
c) Junagadh, Gujarat d) Puttar, Karnataka **Ans : b**

138. Vivekananda Parvatia Krishi Anusandhana Shala is located at
a) Nagpur, Maharashtra b) Indore, M.P.
c) Junagadh, Gujarat d) Almora, U.P **Ans : d**

139. Water Technolocy Centre for Eastern Region is located at
a) Bhubaneswar, Orissa. b) Indore, M.P.
c) Junagadh, Gujarat d) Puttar, Karnataka **Ans : a**

140. Centre for International Forestry Research (CIFOR) is located at
a) Indonesia b) Colombia.
c) Peru d) Mexico **Ans : a**

141. Center International de Agricultural Tropical (CIAT) is located at
a) Indonesia b) Colombia.
c) Peru d) Mexico **Ans : b**

142. Centre International de la Papa (CIP) is located at
a) Indonesia b) Colombia.
c) Peru d) Mexico **Ans : c**

143. Center International de la Mejoramiento de Maizy Trigo (CIMMYT) is located at
a) Indonesia b) Colombia.
c) Peru d) Mexico **Ans : d**

144. International Board for Plant Genetic Resources (IBPGR) is located at
a) Indonesia b) Colombia.
c) Peru d) Rome **Ans : d**

145. International Centre for Agricultural Research in the Dry Area (ICARDA) is located at
a) Syria b) Colombia.
c) Peru d) Mexico **Ans : a**

146. International Centre of Insect Physiology and Ecology (ICIPE) is located at
a) Indonesia b) Colombia.
c) Kenya d) Mexico **Ans : c**

147. International Centre for Integrated Mountain Development (ICIMOD) is located at

a) Indonesia b) Colombia.
c) Peru d) Nepal **Ans : d**

148. International Centre for Research in Agro-forestry (ICRAF) is located at

a) Indonesia b) Colombia.
c) Peru d) Kenya **Ans : d**

149. International Crops Research Institute for Semi-Arid Tropics (ICRISAT) is located at

a) India b) Colombia.
c) Peru d) Mexico **Ans : a**

150. International Institute of Tropical Agriculture (IITA) is located at

a) Indonesia b) Colombia.
c) Nigeria d) Mexico **Ans : c**

151. International Irrigation Management Institute (IIMI) is located at

a) Indonesia b) Srilanka
c) Peru d) Mexico **Ans : b**

152. International Laboratory for Research on Animal Disease (ILRAD) is located at

a) Indonesia b) Colombia.
c) Kenya d) Mexico **Ans : c**

153. International Livestock Centre for Africa (ILCA) is located at

a) Ethiopia b) Colombia.
c) Peru d) Mexico **Ans : a**

154. International Rice Research Institute (IRRI) is located at

a) Indonesia b) Colombia.
c) Peru d) Philippines **Ans : d**

155. International Service for National Agricultural Research (ISNAR) is located at

a) Indonesia b) Netherlands
c) Peru d) Mexico **Ans : b**

156. West Africa Rice Development Association (WARDA) is located at

a) Ivory Coast b) Colombia.
c) Peru d) Mexico **Ans : a**

157. Winrock International Institute for Agricultural Development is located at

a) Indonesia b) Canada
c) Peru d) Mexico **Ans : b**

158. Acre is equivalent to

a) 4,840 sq. yards b) 2.54 ha
c) 0.4047 yards d) 43560 sq.m **Ans : a**

159. Acre is equivalent to

a) 4,840 sq. m b) 2.54 ha
c) 0.4047 ha d) 43560 sq.m **Ans : c**

160. Acre is equivalent to

a) 4,840 sq. m b) 2.54 ha
c) 0.4047 yard d) 43560 sq.ft **Ans : d**

161. One ha is equivalent to

a) 4,840 sq. yards b) 2.54 acre
c) 0.4047 yards d) 43560 sq.m **Ans : b**

162. The unit of crop produced per unit of nutrient added is called

a) Agronomic efficiency b) Nutrient use efficiency
c) Nutrient uptake d) None **Ans : a**

163. Agronomic efficiency is expressed as

a) Kg/kg b) Kg/day
c) Kg/ha d) None **Ans : a**

164. The direct seeded pre-monsoon rice in the deep water is called

a) Aman rice b) Aus rice
c) Boro rice d) None **Ans : a**

165. The ....crop grows with rising flood water and settles at the ground when water subsides during the post monsoon period.

a) Aman rice b) Aus rice
c) Boro rice d) None **Ans : a**

166. The ....rice crop is photosensitive and harvested during October-December.

a) Aman b) Aus
c) Boro d) None **Ans : a**

167. The instrument used in the measurement of force and velocity of wind.

a) Anemometer b) Wind vane
c) Barometer d) Psychrometer **Ans : a**

168. An instrument used for measuring atmospheric pressure

a) Aneroid barometer b) Anemometer
c) Wind vane d) Psychrometer **Ans : a**

169. The main advantage of Aneroid barometer over a mercury barometer is its.

a) Durability b) Speed
c) Accurecy d) Portability **Ans : d**

170. Angstrom is a unit of length equal to

a) 10^{-2} cm b) 10^{-4} cm
c) 10^{-6} cm d) 10^{-8} cm **Ans : d**

❑❑❑

2 Chapter

Agro Meteorology

Multiple Choice Questions (MCQ's)

1. The condition of atmosphere at a given place and time is termed as
   a) Climate b) Weather
   c) Meteorology d) None **Ans : b**

2. Weather pertains to a
   a) Very long period of time b) Small duration
   c) Medium duration d) None **Ans : b**

3. Climate pertains to a
   a) Long period of time b) Small duration
   c) Medium duration d) None **Ans : a**

4. The type of rains on the slopes of the Western Ghats is
   a) Cyclonic b) Convectional
   c) Orographic d) None **Ans : c**

5. The condition favourable for precipitation is
   a) Low moisture content b) High moisture content
   c) Cooling of air d) Descending air **Ans : c**

6. In South Africa the short-grass region is called
   a) Selva b) Veldt
   c) Lans d) Prairie **Ans : b**

7. The xerophytic natural vegetation is found in
   a) The equatorial region   b) The monsoon region
   c) The Sahara region   d) The Tundra region   **Ans : c**

8. Instrument used for measuring wind velocity is known as:
   a) Hydrometer   b) Pyranometer
   c) Altimeter   d) Anemometer   **Ans : d**

9. The term PET was coined by:
   a) Sarkar and Biswas   b) Haigreaves
   c) Thornthwaite   d) Mendel   **Ans : c**

10. Chemical used for the control of transpiration
    a) Kaolin   b) Mobileaf
    c) PMA   d) All   **Ans : d**

11. The study of the relation of agricultural crops and environment is called as
    a) Autoecology   b) Agro-ecology
    c) Agrology   d) Agrometeorology   **Ans : d**

12. A recorded rainfall of ....within a period of 24 hours is considered as rainy day
    a) More than 2.5 mm   b) More than 5.0 mm
    c) More than 3.0 mm   d) More than 1.0 mm   **Ans : d**

13. The maximum CO_2 concentration at normal atmospheric levels occur in relatively low light intensities of________ foot candles for some plants
    a) 2500-5000   b) 6000-7500
    c) 2150-2450   d) 1000-2000   **Ans : d**

14. The estimates postulate that the carbon dioxide content of atmosphere has increased from 280 parts per million to ________ during last three decades.
    a) 300 ppm   b) 350 ppm
    c) 400 ppm   d) 500 ppm   **Ans : b**

15. The instrument used to measure humidity is
    a) Barometer   b) Hydrometer
    c) Hygrometer   d) Altimeter   **Ans : c**

16. The summation of weather conditions over a given region during a comparatively longer period is termed as
    a) Climate   b) Weather
    c) Meteorology   d) None   **Ans : a**

17. The science which deals with laws and principles of atmospheric phenomenon

a) Climatology b) Agricultural meteorology
c) Meteorology d) None **Ans : c**

18. The branch of meteorology that deals with the response of crops to the physical environment

a) Climatology b) Agricultural meteorology
c) Agro climatology d) None **Ans : b**

19. The relationship of climatic regimes and agricultural production is termed as

a) Climatology b) Agricultural meteorology
c) Agro climatology d) None **Ans : c**

20. The gaseous envelop of invisible film of air surrounding the earth is called

a) Climate b) Weather
c) Troposphere d) Atmosphere **Ans : d**

21. Atmosphere extends up to a height of

a) 600 km b) 60 km
c) 6000 km d) 1600 km **Ans : d**

22. Air contains

a) 21% Nitrogen b) 87% Nitrogen
c) 78% Nitrogen d) 0.03% Nitrogen **Ans : c**

23. Air contains

a) 21% Oxygen b) 87% Oxygen
c) 78% Oxygen d) 0.03% Oxygen **Ans : a**

24. Air contains

a) 21% CO_2 b) 87% CO_2
c) 78% CO_2 d) 0.03% CO_2 **Ans : d**

25. Air contains

a) 2.1% Argon b) 0.93% Argon
c) 7.8% Argon d) 0.03% Argon **Ans : b**

26. Which is not a major component of atmosphere

a) Carbon dioxide b) Oxygen
c) Argon d) Carbon monoxide **Ans : d**

27. Based on vertical temperature differences atmosphere is divided into

a) 4 layers b) 3 layers
c) 5 layers d) 6 layers **Ans : a**

28. The lower most layer of atmosphere is
a) Troposphere b) Stratosphere
c) Mesosphere d) Thermosphere **Ans : a**

29. The upper most layer of atmosphere is
a) Troposphere b) Stratosphere
c) Mesosphere d) Thermosphere **Ans : d**

30. The second layer of atmosphere is
a) Troposphere b) Stratosphere
c) Mesosphere d) Thermosphere **Ans : b**

31. The third layer of atmosphere is
a) Troposphere b) Stratosphere
c) Mesosphere d) Thermosphere **Ans : c**

32. The densest part of the atmosphere is
a) Troposphere b) Stratosphere
c) Mesosphere d) Thermosphere **Ans : a**

33. Troposphere is thicker at
a) Poles b) Equator
c) Both d) None **Ans : b**

34. Sunspots are
a) Undulations on the solar surface
b) Huge magnetic storms on the solar surface
c) Places with adequate sunshine
d) Places of tourist importance **Ans : b**

35. The last great ice age existed in
a) Pliocene period b) Pleistocene period
c) Miocene period d) Triassic period **Ans : b**

36. The origin of simplest life is attributed to
a) Proterzoic Era b) Cambrian Era
c) Archaezoic Era d) Carboniferous period **Ans : c**

37. How much time does a polar satellite require for an orbit of the earth?
a) 2 hours b) 50 minutes
c) 100 minutes d) 15 minutes **Ans : c**

38. The height at which the geostationary satellites orbit the earth is approximately
a) 35000 km b) 15 km
c) 35 km d) 1000 km **Ans : a**

39. The continuous lines drawn on a weather map are
a) Isotherms b) Isohyets
c) Isogons d) Isobars **Ans : d**

40. The infrared photograph taken from weather satellites
a) Distinguish one cloud type from another
b) Fails to distinguish one cloud type from another
c) Has nothing to do with clouds
d) Produces confusing picture of clouds. **Ans : a**

41. Troposphere extends up to a height of
a) 8-18 km b) 18-50 km
c) 50-80 km d) >80 km **Ans : a**

42. Stratosphere extends up to a height of
a) 8-18 km b) 18-50 km
c) 50-80 km d) >80 km **Ans : b**

43. Mesosphere extends up to a height of
a) 8-18 km b) 18-50 km
c) 50-80 km d) >80 km **Ans : c**

44. Thermosphere extends up to a height of
a) 8-18 km b) 18-50 km
c) 50-80 km d) >80 km **Ans : d**

45. In troposphere the temperature
a) Increases with altitude b) Decreases with altitude
c) Remains constant d) None **Ans : b**

46. The temperature in the boundary of troposphere is
a) 60°C b) - 60°C
c) 6°C d) -6°C **Ans : b**

47. The thin layer separating stratosphere from troposphere is
a) Tropopause b) Stratopause
c) Mesopause d) Thermopause **Ans : a**

48. The thin layer separating stratosphere from mesosphere is
a) Tropopause b) Stratopause
c) Mesopause d) Thermopause **Ans : b**

49. The thin layer separating mesosphere from thermosphere is
a) Tropopause b) Stratopause
c) Mesopause d) Thermopause **Ans : c**

50. The term ozone layer is synonymous with
   a) Troposphere b) Stratosphere
   c) Mesosphere d) Thermosphere
   Ans : b

51. In thermosphere the temperature
   a) Steadily increases with altitude
   b) Steadily decreases with altitude
   c) Remains constant
   d) None
   Ans : a

52. The amount of solar radiation continuously emitted by sun is
   a) 1.94 cal/cm/ min b) 1.94 cal/cm^2/ min
   c) 1.94 cal/cm^2/ hr d) 1.94 cal/cm^2/ day
   Ans : b

53. Solar radiation is received in the form of
   a) Electromagnetic waves b) Magnetic waves
   c) Thermo magnetic waves d) None
   Ans : a

54. Photo synthetically active radiation (PAR)is more in
   a) Direct radiation b) Diffused radiation
   c) Equal d) None
   Ans : b

55. The solar radiation reflected without any change in its quality is called
   a) Torpedo b) Albedo
   c) Armada d) None
   Ans : b

56. Thermal radiation emitted by the earth is termed as
   a) Solar radiation b) Lunar radiation
   c) Terrestrial radiation d) None
   Ans : c

57. The radiation balance between global and reflected solar radiation is called
   a) Total radiation b) Gross radiation
   c) Net radiation d) None
   Ans : c

58. The instrument used for measurement of total incoming solar radiation is
   a) Solar meter b) Pyranometer
   c) Thermometer d) None
   Ans : b

59. The instrument used for measurement of reflectivity
   a) Torpedo meter b) Albedo meter
   c) Armada meter d) None
   Ans : b

60. Photo synthetically active radiation (PAR)is measured with
   a) Thermo sensor b) Albedo meter
   c) Quantum sensor d) None
   Ans : c

61. Direct solar radiation is measured with
a) Pyrenometer b) Pyrheliometer
c) Polymeter d) None **Ans : b**

62. Heat can flow between substances through
a) Conduction b) Convection
c) Radiation d) All **Ans : d**

63. The transfer of heat by molecular activity is called
a) Conduction b) Convection
c) Radiation d) All **Ans : a**

64. The process of heat transfer within liquids and gases resulting from the motion of fluids is called
a) Conduction b) Convection
c) Radiation d) All **Ans : b**

65. The transfer of energy by electromagnetic waves is called
a) Conduction b) Convection
c) Radiation d) All **Ans : c**

66. Conduction is
a) The transfer of heat by molecular activity
b) The process of heat transfer within liquids and gases resulting from the motion of fluids
c) The transfer of energy by electromagnetic waves
d) All **Ans : a**

67. Convection is
a) The transfer of heat by molecular activity
b) The process of heat transfer within liquids and gases resulting from the motion of fluids
c) The transfer of energy by electromagnetic waves
d) All **Ans : b**

68. Radiation is
a) The transfer of heat by molecular activity
b) The process of heat transfer within liquids and gases resulting from the motion of fluids
c) The transfer of energy by electromagnetic waves
d) All **Ans : c**

69. Which of the following does not require the presence of medium
a) Conduction b) Convection
c) Radiation d) All **Ans : c**

70. The temperature variation between day and night is called
a) Diurnal variation b) Seasonal variation
c) Both d) None
Ans : a

71. The temperature variation between seasons is called
a) Diurnal variation b) Seasonal variation
c) Both d) None
Ans : b

72. The diurnal variation in temperature is less at
a) Coastal areas b) Equator
c) Inlands d) None
Ans : a

73. The Seasonal variation in temperature is less at
a) Coastal areas b) Equator
c) Inlands d) Higher altitudes
Ans : b

74. The lines joining the points of equal temperature
a) Isohytes b) Isotherms
c) Isobar d) None
Ans : b

75. The decrease in temperature in higher altitudes in the air is called
a) Horizontal temperature gradient
b) Vertical temperature gradient
c) Isotherm
d) None
Ans : b

76. The vertical temperature gradient is expressed as
a) Floating rate b) Lapse rate
c) Print rate d) Pulse rate
Ans : b

77. The normal lapse rate is
a) 6.5°C/km b) 65°C/km
c) -6.5°C/km d) -65°C/km
Ans : a

78. The mean heights of cirrus clouds range from
a) 15 to 25 km b) 5 to 13 km
c) 0 to 2 km d) 2 to 7 km
Ans : b

79. Heavy and dense clouds with considerable vertical extent are called
a) Stratocumulus b) Stratus
c) Nimbostratus d) Cumulonimbus
Ans : d

80. The type of cloud that has a fibrous appearance is called
a) Cumulus b) Stratus
c) Cirrus d) Altostratus
Ans : c

81. Ice-crystal theory of precipitation was propounded by
 a) Tor Bergeron b) Evans
 c) Petterssen d) Bjerknes **Ans : a**

82. An air mass with warmer temperature than the underlying surface is called
 a) Stable air mass b) Warm air mass
 c) Cold air mass d) None **Ans : b**

83. Warm and dry air masses originate in
 a) The warm Atlantic b) The warm Pacific
 c) Tropical/subtropical deserts d) Equatorial belt **Ans : c**

84. Summertime continental Polar air masses have their source regions
 a) In the central parts of high-latitude continents
 b) Over the oceans in the higher latitudes
 c) In the subtropical high pressure land areas
 d) In central Canada and Siberia **Ans : a**

85. Air mass concept was developed by
 a) Vilhelm Bjerknes b) Abercromby
 c) Shaw and Lempfert d) Picardy **Ans : a**

86. The carbon dioxide theory of climatic change was advanced by
 a) T.C. Chamberlin b) Arthur Holmes
 c) Sol Berg d) Kennelly **Ans : a**

87. The period called the Little Ice Age is believed to have extended from
 a) 1643 to about 1900 A.D. b) 1550 to about 1850 A.D.
 c) 600 to 900 A.D. d) 1200 to 1500 A.D. **Ans : b**

88. How many degrees westward do the polar satellites drift during each orbit?
 a) 15^0 b) 24^0
 c) 10^0 d) 5^0 **Ans : a**

89. The normal lapse rate is
 a) 3.5°F/1000 feet b) 35°F/1000 feet
 c) -3.5°F/1000 feet d) -35°F/1000 feet **Ans : a**

90. The rate at which temperature changes as air rises or falls is called
 a) Actual lapse rate b) Adiabatic lapse rate
 c) Thermal lapse rate d) None **Ans : b**

91. Inversion is the
a) Abrupt fall instead of rise in temperature
b) Vertical temperature gradient
c) Abrupt rise instead of fall in temperature
d) None **Ans : b**

92. The heat flow into or out of the soil is called
a) Actual lapse rate b) Solar heat flux
c) Thermal lapse rate d) Soil heat flux **Ans : d**

93. The thermal conductivity into the soil depends on
a) Texture b) Moisture
c) Organic matter content d) All **Ans : d**

94. Which of the following can influence the heat flow in soil
a) Irrigation b) Ploughing
c) Both d) None **Ans : c**

95. The melting and boiling point of water are
a) 32 and 210°F b) 32 and 110°F
c) 3.2 and 210°F d) -32 and 210°F **Ans : a**

96. The melting point of water is
a) 273°K b) 100°K
c) 0°K d) -273°K **Ans : a**

97. There is no activity of molecules of a substance at
a) 273°K b) 100°K
c) 0°K d) -273°K **Ans : c**

98. A mechanical device to record air temperature continuously
a) Thermometer b) Thermograph
c) Thermosensor d) None **Ans : b**

99. The pressure exerted by the atmosphere on the earth's surface is called
a) Thermal pressure b) Atmospheric pressure
c) Solar pressure d) Vapour pressure **Ans : b**

100. Pressure is
a) Weight/unit area b) Force/unit area
c) Volume/unit area d) None **Ans : b**

101. The atmospheric pressure
a) Decreases with decrease in elevation
b) Decreases with increase in elevation
c) Remains constant
d) Increases with increase in elevation **Ans : b**

102. Barometer is used to measure
a) Temperature b) Atmospheric pressure
c) Wind d) None **Ans : b**

103. The atmospheric pressure is expressed in
a) Bars b) Millibars
c) Centibars d) None **Ans : b**

104. Wind is air in horizontal motion which travels from a
a) High pressure area to a low pressure area
b) Low pressure area to a high pressure area
c) High tension area to a low tension area
d) Low tension area to a high tension area **Ans : a**

105. The direction from which the wind comes is
a) Leeward b) Windward
c) Seaward d) None **Ans : b**

106. The direction towards which the wind blows is
a) Leeward b) Windward
c) Seaward d) None **Ans : a**

107. Wind direction is indicated by
a) Barometer b) Wind vane
c) Anemometer d) None **Ans : b**

108. Wind velocity is measured by
a) Barometer b) Wind vane
c) Anemometer d) None **Ans : c**

109. Wind velocity near the ground surface is
a) High than at higher elevations
b) Less than at higher elevations
c) Both
d) None **Ans : b**

110. Humidity refers to the
a) Temperature of the atmosphere
b) Water vapour content of the atmosphere
c) Gaseous content of the atmosphere
d) None **Ans : b**

111. The amount of water vapour in the atmosphere depends upon
a) Air and Pressure b) Air and Temperature
c) Pressure and Temperature d) None **Ans : b**

112. When air contains maximum amount of water at a particular temperature it is said to be

a) Saturated b) Unsaturated
c) Both d) None **Ans : a**

113. The quantity of water vapour present in the atmosphere at any particular time is expressed as

a) Absolute humidity b) Specific humidity
c) Relative humidity d) All **Ans : d**

114. Absolute humidity is

a) Weight of water vapour/ volume of air
b) Weight of water vapour/ weight of air including water vapour
c) Water vapour present in the atmosphere/ water vapour required for saturation X 100
d) None **Ans : a**

115. Specific humidity is

a) Weight of water vapour/ volume of air
b) Weight of water vapour/ weight of air including water vapour
c) Water vapour present in the atmosphere/ water vapour required for saturation X 100
d) None **Ans : b**

116. Relative humidity is

a) Weight of water vapour/ volume of air
b) Weight of water vapour/ weight of air including water vapour
c) Water vapour present in the atmosphere/ water vapour required for saturation X 100
d) None **Ans : c**

117. Absolute humidity is expressed in

a) g/m^3 b) g/kg
c) % d) g/cc **Ans : a**

118. Specific humidity is expressed in

a) g/m^2 b) g/kg
c) % d) g/cc **Ans : b**

119. Relative humidity is expressed in

a) g/m^2 b) g/kg
c) % d) g/cc **Ans : c**

120. Vapour pressure deficit is the difference between the

a) Saturation vapour pressure and unsaturated vapour pressure
b) Actual vapour pressure and saturation vapour pressure
c) Saturation vapour pressure and actual vapour pressure
d) None **Ans : c**

121. Atmospheric humidity is measured by

a) Radiometer b) Hydrometer
c) Hygrometer d) Micrometer **Ans : c**

122. Fogs produced in the coastal areas of Japan are

a) Frontal fogs b) Radiation fogs
c) Advection fogs d) Upslope fogs **Ans : c**

123. Humidity refers to

a) Rainfall b) Fog
c) Cloud d) Atmospheric water vapour **Ans : d**

124. The rain shadow effect is related with

a) Drizzling b) Cyclonic precipitation
c) Convectional precipitation d) Orographic precipitation **Ans : d**

125. The type of rainfall occurring in the equatorial region is

a) Convectional b) Frontal
c) Cyclonic d) Advectional **Ans : a**

126. One of the forms of precipitation is

a) Dew b) Frost
c) Sleet d) Sublimation **Ans : c**

127. Distribution of precipitation can be shown in a map by

a) Isotherms b) Isobars
c) Isohyets d) Isohalines **Ans : c**

128. kPa stands for

a) Kelvin Pascals b) Kelvin Pattern
c) Kilo Pascals d) Kilo Pattern **Ans : c**

129. Air is said to be humid when the vapour pressure deficit is

a) <1.5kPa b) >1.5kPa
c) <2.5kPa d) >2.5kPa **Ans : a**

130. Air is said to be dry when the vapour pressure deficit is

a) <1.5kPa b) >1.5kPa
c) <2.5kPa d) >2.5kPa **Ans : d**

131. The instrument used for measuring relative humidity is

a) Psychrometer b) Hygrometer
c) Both d) None **Ans : c**

132. The wet and dry bulb thermometers are kept in
  a) Side screen b) Sight screen
  c) Stevenson's screen d) Sun screen Ans : c

133. RH can be accurately measured in crop canopy with
  a) Assman's psychrometer b) Hair hygrometer
  c) Hygrograph d) None Ans : a

134. RH can be accurately measured inside the room with
  a) Assman's psychrometer b) Hair hygrometer
  c) Hygrograph d) None Ans : b

135. Humidity can be continuously measured with
  a) Assman's psychrometer b) Hair hygrometer
  c) Hygrograph d) None Ans : c

136. The sensitive element in hair hygrometer
  a) Horse hair b) Cattle hair
  c) Human hair d) Thermometer Ans : c

137. The process by which water in a liquid state is converted to vapour and escapes into air
  a) Evaporation b) Transpiration
  c) Evapotranspiration d) Vapourisation Ans : a

138. The equipment used for measurement of evaporation
  a) USWB open pan evaporimeter b) Sunken screen evaporimeter
  c) Can evaporimeter d) All Ans : d

139. The highest clouds are
  a) Cirrus b) Stratus
  c) Cumulus d) Cumulo -nimbus Ans : a

140. The clouds associated with rains in middle and high altitude is
  a) Cirrus b) Stratus
  c) Alto- stratus d) Cumulo- nimbus Ans : c

141. The cloud in the form of uniform grey cloud sheet
  a) Cirrus b) Stratus
  c) Cumulus d) Cumulo nimbus Ans : b

142. The clouds associated with steady precipitation
  a) Cirrus b) Nimbo-Stratus
  c) Cumulus d) Cumulo nimbus Ans : b

143. Clouds prominent during summer

a) Cirrus b) Stratus
c) Cumulus d) Cumulo nimbus **Ans : c**

144. Clouds associated with thunder storms

a) Cirrus b) Stratus
c) Cumulus d) Cumulo nimbus **Ans : d**

145. The outer solid crust of earth is called

a) Atmosphere b) Lithosphere
c) Hydrosphere d) None **Ans : b**

146. The total volume of water contained in oceans is about

a) $1.94 x 10^9$ Km^3 b) $1.64 x 10^9$ Km^3
c) $1.96 x 10^9$ Km^3 d) $1.46 x 10^9$ Km^3 **Ans : d**

147. The average annual world precipitation is

a) 1500mm b) 1200mm
c) 1000mm d) None **Ans : c**

148. Temperature is high near the

a) Poles b) Equator
c) Remains unchanged d) None **Ans : b**

149. The low pressure areas near the equator is called

a) Oldrums b) Doldrums
c) Polar calms d) None **Ans : b**

150. Horse latitude is the belt between

a) 30 to35^0 North latitude b) 30 to35^0 South latitude
c) Both d) None **Ans : c**

151. The decrease in air pressure due to movement of air from polar region is known as

a) Polar storms b) Doldrums
c) Polar calms d) None **Ans : c**

152. The normal date of arrival of monsoon in India is

a) 1st June b) 15th June
c) 20th June d) 30th June **Ans : a**

153. The grand period of rainfall in India is

a) South -west monsoon b) South -east monsoon
c) North -east monsoon d) None **Ans : a**

154. The part of India not receiving rains from south -west monsoon is

a) South -west Tamilnadu b) East coast of Tamilnadu
c) Kerla d) None **Ans : b**

155. South coastal Andhra Pradesh and Tamilnadu receives rainfall from

a) South -west monsoon b) South -east monsoon
c) North -east monsoon d) None **Ans : c**

156. The short range weather forecast is for a period of

a) 1-2 days b) 3-4 days to 2 weeks
c) More than four weeks d) Less than four weeks **Ans : a**

157. The medium range weather forecast is for a period of

a) 1-2 days b) 3-4 days to 2 weeks
c) More than four weeks d) Less than four weeks **Ans : b**

158. The long range weather forecast is for a period of

a) 1-2 days b) 3-4 days to 2 weeks
c) More than four weeks d) Less than four weeks **Ans : c**

159. Which of the following is not a method of weather forccast

a) Synoptic method b) Statistical method
c) Numerical method d) Asynoptic method **Ans : d**

160. Cloud seeding is done with the help of

a) Silver iodide b) Magnesium iodide
c) Strontium chloride d) All **Ans : a**

161. The technique of cloud seeding aims at correcting the

a) Deficiencies of nuclei b) Excess of nuclei
c) Deficiencies of neutrons d) All **Ans : a**

162. Silver iodide is used for

a) Warm cloud seeding b) Cold cloud seeding
c) Both d) None **Ans : b**

163. Sodium chloride is used for

a) Warm cloud seeding b) Cold cloud seeding
c) Both d) None **Ans : a**

164. The adverse effect of high wind velocity can be reduced by

a) Wind breaks b) Shelterbelts
c) Both d) None **Ans : c**

165. India has been divided into

a) 12 agro climatic zones
b) 15 agro climatic zones
c) 18 agro climatic zones
d) None

Ans : b

166. Radiation harmful to the plants comes in the spectral region

a) <0.4μm
b) 0.4μm - 0.7μm
c) 0.7μm – 1.0μm
d) >0.4μm

Ans : a

167. Photosynthetically active radiation comes in the spectral region

a) <0.4μm
b) 0.4μm - 0.7μm
c) 0.7μm – 1.0μm
d) >0.4μm

Ans : b

168. Near infrared radiation comes in the spectral region

a) <0.4μm
b) 0.4μm - 0.7μm
c) 0.7μm – 1.0μm
d) >0.4μm

Ans : c

169. Radiation having influence on seed germination and dormancy of seeds

a) Near infrared
b) Photosynthetically active radiation
c) Ultraviolet
d) None

Ans : a

170. Chlorophyll formation is promoted by light in the spectral region

a) <0.25μm
b) 0.3μm - 0.338μm
c) 0.4μm – 1.0μm
d) >1.0μm

Ans : b

171. Phytochrome is excited by

a) Green light
b) Red light
c) White light
d) All

Ans : a

172. The photosynthetic electron transport chain is formed during subsequent development of chloroplast under

a) Green light
b) Red light
c) White light
d) All

Ans : c

173. Formation of protein –pigment system of the chloroplast and their resistance to senescence is due to

a) Green light
b) Red light
c) White light
d) All

Ans : b

174. The amount of solar radiation present in any layer of the crop can be calculated by

a) $I = I_0 e^{-KLAI}$
b) $I = I_0 e^{-K}$
c) $I_0 = I e^{-KLAI}$
d) $I = I_0 e^{-LAI}$

Ans : a

175. The effect of light on plants is known as

a) Thermoperiodism b) Photoperiodism
c) Both d) None **Ans : b**

176. Long day plants flower when the length of light is

a) <14 hrs b) >14 hrs
c) <10 hrs d) >10 hrs **Ans : b**

177. Short day plants flower when the length of light is

a) <14 hrs b) >14 hrs
c) <10 hrs d) >10 hrs **Ans : c**

178. Cockelbur is an example of

a) Qualitative short day plant b) Quantitative short day plant
c) Qualitative long day plant d) Quantitative short day plant **Ans : a**

179. At the time of equinoxes

a) The length of day and night is equal
b) The length of day and night is not equal
c) The length of light is equal
d) None **Ans : a**

180. The rate of increase in reaction for every 10^0C in temperature is called

a) Q_{10} b) K_{10}
c) C_{10} d) T_{10} **Ans : a**

181. The relationship between growth duration and temperature is termed as

a) Base temperature b) Growing degree days
c) Photothermal unit d) None **Ans : b**

182. The lowest temperature below which there is no growth is called as

a) Q_{10} b) Base temperature
c) Photothermal unit d) T_{10} **Ans : b**

183. Base temperature for wheat is

a) 4.5^0C b) 45^0C
c) 10^0C d) 1.0^0C **Ans : a**

184. Base temperature for rice is

a) 4.5^0C b) 5^0C
c) 10^0C d) 1.0^0C **Ans : c**

185. Coefficient of variation is given by

a) SD/Mean x100
b) S.D + Mean / 100
c) $\frac{\text{SD x Mean}}{100}$
d) None

Ans : a

186. Q_{10} can be calculated by

a) $Q10 = \frac{\text{Rate of reaction at (t+10)°C}}{\text{Rate of reaction at t°C}}$
b) $Q10 = \frac{\text{Rate of reaction at t°C}}{\text{Rate of reaction at (t+10)°C}}$
c) $Q10 = \frac{\text{Rate of reaction at (t-10)°C}}{\text{Rate of reaction at t°C}}$
d) $Q10 = \frac{\text{Rate of reaction at t°C}}{\text{Rate of reaction at (t-10)°C}}$

Ans : a

187. Composition of the atmosphere

a) Varies with seasons
b) Varies with altitude
c) Remains constant in the lower layers
d) Varies from place to place

Ans : c

188. The atmospheric layer characterized by selective absorption of ultraviolet radiation is called

a) Ionosphere
b) Exosphere
c) Ozonosphere
d) Stratosphere

Ans : c

189. The upper boundary of the stratosphere is called

a) Tropopause
b) Thermosphere
c) Stratopause
d) Mesopause

Ans : c

190. The atmospheric layer reflecting radio waves is called

a) Homosphere
b) Ionosphere
c) Stratosphere
d) Ozonosphere

Ans : b

191. About 50% of the atmosphere lies below an altitude of

a) 5.6 km
b) 10 km
c) 15 km
d) 30 km

Ans : a

192. Varied colours of red and orange at sunrise and sunset are cased by

a) Water vapour
b) Dust particles
c) Gases
d) Insolation

Ans : b

193. To reach the earth's surface the rays of the sun take about

a) 18 minutes
b) 12 minutes
c) 8 minutes
d) 30 minutes

Ans : c

194. The insolation reaching the earth's surface is equal to
a) 23 billion horse power b) 15 billion horse power
c) 5 billion horse power d) 10 billion horse power **Ans : a**

195. The highest temperatures are recorded
a) In the late evening b) At midday
c) In the afternoon d) In the morning **Ans : c**

196. The total annual insolation is greatest at
a) The tropic of Cancer b) The Tropic of Capricorn
c) The equator d) The Arctic circle **Ans : c**

197. The process in which incident radiation is retained by a substance is called
a) Absorption b) Scattering
c) Diffusion d) Reflection **Ans : a**

198. In the direct absorption of insolation the most significant part is played by
a) Dust particles b) Water vapour
c) Gases d) Clouds **Ans : b**

199. The albedo of the earth as a whole is
a) 25% b) 50 %
c) 35% d) 10% **Ans : c**

200. The process of the transfer of heat through movement of a substance from one place to another is called
a) Radiation b) Conduction
c) Diffusion d) Convection **Ans : d**

201. The most important source for the latent heat of evaporation is
a) Cloud b) Fog
c) Ocean d) Atmosphere **Ans : c**

202. The intensity of insolation depends on
a) Altitude b) Latitude
c) Nature of the earth's surface d) Winds **Ans : b**

203. The specific heat of water is
a) 1 b) 3
c) 5 d) 4 **Ans : a**

204. The thermal equator is found
   a) At the equator
   b) South of the geographical equator
   c) North of the geographical equator
   d) At the Tropic of Cancer
   Ans : c

205. Isotherms are used to show the horizontal distribution of
   a) Salinity
   b) Temperature
   c) Pressure
   d) Rainfall
   Ans : b

206. The height of midday sun at the equator is never less than
   a) 66 ½
   b) 23 ½
   c) 90^0
   d) 30^0
   Ans : a

207. The cold Pole of the earth is
   a) The North Pole
   b) The South Pole
   c) Verkhoyansk
   d) Oslo
   Ans : c

208. Inversion of temperature represents
   a) Decrease of temperature with altitude
   b) Increase of temperature with altitude
   c) Uniform temperature
   d) Decrease of temperature with latitude
   Ans : b

209. The annual range of temperature in the equatorial region hardly exceeds
   a) 1^0C
   b) 2^0C
   c) 15^0C
   d) 2^0C
   Ans : b

210. The dry adiabatic lapse rate is
   a) 6.4^0C
   b) 10^0C
   c) 5^0C
   d) 2^0C
   Ans : b

211. The normal lapse rate of temperature per kilometer is
   a) 6.4^0C
   b) 4.6^0C
   c) 10^0C
   d) 8^0C
   Ans : a

212. An adiabatic change of temperature is caused by
   a) Increase in the moisture content of an air parcel
   b) Increasing amount of dust particles in an air parcel
   c) Expansion or compression of an air parcel
   d) Horizontal movement of an air parcel
   Ans : c

213. The process involved in adiabatic change of temperature is
   a) Chemical
   b) Physical
   c) Advectional
   d) Sublimation
   Ans : b

214. The highest annual range of temperature i.e. 65^0C is recorded at

a) Paris b) Bombay
c) Verkhoyansk d) North Pole Ans : c

215. The immediate cause of wind is

a) Humidity differences b) Temperature differences
c) Rotation of the earth d) Pressure differences Ans : d

216. The standard sea-level pressure at 45^0 latitude at a temperature of 15^0C is

a) 1013.25 mb b) 998 mb
c) 1031 mb d) 1034 mb Ans : a

217. Extremely sensitive atmospheric pressure recording instrument is called

a) Hygrometer b) Microbarovariograph
c) Aneroid barometer d) Thermograph Ans : b

218. All the weather changes are closely related to

a) Pressure variations b) Microbarovariograph
c) Humidity variations d) Cyclonic activity Ans : a

219. Wind velocity is measured by

a) Barograph b) Wind vane
c) Cup anemometer d) Aneroid activity Ans : c

220. Pressure distribution is shown on a map by

a) Isogonic lines b) Isohyets
c) Isobars d) Isonephs Ans : c

221. Continuous recording of air pressure is done by

a) hygrograph b) seismograph
c) thermograph d) barograph Ans : d

222. Horse latitudes lie within

a) Polar high b) Equatorial low
c) Subtropical high d) Subpolar low Ans : c

223. The pressure system with higher pressure at the centre is called

a) Anticyclone b) Front
c) Cyclone d) Depression Ans : a

224. Jet stream is a

a) Warm current b) Tributary of a river
c) Upper air westerlies d) Local wind Ans : c

225. Doldrum is a zone of
a) Intertropical divergence
b) Intertropical convergence
c) Frontolysis
d) Local wind
Ans : b

226. Chinook is
a) A very warm and dry wind on the eastern slopes of the Rockies
b) A violent and extremely cold wind
c) An extremely cold wind in central Siberia
d) A dry and dusty wind of the west coast of Africa blowing from the deserts.
Ans : a

227. Relatively narrow bands of high velocity winds in the upper troposphere are called
a) Hurricanes
b) Atmospheric disturbances
c) Jet stream
d) Antitrades
Ans : c

228. Sirocco is
a) An island
b) A bird
c) A local wind
d) A volcano
Ans : c

229. Absolute humidity refers to
a) The weight of water vapour per weight of a given mass of air
b) The weight of water vapour per unit volume of air
c) Vapour pressure
d) The weight of water vapour per unit weight of dry air
Ans : b

230. Collision coalescence theory was propounded to explain
a) Tornado
b) Precipitation
c) Clouds
d) Cyclones
Ans : b

231. The average slope of a warm front is
a) 1:100 to 1: 200
b) 1: 50 to 1: 100
c) 1: 20 to 1: 30
d) 1: 10 to 1: 20
Ans : a

232. The term front genesis refers to
a) The decay of a front
b) The creation of a front
c) The migration of a front
d) Occlusion
Ans : b

233. Extra tropical cyclones have
a) Eye
b) Diverging winds
c) Fronts
d) No fronts
Ans : c

234. The name 'hurricane' is given to tropical cyclones in
a) The north Pacific Ocean
b) The north Atlantic Ocean
c) Australia
d) The Bay of Bengal
Ans : b

235. The typhoons derive their energy from
- a) The falling rain
- b) The rising air
- c) Density contrast
- d) Latent heat of condensation

Ans : d

236. Close to the equator the tropical cyclones do not occur because of
- a) Weak Coriolis force
- b) Light and variable winds
- c) Excessive humidity
- d) Convective activity

Ans : a

237. Tornadoes may be called
- a) Local winds
- b) Whirl winds
- c) The most violent storms
- d) Convective activity

Ans : c

238. The duration of a waterspout is about
- a) 5 to 8 hrs
- b) 30 to 60 minutes
- c) 2 to 3 days
- d) 2 to 4 hours

Ans : b

239. The diameter of a hurricane ranges from
- a) 50 to 100 km
- b) 100 to 200 km
- c) 10 to 50 km
- d) 500 to 600 km

Ans : d

240. The central low pressure core of a tropical cyclone is called
- a) Col
- b) Trough
- c) Eye
- d) Depression

Ans : c

241. The diameter of hurricane eye varies from
- a) 20 to 40 km
- b) 100 to 200 km
- c) 10 to 50 meters
- d) 200 to 300 km

Ans : a

242. The maximum frequency of hurricanes is found in between latitudes
- a) 5^0 and 10^0
- b) 10^0 and 15^0
- c) 0^0 and 5^0
- d) 20^0 and 25^0

Ans : b

243. Thunderstorms originate from
- a) Altocumulus clouds
- b) Stratus clouds
- c) Cumulonimbus clouds
- d) Nimbostratus clouds

Ans : c

244. The height of thunderstorms may vary from
- a) 4 to 20 km
- b) 100 to 500 meters
- c) 20 to 40 km
- d) 10 to 50 meters

Ans : a

245. The most important factor in thunderstorms development is
- a) Atmospheric stability
- b) Atmospheric instability
- c) Abundance of moisture
- d) Temperature inversion

Ans : b

246. In a tropical cyclone the pattern of isobars is
a) Elliptical b) Semi-circular
c) Circular d) Rectangular **Ans : c**

247. The slope of a cold front varies from
a) 1: 50 to 1: 100 b) 1: 100 to 1: 200
c) 1: 200 to 1: 400 d) 1: 400 to 1: 600 **Ans : a**

248. Zones separating different air masses are called
a) Anticyclones b) Cyclones
c) Fronts d) Cols **Ans : c**

249. The largest number of temperate cyclones originate mostly over the
a) Indian ocean b) North Atlantic ocean
c) North Pacific ocean d) Arctic ocean **Ans : b**

250. The Mediterranean type of climate receives most of its precipitation during
a) Summer b) Winter
c) Spring d) The whole year **Ans : b**

251. In the equatorial type of climate the diurnal range of temperature varies from
a) 5^0 to 14^0 C b) 0^0 to 2 C
c) 15^0 to 25 C d) 30^0 to 40^0 C **Ans : a**

252. The equatorial rain forest is also known as
a) Savanna b) Campos
d) Llanos d) Selva **Ans : d**

253. Sclerophyllus natural vegetation is found in
a) Mediterranean regions b) Equatorial regions
c) Monsoon regions d) Tundra regions **Ans : a**

254. The average annual temperature in tundra climate is
a) -20^0C b) -12^0C
c) -5^0C d) 0^0C **Ans : b**

255. In the Koppen system polar climates are designated as
a) F b) E
c) T d) B **Ans : b**

256. Zerophytes represent the class of vegetation found in
a) Tropical rain forest b) Humid microthermal climates
c) Semi-arid steppes d) Tundra **Ans : c**

257. Temperate marine type of climate (Do) is also called
a) China type of climate
b) Mediterranean type climate
c) Marine west coast type of climate
d) Monsoon type of climate **Ans : c**

258. The tropical/subtropical hot deserts are found between
a) 15^0 and 35^0 N and S b) 30^0 and 40^0 N and S
c) 50^0 and 60^0 N d) 55^0 and 70^0 N **Ans : a**

259. In the tropical hot deserts
a) Precipitation exceeds evaporation
b) Evaporation exceeds precipitation
c) Precipitation equals evaporation
d) Precipitation is abundant **Ans : b**

260. Precipitation in the Mediterranean type of climate is
a) Reliable b) Unreliable
c) Highly unreliable d) Highly reliable **Ans : b**

261. In taiga region the average temperature of the coldest month is
a) Below 0^0C b) Above 0^0C
c) 5^0C d) 3^0C **Ans : a**

262. Verkhoyansk has recorded the lowest temperature of
a) -10^0C b) -40^0C
c) -66.8^0C d) -45^0C **Ans : c**

263. Tundra climate is characterized by
a) Abundant precipitation b) Meager precipitation
c) Moderate precipitation d) Droughts **Ans : b**

264. Extreme continentality is the characteristic feature of the
a) Continental steppe climate b) Middle-latitude desert climate
c) Tropical monsoon climate d) Savanna climate **Ans : a**

265. Smog is a type of
a) Storm b) Soil
c) Precipitation d) Fog which is mixed with smoke **Ans : d**

266. Smaze is a type of
a) Atmospheric disturbance b) Frost
c) Precipitation d) Fog which is an admixture **Ans : d**

267. Sounding balloons are used for
a) exploring the atmosphere b) exploring the space
c) exploring the sound d) exploring the oceans **Ans : a**

268. The barometer was invented by Torrecelli in

a) 1543 b) 1680
c) 1643 d) 1810 **Ans : c**

269. The number of World Meteorological Centres are

a) 3 b) 5
c) 4 d) 2 **Ans : a**

270. TIROS I was launched in 1960 by

a) France b) Germany
c) USA d) Japan **Ans : c**

271. The last TIROS was launched in

a) 1965 b) 1985
c) 1970 d) 1975 **Ans : a**

272. On surface weather maps isobars are usually plotted at intervals of

a) 6 millibars b) 1 millibar
c) 4 millibars d) 3 millibars **Ans : c**

273. World Meteorological Organization (WMO) was established in

a) 1950 b) 1873
c) 1945 d) 1860 **Ans : a**

274. The World Meteorological Organization is located at

a) Washington b) Geneva
c) Moscow d) London **Ans : b**

275. India gets most of its rains from

a) Tropical cyclones b) The summer monsoon
c) The retreating monsoon d) The winter monsoon **Ans : b**

276. The main cause of monsoon climate is

a) Low temperature b) High temperature
c) Undulating terrain d) Seasonal reversal of winds **Ans : d**

277. The Coriolis force is

a) Nil at the equator b) Maximum at the equator
c) Nil at the poles d) Maximum at 66½ N and S **Ans : a**

278. The equipment not used for measurement of evaporation

a) USWB open pan evaporimeter b) Sunscreen evaporimeter
c) Can evaporimeter d) Portable evaporimeter **Ans : b**

279. The height of cylinder in USWB open pan evaporimeter is
a) 25 cm b) 250cm
c) 122 cm d) 122 mm Ans : a

280. The diameter of cylinder in USWB open pan evaporimeter is
a) 25 cm b) 250cm
c) 122 cm d) 122 mm Ans : c

281. Evaporation is measured in
a) mm/day b) cm/day
c) kg/day d) cm/min Ans : a

282. The height of cylindrical vessel in sunken screen evaporimeter is
a) 4.5 cm b) 45cm
c) 450 cm d) 122 mm Ans : b

283. The diameter of cylindrical vessel in sunken screen evaporimeter is
a) 4.5 cm b) 45cm
c) 60 cm d) 60 mm Ans : c

284. Equipment used to measure evapotranspiration under field condition is
a) USWB open pan evaporimeter
b) Sunken screen evaporimeter
c) Can evaporimeter
d) Both b and c Ans : d

285. Portable evaporimeter was developed in
a) India b) Indenosia
c) Israel d) USA Ans : c

286. Portable evaporimeter consists of
a) Glass tube b) Iron tube
c) Plastic tube d) Aluminium tube Ans : b

287. The falling down of condensed moisture to ground surface is called
a) Mist b) Precipitation
c) Fog d) Hail storm Ans : b

288. When the water droplets evaporate before it reaches the ground it is called
a) Mist b) Precipitation
c) Fog d) Hail storm Ans : a

289. When rain falls on the ground and forms a sheet or coating of ice it is called

a) Mist b) Glaze
c) Fog d) Hail storm **Ans : b**

290. Rime is a

a) Fog b) Glaze
c) Freezing fog d) Hail storm **Ans : c**

291. The solid precipitation on the ground in the form of ice crystals it is called

a) Snow b) Glaze
c) Freezing fog d) Hail storm **Ans : a**

292. The solid precipitation on the ground in the form of small particles of clear ice it is called

a) Snow b) Glaze
c) Sleet d) Hail storm **Ans : c**

293. The solid precipitation in the form of large, hard rounded pallets of ice and compact snow it is called

a) Snow b) Glaze
c) Sleet d) Hail storm **Ans : d**

294. Instrument used for measurement of rainfall continuously

a) Rain guage b) Automatic rain guage
c) Screw guage d) None **Ans : b**

295. The aggregate of minute drops of water suspended in air at higher altitude is called

a) Snow b) Glaze
c) Sleet d) Cloud **Ans : d**

296. Which of the following is not a basic form of cloud

a) Cirrus b) Stratus
c) Cumulus d) Cumulo-nimbus **Ans : d**

297. Is the clouds associated with rains, the word prefixed or suffixed to the basic form is

a) Alto b) Nimbus
c) Cumulus d) Cumulo- nimbus **Ans : b**

298. The clouds are classified in to .... characteristic forms

a) 7 b) 8
c) 9 d) 10 **Ans : d**

299. The rate of evapo-transpiration equal to all smaller than potential evapo-transpiration as affected by the level of the available soil water, salinity, filed size, or other causes is termed as

a) Actual crop evapo-transpiration
b) PET
c) ET
d) None

Ans : a

300. The rate of actual evapo-transpiration is expressed in terms of

a) cm/hr b) mm/day
c) m/day d) None

Ans : b

301. The rate at which temperature changes as air rise or falls is called

a) Adiabatic lapse rate b) Altitudinal change
c) Temperature inversion d) None

Ans : a

302. ....is constant for dry air.

a) Adiabatic lapse rate b) Altitudinal change
c) Temperature inversion d) None

Ans : a

303. The instrument used in the measurement of force and velocity of wind

a) Anemometer b) Wind vane
c) Barometer d) Psychrometer

Ans : a

304. An instrument used for measuring atmospheric pressure

a) Aneroid barometer b) Anemometer
c) Wind vane d) Psychrometer

Ans : a

305. The main advantage of Aneroid barometer over a mercury barometer is its.

a) Durability b) Speed
c) Accuracy d) Portability

Ans : d

306. An extremely dry climate with an annual average precipitation usually less than 250 mm and generally rainfall is well short of evapo-transpiration demand of the atmosphere is

a) Arid b) Sem arid
c) Tropical d) None

Ans : a

❑❑❑

3 Chapter

Soil and Tillage

Multiple Choice Questions (MCQ's)

1. Most important soil forming processes for the development of the order Aridisols are
   a) Calcification and Gypsification
   b) Calcification and Salinization
   c) Salinization and Alkalization
   d) Salinization and Gypsification **Ans : b**

2. Prominent secondary clay mineral in Ultisols is
   a) Kaolinite b) Illite
   c) Chlorite d) Vermiculite **Ans : a**

3. Soils formed due to alternate wet and dry conditions are called as
   a) Rendzenas b) Lithosols
   c) Plannosols d) Regosols **Ans : c**

4. Weakly developed coarse textured soil is
   a) Acrisols b) Alisols
   c) Arenosols d) Plannosols **Ans : c**

5. Saline sodic soils has the value of
   a) EC More than 4 mmhos/cm. ESP more than 15, pH more than 8.5
   b) EC less than 4 mmhos/cm. ESP more than 15, pH less than 8.5
   c) EC more than 4 mmhos/cm. ESP less than 15. pH less than 8.5
   d) EC less than 4 mmhos/cm. ESP 'more than 15: pH more than 8.5 **Ans : a**

6. Sodic soils contain sufficient amount of .... to interfere with plant growth
   a) Exchangeable calcium   b) Changeable sodium
   c) Exchangeable potassium   d) Exchangeable magnesium   **Ans : b**

7. In degraded acid soils root extension into anaerobic and highly reduced layer is reduced due to injurious concentration of
   a) Calcium   b) Iron
   c) Aluminium   d) Magnesium   **Ans : b**

8. Barley and rice crops are_________ salt tolerant
   a) High   b) Low
   c) Medium   d) Very low   **Ans : a**

9. Gypsum is commonly used for reclamation of
   a) Saline soil   b) Sodic soil
   c) Acidic soil   d) All of these   **Ans : b**

10. In alkali soils crop growth is affected due to ....
   a) Higher salt concentration   b) Poisonous ion concentration
   c) Poor physical properties   d) Microorganisms   **Ans : a**

11. Soluble salt content of soil is determined by measuring its
   a) Electrical conductivity
   b) pH value
   c) Organic matter content carbon
   d) None of these   **Ans : a**

12. Plants tolerant to high soil salinity conditions
   a) Heliophytes   b) Halophytes
   c) Heliotropic   d) None of these   **Ans : b**

13. Degree of soil salinity is indicated by its
   a) Total soluble salt content   b) Organic matter content
   c) Nitrogen content   d) Phosphorus content   **Ans : a**

14. Unit of measuring electrical conductivity is
   a) Micro - gram/sq.cm.   b) Angstran/cm.
   c) Millimhos/cm   d) None of these   **Ans : c**

15. The negative logarithm of H+ ion concentration (activity) in the soil solution
   a) Acidity   b) Salinity
   c) Alkalinity   d) pH   **Ans : d**

16. On the basis of pH measurements, the nearly neutral
    a) 6.0 to 6.5 b) 6.6 to 7.5
    c) 7.5 to 8.5 d) 5.0 to 6.0 **Ans : b**

17. Saline sodic soils has the value of
    a) EC More than 4 mmhos/cm. ESP more than 15: pH more than 8.5
    b) EC less than 4 mmhos/cm. ESP more than 15: pH less than 8.5
    c) EC more than 4 mmhos/cm. ESP less than 15: pH less than 8.5
    d) EC less than 4 mmhos/cm. ESP 'more than 15: pH more than 8.5 **Ans : a**

18. Sodic sons contain sufficient amount of .... to interfere with plant growth
    a) Exchangeable calcium b) Changeable sodium
    c) Exchangeable potassium d) Exchangeable magnesium **Ans : b**

19. Lime is used for reclamation of
    a) Acidic soils b) Alkaline soils
    c) Calcareous soils d) None of these **Ans : a**

20. Tree species used for the reclamation of salt affected soils is/are
    a) Casuarinas b) Prosopis
    c) Eucalyptus d) All **Ans : d**

21. What is the total area of acid soils in India?
    a) 28 m ha b) 25 m ha
    c) 43 m ha d) 48 m ha **Ans : d**

22. What is the total area of acid soils with pH below 5.5 in India?
    a) 23 m ha b) 25 m ha
    c) 28 m ha d) 32 m hass **Ans : b**

23. What is the total area of acid soils with pH between 5.6 and 6.5 in India?
    a) 23 m ha b) 25 m ha
    c) 28 m ha d) 32 m ha

24. In lime requirement, liming materials are added to raise the soil pH to
    a) 6.0 – 7.0 b) 6.0 – 8.0
    c) 7.0 – 8.0 d) 7.0 – 9.0 **Ans : a**

25. 'Soil is a nutrient bin'- this statement was given by
    a) Ruffin b) Whitney
    c) Hilgard d) Dokuchaiev **Ans : b**

26. 'Soil is mixture of earth's upper most mantles of weathered rocks and organic matter'- this statement given by
    a) Ruffin b) Whitney
    c) Hilgard d) Dokuchaiev **Ans : a**

27. Who is the father of soil science?
    a) Dokuchaiev b) Justin von Liebig
    c) J.W. Leather d) Marbut **Ans : a**

28. Who is the father of Indian soil science?
    a) Dokuchaiev b) Justin von Liebig
    c) J.W. Leather d) Marbut **Ans : c**

29. A loose, unconsolidated inorganic material of weathered rock on the earth surface is called as
    a) Regolith b) Parent material
    c) Solum d) Both a and b **Ans : d**

30. What is the percentage of nitrogen in atmospheric air?
    a) 78.09 b) 79.0
    c) 77.0 d) 80.1 **Ans : a**

31. What is the percentage of carbon dioxide in atmospheric air?
    a) 0.3 b) 0.03
    c) 0.003 d) 3.0 **Ans : b**

32. What is the density of earth's crust?
    a) 2.3 - 2.6 g cm^{-3} b) 3.0 - 4.5 g cm^{-3}
    c) 9.0 - 12.0 g cm^{-3} d) 2.6 - 3.0 g cm^{-3} **Ans : d**

33. What is the density of earth's core?
    a) 2.6 - 2.7 g cm^{-3} b) 3.0 - 4.5 g cm^{-3}
    c) 9.0 - 12.0 g cm^{-3} d) 2.6 - 3.0 g cm^{-3} **Ans : c**

34. What is the density of earth's mantle?
    a) 2.6 - 2.7 g cm^{-3} b) 3.0 - 4.5 g cm^{-3}
    c) 9.0 - 12.0 g cm^{-3} d) 2.6 - 3.0 g cm^{-3} **Ans : b**

35. Two elements occurring in greatest abundance in the earth's crust are
    a) Oxygen and Aluminium b) Silicon and Aluminium
    c) Silicon and Iron d) Oxygen and Silicon **Ans : d**

36. What is the percentage of silicon (Si^{4+}) in the earth's crust?
    a) 46.60 b) 27.72
    c) 8.13 d) 5.0 **Ans : b**

37. Honey-comb structure is found in
a) Orthosilicates b) Inosilicates
c) Phyllosilicates d) Tectosilicates **Ans : c**

38. What is the common name of $Fe_2 O_3 . 3H_2O$?
a) Haematite b) Limonite
c) Goethite d) Magnetite **Ans : b**

39. What is the common name of $Al_2 O_3 . 3H_2O$?
a) Haematite b) Limonite
c) Goethite d) Gibbsite **Ans : d**

40. The metamorphic rock, slate is formed from
a) Shale b) Sandstone
c) Limestone d) Coal **Ans : a**

41. The metamorphic rock, quartzite is formed from
a) Shale b) Sandstone
c) Limestone d) Coal **Ans : b**

42. The metamorphic rock, marble is formed from
a) Shale b) Sandstone
c) Limestone d) Coal **Ans : c**

43. The metamorphic rock, graphite is formed from
a) Shale b) Sandstone
c) Limestone d) Coal **Ans : d**

44. Crystalline metamorphic rock is
a) Schist b) Gneiss
c) Marble d) Slate **Ans : b**

45. When water freezes in to ice, its volume is increased by
a) 6 per cent b) 7 per cent
c) 8 per cent d) 9 per cent **Ans : d**

46. Most important weathering for soil formation is/are
a) Physical weathering b) Chemical weathering
c) Biological weathering d) All **Ans : b**

47. The mineral bauxite is formed from Al_2O_3 by
a) Hydration b) Hydrolysis
c) Oxidation d) Reduction **Ans : a**

48. Cleavage occur in only one plane during the weathering of
a) Olivine b) Feldspar
c) Hornblende d) Mica **Ans : d**

49. Cleavage occur in two planes during the weathering of
    a) Olivine and Augite b) Feldspar
    c) Hornblende d) All **Ans : d**

50. Minerals formed during the intermediate stage of weathering are
    a) Kaolinite and Halloysite
    b) Haematite and Gibbsite
    c) Vermiculite and Montmorillonite
    d) All **Ans : c**

51. Minerals formed during the advanced stage of weathering are
    a) Kaolinite and Halloysite b) Quartz and Illite
    c) Muscovite and Hornblende d) All **Ans : a**

52. The equation, S = f (cl, b, r, p, t....) was first formulated by
    a) Dokuchaiev b) Jenny
    c) Hilgard d) Whitney **Ans : b**

53. Amongst all the soil forming factors, the most influential factor in soil development is
    a) Parent material b) Vegetation
    c) Climate d) Time **Ans : c**

54. For every 10 °C rise in temperature, the speed of chemical reactions increases by a factor of
    a) Two b) Three
    c) Four d) Both a and b **Ans : d**

55. Soils formed by the process of gleization is called as
    a) Calcimorphic soils b) Hydromorphic soils
    c) Halomorphic soils d) Dynamomorphic soils **Ans : b**

56. Pedogenic process responsible for the development of bluish to grayish horizons in soils is
    a) Eluviation b) Laterization
    c) Gleization d) Podzolization **Ans : c**

57. Process of accumulation of salts, such as sulphate, chlorides of calcium, magnesium, sodium and potassium, in soils is
    a) Salinization b) Alkalization
    c) Solonization d) Solodization **Ans : a**

58. Saline soils are also called as
    a) Solonetz b) Solonchalks
    c) Soloth d) Both b and c **Ans : b**

59. Process that involves the accumulation of sodium ions on the exchange complex of the clay is/are
a) Solonization b) Alkalization
c) Solodization d) Both a and b **Ans : d**

60. Sodic soils are also called as
a) Soloth b) Solonetz
c) Solonchalks d) Both a and b **Ans : b**

61. A typical columnar structure is a characteristic feature of
a) Solonchalks b) Solonetz
c) Soloth d) All **Ans : b**

62. Soils formed as the result of dealkalization is
a) Saline soils b) Sodic soils
c) Soloth d) Neutral soils **Ans : c**

63. Moor soils are
a) Normal soils b) Transitional soil
c) Abnormal soils d) None **Ans : c**

64. Tundra soils are
a) Normal soils b) Transitional soils
c) Abnormal soils d) None **Ans : a**

65. What is the total number of soil order?
a) 10 b) 11
c) 12 d) 13 **Ans : c**

66. Best agricultural soils of the world is
a) Entisols b) Inceptisols
c) Mollisols d) Vertisols **Ans : c**

67. Weakly developed shallow soil is
a) Acrisols b) Alisols
c) Arenosols d) Plannosols **Ans : d**

68. Land capability class was given by
a) Sys b) Klingebiel and Montgomery
c) Sys and Verhegs d) Storie an Ricquier **Ans : b**

69. Total number of categories in the land capability classification is
a) Three b) Five
c) Eight d) Twelve **Ans : a**

70. Colour of the land capability class-I on LCC maps is
a) Yellow b) Pink
c) Green d) Red Ans : c

71. Upland paddy crop is the one which is grown
a) on raised seed beds
b) as rainfed, direct seeded crop
c) along the hills on terraces with assured supplemental irrigation
d) where seedlings are grown on raised seed bed as in Japanese method of nursery raising Ans : b

72. Science that deals with soil genesis and classification is called as
a) Petrology b) Pedology
c) Pedagogy d) Petrography Ans : b

73. Rocks formed by cooling of the molten materials (magma) from the sun is called as
a) Metamorphic rocks b) Igneous rocks
c) Sedimentary rocks d) All Ans : b

74. Of the followings, which is/are the examples of igneous rocks
a) Granites b) Basalts
c) Syenites d) All Ans : d

75. Rocks formed from igneous rocks by the influence of high pressure and temperature are called as
a) Metamorphic rocks b) Primary rocks
c) Sedimentary rocks d) All Ans : c

76. Gneiss, Schist and Slate are the examples of
a) Metamorphic rocks b) Sedimentary rocks
c) Primary rocks d) Igneous rocks Ans : a

77. Sandstone and limestone are
a) Metamorphic rocks b) Sedimentary rocks
c) Primary rocks d) Igneous rocks Ans : b

78. Major component of igneous rocks are
a) Quartzs b) Amphiboles and pyroxenes
c) Feldspars d) Micas Ans : c

79. Percentage of quartz in igneous rocks is
a) 4 b) 12
c) 18 d) 60 Ans : b

80. Percentage of feldspar in igneous rocks is
a) 4 b) 12
c) 18 d) 60 **Ans : d**

81. Percentage of mica in igneous rocks is
a) 4 b) 12
c) 18 d) 60 **Ans : a**

82. The process of disintegration and decomposition of minerals within the soil is called as
a) Geochemical weathering b) Pedochemical weathering
c) Physical weathering d) Chemical weathering **Ans : b**

83. The process of weathering that takes place before soil development and continues beneath the soil cover is called as
a) Geochemical weathering b) Pedochemical weathering
c) Physical weathering d) Chemical weathering **Ans : a**

84. Of the following, which is an important chemical weathering process
a) Carbonation b) Oxidation
c) Hydrolysis d) Hydration **Ans : c**

85. Which is/are the chemical weathering processes
a) Carbonation b) Hydrolysis
d) Hydration d) All **Ans : d**

86. Which of the following processes is responsible for the formation of gypsum from calcium sulphate?
a) Hydration b) Carbonation
c) Oxidation d) Hydrolysis **Ans : a**

87. Which of the following processes is responsible for the formation of allophane from orthoclase feldspar?
a) Hydration b) Carbonation
c) Oxidation d) Hydrolysis **Ans : d**

88. Transformation of biotite to kaolinite takes place by the process of
a) Physical weathering b) Chemical weathering
c) Biological weathering d) Both b and c **Ans : d**

89. Weathered products of rocks and rock minerals are called as
a) Pedon b) Parent materials
c) Solum d) Polypedons **Ans : b**

90. Most easily weatherable mineral is

a) Gypsum b) Biotite
c) Olivine d) Calcite **Ans : c**

91. The secondary clay mineral kaolinite is mainly derived from

a) Igneous rocks b) Sedimentary rock
c) Metamorphic rocks d) All **Ans : b**

92. The secondary clay mineral illite is mainly derived from

a) Granite b) Basalt
c) Olivines d) Pyroxenes and amphiboles **Ans : a**

93. The secondary clay mineral montmorillonite is formed from

a) Basalt b) Gabbro
c) Phonolite d) All **Ans : d**

94. Materials transported and deposited by water are called as

a) Alluvium b) Colluvium
c) Loess d) Aeolian **Ans : a**

95. Materials transported by the action of gravity are called as

a) Alluvium b) Lacustrine
c) Glacial drift d) Colluvium **Ans : d**

96. Materials transported by water and settled out in lakes are called as

a) Alluvium b) Lacustrine
c) Glacial drift d) Colluvium **Ans : b**

97. Silt materials transported by wind are called as

a) Delta b) Meander
c) Loess d) Aeolian **Ans : c**

98. Sand materials transported by wind are called as

a) Delta b) Meander
c) Loess d) Aeolian **Ans : d**

99. Property of parent material, to a large extent, determines the depth of soil profile formation is

a) Texture b) Structure
c) Temperature d) Porosity **Ans : a**

100. Soils are deeper when the texture of the parent materials is

a) Light b) Heavy
c) Medium d) None **Ans : a**

101. The local name Kanhar means what?
a) Red soils b) Black soils
c) Yellow soils d) Brown soils **Ans : b**

102. The local name pokkali or cat clays means what?
a) Red soils b) Black soils
c) Laterite soils d) Alluvial soils **Ans : c**

103. Azonal soils are
a) Black soils b) Red soils
c) Laterite soils d) Alluvial soils **Ans : d**

104. Thornthwait's formula is used for the calculation of
a) Evaporation b) Transpiration
c) Evapotranspiration d) Potential evapotranspiration **Ans : d**

105. Microorganism involved in soil granulation
a) Earthworms b) Moles
c) Ants d) Fungi **Ans : a**

106. Immobilization and accumulation of materials from the upper horizons at a depth beneath the soil surface is called as
a) Illuviation b) Eluviation
c) Laterization d) Podzolization **Ans : a**

107. Which is the fundamental soil forming process?
a) Laterization b) Podzolization
c) Gypsification d) Humification **Ans : d**

108. Which is/are the fundamental soil forming processes?
a) Illuviation b) Eluviation
c) Humification d) All **Ans : d**

109. Formation of 1: 1 type clay minerals of kaolinitic group is a characteristic process of
a) Argillization b) Podzolization
c) Laterization d) All **Ans : c**

110. Cool and humid climatic conditions are favourable for
a) Podzolization b) Laterization
c) Humification d) Pedoturbation **Ans : a**

111. Silicious parent materials are favourable for
a) Podzolization b) Laterization
c) Humification d) Pedoturbation **Ans : a**

112. Warm and humid climatic conditions are favourable for

a) Podzolization	b) Laterization	
c) Humification	d) Pedoturbation	**Ans : b**

113. Leaching of soluble salts from soil either by irrigation water or rain water is called as

a) Alkalization	b) Salinization	
c) Desalinization	d) Solodization	**Ans : c**

114. Soil horizons A and B (A+B) are collectively called as

a) Pedon	b) Soil profile	
c) Soil regolith	d) Soil solum	**Ans : d**

115. Soil horizons A, B and C (A+B+C) are collectively called as

a) Pedon	b) Soil profile	
c) Soil regolith	d) Soil solum	**Ans : c**

116. Bed rocks are obviously absent in

a) Red soils	b) Black soils	
c) Laterite soils	d) Alluvial soils	**Ans : d**

117. Master horizons are

a) and C	b) O, A, E, B and C	
c) O, A, B and C	d) A, B and C	**Ans : b**

118. Soil horizon that exhibits obliteration of all or most of the original rock structure is

a) E horizon	b) A horizon	
c) B horizon	d) All	**Ans : a**

119. B horizon is a

a) Illuvial horizon	b) Elluvial horizon	
c) Organic horizon	d) Bed rock	**Ans : a**

120. Bleached or ashy appearance is a characteristic feature of

a) O horizon	b) A horizon	
c) E horizon	d) B horizon	**Ans : c**

121. Elluvial horizon is

a) A horizon	b) E horizon	
c) B horizon	d) O horizon	**Ans : b**

122. Prominent structure in black soils is

a) Subangular blocky	b) Angular blocky	
c) Prismatic	d) Columnar	**Ans : b**

123. Typical soils for the temperate humid regions are
a) Lateritic and laterite soils
b) Chernozems
c) Podzols
d) Chesnut soils
Ans : c

124. Soils of semi-arid and sub humid regions are
a) Lateritic and laterite soils
b) Chernozems
c) Podzols
d) Sierozem soils
Ans : b

125. Soils of subtropics and tropical regions are
a) Lateritic and laterite soils
b) Chernozems
c) Podzols
d) Tundra soils
Ans : a

126. Soils of semi-arid and arid regions are
a) Lateritic and laterite soils
b) Chernozems
c) Podzols
d) Desert soils
Ans : d

127. Soils associated with permanently cold climate in the arctic regions are
a) Podzols
b) Priarie soils
c) Tundra soils
d) Chesnut soils
Ans : c

128. Soils formed by physical weathering are
a) Podzols
b) Priarie soils
c) Tundra soils
d) Chesnut soils
Ans : c

129. Coniferous forest is necessary for the formation of
a) Lateritic and laterite soils
b) Podzols
c) Priarie soils
d) Brown earth soils
Ans : b

130. Deciduous forest vegetation is necessary for the formation of
a) Lateritic and laterite soils
b) Podzols
c) Priarie soils
d) Brown earth soils
Ans : d

131. The mixture that consists of well decomposed humus and minerals made by earthworms is called as
a) Mur
b) Mull
c) Peat
d) Muck
Ans : a

132. Bog soils are
a) Zonal soils
b) Intrazonal soils
c) Azonal soils
d) None
Ans : b

133. Which of the following is a fundamental unit in USDA soil classification?
a) Soil solum
b) Soil pedon
c) Soil profile
d) All
Ans : c

134. Gley soils and Planosols are

a) Zonal soils b) Intrazonal soils
c) Azonal soils d) None **Ans : b**

135. Podzolic soils and Sierozems are

a) Zonal soils b) Intrazonal soils
c) Azonal soils d) None **Ans : a**

136. Lithosols and regosols are

a) Zonal soils b) Intrazonal soils
c) Azonal soils d) None **Ans : c**

137. Which of the following soil classification is followed in India?

a) USSR soil classification b) USDA soil classification
c) Marbut's soil classification d) Coffey's soil classification **Ans : b**

138. The soil moisture regimes Udic means

a) Burnt b) Hot and dry
c) Humid d) Water **Ans : c**

139. The soil moisture regimes Udus means

a) Burnt b) Hot and dry
c) Humid d) Water **Ans : a**

140. Soils under the soil order Alfisols is/are

a) Red soils b) Laterite soils
c) Black soils d) Both a and b **Ans : d**

141. What is the depth of cracks formed in Vertisols?

a) Up to 25 cm b) Up to 50 cm
c) Up to 75 cm d) Up to 1 m **Ans : b**

142. Total geographical area of India is

a) 327 mha b) 328 mha
c) 329 mha d) 325 mha **Ans : b**

143. Fertile soils of India are locally called as

a) Urvara b) Anurvara
c) Usara d) Kari **Ans : a**

144. Barren soils of India are locally called as

a) Usara b) Anurvara
c) Urvara d) Kallar
e) Both a and b **Ans : e**

145. Sodic or alkali soils of India are commonly called as
a) Reh b) Kallar
c) Kari d) Matarsi Ans : b

146. Acid sulphate soils of India are locally called as
a) Kari soils b) Pokkali soils
c) Nullah regadi d) Both a and b Ans : d

147. Which is the soil order of shallow black soils?
a) Entisols b) Inceptisols
c) Vertisols d) Both a and b Ans : d

148. Which is the soil order of Tarai soil group?
a) Mollisols b) Histosols
c) Podzols d) Ultisols Ans : a

149. Which is the soil order of medium and deep black soils?
a) Entisols b) Inceptisols
c) Vertisols d) Both b and c Ans : c

150. Total area of black soils in India is
a) 72 mha b) 74 mha
c) 75 mha d) 70 mha Ans : b

151. Total area of red soils in India is
a) 72 mha b) 74 mha
c) 75 mha d) 70 mha Ans : a

152. Total area of laterite and lateritic soils in India is
a) 28 mha b) 25 mha
c) 32 mha d) 29 mha Ans : b

153. Total area of desert soils in India is
a) 28 mha b) 25 mha
c) 32 mha d) 29 mha Ans : d

154. Total area of alluvial soils in India is
a) 72 mha b) 74 mha
c) 75 mha d) 70 mha Ans : c

155. Honey-comb structure is diagnostic feature of
a) Red soils b) Laterite soils
c) Lateritic soils d) Both b and c Ans : c

156. The most important soils for agriculture are
a) Alluvial soils b) Black soils
c) Red soils d) Laterite and lateritic soils Ans : a

157. Newer or recent alluvium is locally called as

a) Bhangar b) Khadar
c) Bhata d) Matarsi **Ans : b**

158. Older alluvium is locally called as

a) Bhangar b) Khadar
c) Bhata d) Matarsi **Ans : a**

159. Pan formation or compact soil layer is a common feature of

a) Black soils b) Red soils
c) Laterite and lateritic soils d) Alluvial soils **Ans : d**

160. Percentage of organic matter in peaty and marshy soils is

a) 10 – 30 % b) 10 – 40 %
c) 10 – 50 % d) 20 – 60 % **Ans : d**

161. Agro-ecological zones in India are

a) 8 b) 15
c) 13 d) 21 **Ans : a**

162. Total area of salt affected soils in India is

a) 25 mha b) 28 mha
c) 29 mha d) 10 mha **Ans : b**

163. Total area of saline soils in India is

a) 10 mha b) 7.2 mha
c) 2.8 mha d) 6.1 mha **Ans : b**

164. Total area of sodic or alkali soils in India is

a) 10 mha b) 7.2 mha
c) 2.8 mha d) 6.1 mha **Ans : c**

165. Primary source of salts in salt affected soils is

a) Rock weathering b) Ground water fluctuation
c) Irrigation and fertilizers d) Sea inundation **Ans : a**

166. What is the limit of electrical conductivity for the categorization of saline soils?

a) 2 dS m^{-1} b) 4 dS m^{-1}
c) 6 dS m^{-1} d) 8 dS m^{-1} **Ans : b**

167. What is the pH of saline-sodic soils?

a) Less than 8.5 b) More than 8.5
c) 8.5 – 10.0 d) 8.5 **Ans : a**

168. What is the electrical conductivity of saline-sodic soils?
a) Less than 4 dS m^{-1} b) More than 4 dS m^{-1}
c) 4 dS m^{-1} d) None Ans : b

169. Most of the coastal soils are rich in
a) Nitrogen b) Phosphorous
c) Potassium d) Calcium and Magnesium Ans : c

170. High salt tolerant crops is/are
a) Barley b) Rice and sugarcane
c) Oats d) All Ans : d

171. Medium salt tolerant crops is/are
a) Cotton b) Sorghum and pearl millet
c) Wheat and maize d) All Ans : d

172. Low salt tolerant crops
a) Pulses b) Pea
c) Sesamum d) All Ans : d

173. A soil may be called as alkaline when the pH of the soil is more than
a) 6.5 b) 7.0
c) 7.5 d) 8.0 Ans : b

174. Laboratory method that is widely used for the calculation of gypsum requirement is
a) Schoonover's method b) Schofield's method
c) SMP method d) Puri's method Ans : a

175. What is the gypsum equivalent of iron pyrites?
a) 0.19 b) 0.63
c) 0.85 d) 1.62 Ans : b

176. What is the gypsum equivalent of sulphur?
a) 0.19 b) 0.63
c) 0.85 d) 1.62 Ans : a

177. What is the gypsum equivalent of $CaCl_2 . 2H_2O$?
a) 0.19 b) 0.63
c) 0.85 d) 1.62 Ans : c

178. Which is high sodium tolerant crop?
a) Rice b) Wheat
c) Barley d) Cotton Ans : a

179. Which is/are the medium sodium tolerant crops?
a) Wheat b) Barley
c) Cotton d) All Ans : d

180. Colour of the land capability class-II on LCC maps is
a) Yellow b) Pink
c) Green d) Red Ans : a

181. Colour of the land capability class-III on LCC maps is
a) Purple b) Brown
c) Dark grey d) Orange Ans : b

182. Colour of the land capability class-IV on LCC maps is
a) Yellow b) Pink
c) Green d) Red Ans : b

183. Colour of the land capability class-V on LCC maps is
a) Yellow b) Pink
c) Green d) Red Ans : c

184. Colour of the land capability class-VI on LCC maps is
a) Yellow b) Pink
c) Green d) Red Ans : d

185. Colour of the land capability class-VII on LCC maps is
a) Yellow b) Pink
c) Green d) Red Ans : d

186. Colour of the land capability class-VIII on LCC maps is
a) Purple b) Brown
c) Dark grey d) Orange Ans : a

187. What is the total number of land suitability classification for irrigation?
a) 4 b) 6
c) 8 d) 10 Ans : b

188. Land suitable for irrigation with nil or moderate limitations are
a) Class I and II b) Class I, II and III
c) Class I, II, III and IV d) All Ans : b

189. The size range of gravels in diameter is
a) 2-50 mm b) 2-75 mm
c) 2-100 mm d) 5-75 mm Ans : b

190. What is the total area of soil degraded by water erosion?
a) 148.9 mha b) 132.5 mha
c) 161.2 mha d) 171.9 mha **Ans : d**

191. Total area of soil degraded by wind erosion is
a) 13.5 mha b) 13.8 mha
c) 10.1 mha d) 11.6 mha **Ans : a**

192. Total area of soil degraded by physical deterioration is
a) 13.5 mha b) 13.8 mha
c) 10.1 mha d) 11.6 mha **Ans : d**

193. Total area of soil degraded by chemical deterioration is
a) 13.5 mha b) 13.8 mha
c) 10.1 mha d) 11.6 mha **Ans : b**

194. Total area of soil degraded by different degradation types is
a) 185.8 mha b) 187.8 mha
c) 180.8 mha d) 183.8 mha **Ans : b**

195. In lateritic soils, soil loss is primarily caused by
a) Sheet erosion b) Rill erosion
c) Splash erosion d) Gully erosion **Ans : b**

196. Contour bunding may be adopted for the soil water conservation on slopes ranging between
a) 10-12% b) 3-6%
c) 15% d) 6-12% **Ans : b**

197. Contour terracing or bench terracing may be adopted for the soil water conservation on slopes ranging between
a) 10-12% b) 6-33%
c) 12-22% d) 22-30% **Ans : b**

198. Contour bunds or graded bunds may be adopted for the soil water conservation on slopes ranging between
a) Upto 4% b) 3-6%
c) 10-12% d) 12-24% **Ans : a**

199. Total number of physiographic units in India is
a) 3 b) 5
c) 7 d) 9 **Ans : a**

200. Prominent soils in Peninsular regions of India are
a) Black soils b) Red soils
c) Alluvial soils d) Laterite and Lateritic soils **Ans : b**

201. Total Agro Ecological Zones (AEZ) in India given by NBSS and LUP is

a) 8 b) 15
c) 20 d) 21 **Ans : c**

202. What is the length of growing period in semi-arid regions?

a) 90-150 days b) 150-210 days
c) 210-270 days d) >270 days **Ans : a**

203. Total area of Alfisols in India is

a) 79.7 mha b) 95.8 mha
c) 80.1 mha d) 26.3 mha **Ans : a**

204. Total area of Vertisols in India is

a) 79.7 mha b) 95.8 mha
c) 80.1 mha d) 26.3 mha **Ans : d**

205. Total area of Entisols in India is

a) 79.7 mha b) 95.8 mha
c) 80.1 mha d) 26.3 mha **Ans : c**

206. Total area of Inceptisols in India is

a) 79.7 mha b) 95.8 mha
c) 80.1 mha d) 26.3 mha **Ans : b**

207. Total area of Aridisols in India is

a) 26.3 mha b) 14.6 mha
c) 8.0 mha d) 23.1 mha **Ans : b**

208. Total area of Mollisols in India is

a) 26.3 mha b) 14.6 mha
c) 8.0 mha d) 23.1 mha **Ans : c**

209. The dominant soil order in India is

a) Vertisols b) Inceptisols
c) Alfisols d) Entisols **Ans : b**

210. Hydrolysis and oxidation are examples of

a) Chemical weathering processes
b) Metamorphic processes
c) Lithification processes
d) Mechanical weathering processes **Ans : a**

211. The water sphere of the earth is known as the

a) Atmosphere b) Troposphere
c) Lithosphere d) Hydrosphere **Ans : d**

212. Particles ranging from 0.01 - 0.03 mm diameter would be considered as

a) Sand b) Silt
c) Clay d) Gravel Ans : b

213s. The metal ion associated with hemoglobin is

a) Copper b) Manganese
c) Iron d) Magnesium Ans : c

214. The only metal which exists as liquid at room temperature is

a) Silver b) Mercury
c) Tungsten d) Nicrome Ans : b

215. Which of the following fertilizers are not produced in India?

a) Nitrogen b) Phosphorus
c) Potassium d) Gypsum Ans : c

216. Which of the following gas contributes the least in green house effect?

a) CO_2 b) CFC
c) Methane d) Nitrous oxide Ans : d

217. Which of the following soil control measures is a mechanical measure?

a) Stubble cropping b) Green manuring
c) Strip cropping d) Basin listing Ans : d

218. Soil erosion is mostly affected by

a) Rainfall intensity b) Rainfall variability
c) Rainfall erosivity d) Total annual rainfall Ans : a

219. The main effect of raindrop erosion is

a) Transport of soil particles by fast flowing water
b) Detachment of soil particles thus destruction of soil structure
c) Beginning of gully erosion
d) Reducing soil moisture content Ans : b

220. The parent rock which gives rise to fertile heavy black clay soil is usually

a) Quartz b) Basalt
c) Limestone d) Granite Ans : b

221. Which of these will add calcium to the soil without changing the soil pH?

a) Lime b) Dolomite
c) Super phosphate d) Gypsum Ans : c

222. The correct order of soil classification is
a) Soil orders, Soil suborders, Soil families, Soil types
b) Soil orders, Soil suborders, Soil types, Soil families
c) Great soil groups, Soil families, Soil series, Soil types
d) Great soil groups, Soil types, Soil orders, Soil series
Ans : c

223. Tillage is most important to crop production because it
a) Destroy soil structure
b) Prepares a good seed bed
c) Improves soil colour
d) Increases the acidity of the soil
Ans : b

224. Micro organisms such as bacteria and fungi are useful in the soil because they
a) Are fed by plants
b) Remove carbon dioxide
c) Increase pore space
d) Decay organic matter
Ans : d

225. Nitrification
a) Does not involve bacterial activity
b) Is associated with ammonification
c) Is the application of nitrogen fertilizer
d) Includes the conversion of organic matter to nitrates
Ans : b

226. Legumes are planted in a pasture to
a) Remove soil minerals
b) Build up soil nitrate levels
c) Bring moisture to the surface
d) Provide large amounts of dry matter
Ans : b

227. The property that is not affected by the amount of organic matter in the soil
a) Cation exchange capacity
b) Soil texture
c) Biological activity
d) Soil structure
Ans : b

228. A green manure crop would
a) Lower water penetration
b) Decrease aeration
c) Increase nitrogen content
d) None of the above
Ans : c

229. The role of earthworms in soil could be described as
a) Causing plant diseases
b) Breaking down organic matter
c) Destroying soil structure
d) Raising little earthworms
Ans : b

230. The definition 'Soil is a dynamic natural body' was given by
a) Buckman and Brady
b) Hilgard
c) Dokuchaiev
d) Joffe (1949)
Ans : c

231. The size of the clay fraction is less than
a) 0.02mm
b) 0.002mm
c) 0.2mm
d) 0.0002mm
Ans : b

232. Clay minerals are called as __________ minerals.
a) Primary minerals
b) Secondary minerals
c) Tertiary minerals
d) Accessory minerals
Ans : b

233. What is the unit of electrical conductivity (a mean of salinity determination)?
a) dSm^{-1}
b) mmho cm^{-1}
c) cmho cm^{-1}
d) Sm^{-1}
e) Both a and b
Ans : e

234. Mineral formed by exfoliation, a type of physical weathering, is
a) Granite
b) Basalt
c) Gypsum
d) Calcite
Ans : a

235. When bases are more, the stability of minerals will
a) Decrease
b) Increase
c) Unaltered
d) None
Ans : b

236. The main products of weathering is
a) Silica
b) Alumina
c) Soluble cations
d) All
Ans : d

237. Secondary minerals that act as a primary minerals are
a) Biotite and muscovite
b) Muscovite and montmorillonite
c) Mica and biotite
d) Mica and muscovite
Ans : c

238. Which is the product of exhaustive hydrolysis and leaching?
a) Vermiculite
b) Montmorillonite
c) Illite
d) Kaolinite
Ans : d

239. Which is the dominant clay mineral in humid tropics?
a) Kaolinite
b) Montmorillonite
c) Vermiculite
d) Illite
Ans : a

240. Clay minerals predominate in the soils of arid regions are
a) Montmorillonite and vermiculite
b) Illite and montmorillonite
c) Illite and chlorite
d) Illite and vermiculite
Ans : c

241. Clay minerals present in humid regions with alkaline environment are
a) Montmorillonite and vermiculite
b) Illite and montmorillonite
c) Illite and chlorite
d) Illite and vermiculite
Ans : d

242. General formula for quartz is

a) $R_2(Si_2O_6)$ b) $RSiO_3$
c) R_2SiO_4 d) SiO_2 **Ans : d**

243. What is the general formula for amphiboles?

a) $R_2(Si_2O_6)$ b) $RSiO_3$
c) R_2SiO_4 d) SiO_2 **Ans : b**

244. Apatite is a __________ mineral.

a) Primary b) Secondary
c) Tertiary d) Accessory **Ans : b**

245. What is the percentage of SiO_2 in quartz?

a) 80 b) 90
c) 100 d) 95 **Ans : c**

246. Which is an amorphous clay mineral?

a) Kaolinite b) Smectite
c) Allophane d) Chlorite **Ans : c**

247. The most widely occurring 1:1 type clay minerals is

a) Kaolinite b) Halloysite
c) Serpentine d) Greenalite **Ans : a**

248. Crystal units of montmorillonite are held together by

a) Potassium b) Water
c) Sodium d) Brucite **Ans : b**

249. __________ is an example for expanding group of clay minerals.

a) Kaoline b) Smectite
c) Mica d) Chlorite **Ans : b**

250. __________ is an example for 2:1 type of non-expanding clay minerals.

a) Mica b) Smectite
c) Chlorine d) Kaoline **Ans : a**

251. __________ is an example for 2:1 type of non-expanding clay mineral.

a) Montmorillonite b) Vermiculite
c) Beidellite d) Nontronite **Ans : b**

252. Crystal units of mica group minerals are held together by

a) Water b) Potassium
c) Magnesium d) Both a and b
e) All the above **Ans : e**

253. Pedology is the study of:
a) Rocks b) Soils
c) Nutrient management d) Crop production Ans : b

254. Sedimentary rocks are formed by
a) Solidification of lava b) Increase in., pressure
c) Increase in temperature d) Cementation of broken parts Ans : d

255. Refers to the study of the origin of; soils in there natural environment, its classification and description.
a) Edaphology b) Pedology
c) Geology d) None, of the above Ans : b

256. ______________Refers the study of the soil properties in relation crop production-
a) Soil taxonomy b) Geology
c) Pedology d) Edaphology Ans : d

257. Refers to the upper most biochemically weathered portion of the regolith
a) Soil b) Earth
c) Rock d) None of the above Ans : a

258. The arrangement of soil particles (sand, silt and clay) within the soil is termed *as*-
a) Soil structure b) Soil texture
c) Soil colloids d) None of the above Ans : a

259. Alfisol soils *are* mostly ____________ in respect of their fertility and are used either as crop land Of forest or range land)
a) High b) Medium
c) Low d) Extremely low Ans : a

260. Potassium ammonium ions *are* apparently just the right size to lift in to the spaces between the crystal units of –
a) Vermiculite b) Montmorillonite
c) Kaolinite d) None of the above Ans : a

261. The organic matter of soils is ordinarily obtained by multiplying the carbon content by -
a) 1.42 b) 1.72
c) 1.82 d) 1.92 Ans : b

262. Ammonia lost through volatilization in significant quantities from:

a) Alkaline soils b) Acidic soils
c) Saline soils d) None of these

Ans : a

263. The diameter of most erodible soils particle is

a) 0.2 mm b) 0.1 mm
c) 0.3 mm d) 0.4 mm

Ans : a

264. According to land' capability classification, the soils which are not suitable for crop cultivation belong to:

a) Class I b) Class III
c) Class IV d) Class VIII

Ans : d

265. Silts have particle size range between

a) 2.00 to 0.20 mm b) 0.02 to 0.002 mm
c) 0.20 to 0.02 mm d) 0.002 mm

Ans : b

266. The capacity of soils to be moulded, that is to change shape in response to, stress and peep that shape when stress is removed is termed as:

a) Plasticity b) Stickiness
c) Consistency d) Cementation

Ans : a

267. Occurrence of quartz, in high amounts makes a rock

a) Hard b) Soft
c) Alkaline d) Acidic

Ans : d

268. Silica content of an igneous rock is

a) 80% b) 5-10%
c) 40-50 % d) 60-7S%

Ans : d

269. Formation of soil is the result of

a) Action of wind b) Chemical processes
c) Bio-chemical processes d) Physical processes

Ans : c

270. Standard way of collecting a soil sample is to take

a) One sample from centre of the field
b) One sample each from every corner of the field and mix
c) Several samples form all over the field and mix
d) None of these

Ans : c

271. Soil samples are collected in

a) Plastic bags b) Metal containers
c) Gunny bags d) Wooden boxes

Ans : a

272. Diameter of clay particle is

a) Less than 0.002 mm b) O.OS-O.S mm
c) 0.S-1.0 mm d) 0.7 - 0.9 mm **Ans : a**

273. Adsobed Ca++, Mg++ or AF+ may encourage soil aggregate formation starting with a process called

a) Flocculation b) Exfoliation
c) Soil solarization d) None of above **Ans : a**

274. Most of the. plants are likely to suffer if the soil volume occupied by air is less than ________ percent

a) 20-25 b) 15-20
c) 12-15 d) 10-12 **Ans : a**

275. The colloidal surface area in the upper 15 cm of a hectare of a clay soil could be as high as Km.

a) 700 b) 7000
c) 70000 d) 700,000 **Ans : d**

276. Soil particles get deflocculated due to:

a) Calcium b) Sodium
c) Magnesium d) Sulphur **Ans : b**

277. Arrangement of soil particles is referred to as -

a) Soil structure b) Soil texture
c) Soil organization d) None of the above **Ans : a**

278. A pH value of 6.0 indicates that the soil reaction is

a) Acidic b) Alkaline
c) Neutral d) Highly alkaline **Ans : a**

279. Which of the biochemically withered constitutes soil?

a) Upper most portion b) Middle portion
c) Lower portion c) Portion adjacent to bedrock. **Ans : a**

280. Soil is biochemically-withered part of

a) Bedrock b) Patent material
c) Regolith d) Organic matter **Ans : c**

281. Soils are surveyed and classified on the basis of distinctive features of

a) Parent materials b) Soil horizons
c) Soil profiles d) Vegetation **Ans : c**

282. A vertical section cut downwards through the soil is called as

a) Oil profile b) Soil column
c) Soil horizon d) Soil sample **Ans : a**

283. Horizontal layers in the soil profile are known as

a) Soil columns b) Horizontal zones
c) Soil horizon d) Rhizospheres **Ans : b**

284. Depth of cultivated soils is upto

a) 5 cm b) 5 cm
c) 30 cm d) 45 cm **Ans : b**

285. First 15 cm layer of land which is ploughed for cultivation of crop is called

a) Furrlow slice b) Rhizosphere soil
c) Soil d) Soil profile **Ans : a**

286. The ideal soil type for growing groundnut is black clay loam.

a) True b) False
c) Don't know **Ans : a**

287. Rice can be grown under very wide range of soil acidity and alkalinity with soil pH ranging from

a) 4.5 to 7.5 b) 5.5 to 8.5
c) 4.5 to 8.5 d) 6.5 to 9.5 **Ans : b**

288. During submergence in rice fields, iron availability is

a) Reduced b) Increased
c) Not affected d) All **Ans : a**

289. As the clay content in the soil increases, the rate of infiltration decreases.

a) True b) False
c) Don't know **Ans : a**

290. There is a positive relationship between texture and pore size.

a) True b) False
c) Don't know **Ans : b**

291. Crusting is a problem in red soils,

a) True b) False
c) Don't know **Ans : a**

292. Bulk density of sandy soils is more than clay soil.

a) True b) False
c) Don't know **Ans : a**

293. Bulk density of a soil is always higher than particle density
a) True b) False
c) Don't know
Ans : b

294. The soil pH range for successful cotton cultivation is
a) 3.5 to 5.5 b) 8.5 to 9.0
c) 5.5 to 8.5 d) 4.0 to 5.0
Ans : c

295. Nicotine content in tobacco is directly related with which nutrient content in the soil.
a) Nitrogen b) Phosphorus
c) Potassium d) Sulphur
Ans : a

296. Macro pores of the soil is filled with
a) Water b) Air
c) Water and air both d) Solution of nutrient
Ans : b

297. Texture of the soil refers to the relative proportions grouped according to their size or soil separates. Therefore, the particle size of 0.11 to 0.25 mm falls in soil separate group of
a) Clay b) Sand
c) Fine sand d) Silt
Ans : c

298. Indiscriminate cultivation of wet soil
a) Improve soil structure
b) Spoils soil structure
c) Makes no effect on soil structure
d) None
Ans : b

299. The diameter of the silt particle is
a) Less than 0.002 mm
b) Greater then 2.00 mm
c) In the range of 0.02 to 2.00 mm
d) In the range of 0.002 to 0.02 mm
Ans : d

300. If micro and macro pores in the soils are in the ratio of 1:1, it helps in obtaining
a) Retention of water b) Good drainage
c) Good aeration d) All of them
Ans : d

301. Dispersion of finely divided solid particles is known as ______
a) Solution b) Suspension
c) Emulsion d) None
Ans : b

302. The bulk density of sandy soil is always __________ than clay soil.
a) Equal b) Higher
c) Less d) None
Ans : b

303. The arrangement of soil particles in groups or aggregates is termed as __________.

a) Structure b) Texture
c) Both d) None Ans : a

304. The soil _______ refers to the relative proportion of the various size groups of mineral particles in a given soil.

a) Structure b) Texture
c) Both d) None Ans : b

305. At initial stages of reclamation of sodic soils, the cropping sequence which is more remunerative is ....

a) Rice-wheat-dhaincha b) Rice-sunflower-green gram
c) Rice-cotton-soybean d) Rapeseed-wheat-djhaincha Ans : a

306. In acidic soils the plant growth is affected due to ....ions

a) Fe and Cl b) Al and Cu
c) Fe and Al d) Cl and Cu Ans : c

307. Plants tolerant to high soil salinity conditions

a) Heliophytes b) Hylophytes
c) Heliotropic d) None of these Ans : d

308. Saline soils occur mostly in

a) Utter Pradesh b) Haryana
c) Punjab d) Madhya Pradesh Ans : b

309. Addition of the following material makes it possible to take good crop in Sadic soil

a) FYM b) Green manure
c) Gypsum d) All of these Ans : c

310. pH value of 7.0 denotes that the soil reaction is

a) Acidic b) Alkaline
c) Neutral d) Highly alkaline Ans : c

311. The pore______ is important than porosity to determine the function of pore in soil water relationships.

a) Size b) Shape
c) Number d) None Ans : a

312. The hardening of surface soil is technically termed as __________.

a) Compaction b) Hard pan
c) Pulverization d) None Ans : a

313. Tillage refers to the physical condition of the soil in relation to plant growth.
a) True b) False
c) Don't know Ans : a

314. Breaking of clods with blade harrow is known as secondary tillage operation.
a) True b) False
c) Don't know Ans : a

315. Blind hoeing is possible in
a) Rice b) Bengal gram
c) Sugarcane d) Potato Ans : c

316. Inter-cultivation operation are meant for
a) Rouging b) Destroying weeds
c) Thinning d) Maintaining plant stand Ans : b

317. Under ideal tillage conditions the micro and macro pores should be in the ratio of
a) 1: 1 b) 2:1
c) 3:2 d) 1:2 Ans : a

318. Deep ploughing is recommended
a) Every season b) Once in a year
c) Once in two to three years d) Once in 5 to 6 years Ans : c

319. Deep ploughing is done to
a) Improve the moisture retention
b) Improve soil structure
c) Improve soil texture
d) Bring the nutrients to upper layers Ans : b

320. Frequent harrowings results in............... of soil structure
a) Improves b) Destruction
c) No effect d) None Ans : b

321. The depth of penetration of M. B Plough is
a) 10 cm b) 25 cm
c) 40 cm d) 50 cm Ans : d

322. For taking up tillage operation. we should consider
a) The texture of the soil b) Structure of the soil
c) Type of soil d) Moisture content of the soil Ans : d

323. The main objective of intercutivation is to
a) Conserve soil moisture
b) Increase aeration
c) Control weeds
d) Increase water holding capacity

Ans : c

324. Tillage operations done in between crop rows is known as
a) Harvesting
b) Weeding
c) Preporatory cultivation
d) Inter-cultivation

Ans : d

325. The preliminary tillage under wet land condition for rice is aimed at
a) Just suspending the soil particles in water only
b) Suspending the soil particles in water and providing permeable conditions for downward movement of water
c) Suspending the soil particles in water and providing impermeable conditions for the downward movement of water
d) Creating hard pan in the sub soil

Ans : c

326. Loose, roughly powdery, granularand crumbly condition of soil, favourable for crop production is termed as
a) Tillage
b) Tilth
c) Zero tillage
d) Minimum tillage

Ans : b

327. If the tilth is a condition attainted by the soil by the use of tillage implements then the
a) Non-capillary pores should be large
b) Capillary pores should be less
c) Capillary and non-capillary pores should be fairly large pores.
d) Capillary and non-capillary pores should be fairly large and in equal proportions

Ans : d

328. The process of mechanical manipulation of soil to make it suitable for crop production is termed as
a) Tillage
b) Tilth
c) Preparatory tillage
d) Primary tillage

Ans : a

329. The tillage operations done in field from the time of harvest of the crop to the sowing of the next one, go by the name of
a) Primary tillage
b) Secondary tillage Post-sowing tillage
c) pre-sowing tillage
d) All

Ans : d

330. The stage at which, the soil is too moist, the particles do not fall apart and the, mass of soil gets flattened, then the ploughing
a) Can be done with least expenditure of energy
b) Cannot be done as the expenditure of energy is more
c) Can be done only when some moisture gets evaporated from soil
d) None of these **Ans : c**

331. Gram for good germination requires a:
a) Very fine and firm seed bed
b) Very fine powdery but loose seed bed
c) Firm and cloddy seed bed
d) Cloddy and rough seed bed **Ans : d**

332. Ridges are formed for
a) Broadcast crop b) Closer spaced crop
c) Wider spaced crop d) None **Ans : c**

333. Chlorite is a ____________ type clay mineral.
a) 2 : 1 : 1 b) 2 : 1
c) 1 : 1 d) Both a and b **Ans : a**

334. Potassium fixation is more in
a) Vermiculite b) Illite
c) Montmorillonite d) Chlorite **Ans : b**

335. The last layer in the soil profile is the
a) A Horizon b) B Horizon
c) C Horizon d) R Horizon **Ans : d**

336. Garnet is a
a) Igneous rock b) Sedimentary rock
c) Metamorphic rock d) None **Ans : c**

337. The soil orders which have highest suborder (7) are
a) Mollisols and Andosols b) Mollisols and vertisols
c) Vertisols and Andosols d) Inceptisols and Alfisols **Ans : a**

338. What is the C: N ratio of agric horizon?
a) $< 4:1$ b) $< 6:1$
c) $< 8:1$ d) $< 10:1$ **Ans : c**

339. Land capability classification was first given by
a) Schokalskaya (1932)
b) Govindarajan (1971)
c) Krishnamoorthy *et al.*
d) Klingebiel and Montgomery (1961) **Ans : d**

340. The dominant soil order in India is
a) Entisols b) Alfisols
c) Inceptisols d) Vertisols **Ans : c**

341. An example of physical weathering is
a) Minerals dissolved by water b) Oxidation
c) Frost wedging d) Carbonation **Ans : c**

342. Pedology refers to
a) Study of soils as natural bodies in the landscape
b) Using soils to grow plants
c) Adaptation of plants to different soil types
d) All the above **Ans : a**

343. Which process of horizon differentiation is illustrated by illuviation?
a) Translocation b) Transformation
c) Pedoturbation d) None **Ans : a**

344. Which process of horizon differentiation is illustrated by formation of clay minerals from primary minerals?
a) Translocation b) Transformation
c) Pedoturbation d) None **Ans : b**

345. Sand grains consist primarily of
a) Quartz b) Iron oxides
c) Gneiss d) Al oxides **Ans : a**

346. Rocks formed from pre-existing rocks by heat and pressure is
a) Igneous b) Metamorphic
c) Sedimentary d) None **Ans : b**

347. A deposit formed where a river flows into a larger body of water is
a) Colluvium b) Aeolian
c) A delta d) Loess **Ans : c**

348. Decomposition of plant residue by soil microorganisms may
a) Improve soil texture b) Improve soil structure
c) Both of the above d) None **Ans : c**

349. Soil texture could be changed in a flower bed by
a) Adding sand, silt or clay b) Adding peat
c) Either of the above d) None **Ans : a**

350. The surface area of the particles in a given volume is greater for

a) Sand b) Silt
c) Clay d) None **Ans : c**

351. Which textural class has the capacity to hold the greatest amount of water and nutrients?

a) Sandy loam b) Clay
c) Sandy clay loam d) Silt **Ans : b**

352. The natural aggregation of soil separates into peds is called as

a) Soil textural class b) Soil separates
c) Soil structure d) Soil consistency **Ans : c**

353. Soil permeability refers to

a) Stickiness and plasticity of soil
b) Total pore space
c) Movement of air and water through soil
d) None **Ans : c**

354. Soil consistency refers to

a) The degree of plasticity and stickiness of a soil
b) Movement of air and water through soil
c) The occurrence of similar soil profiles in similar landscape positions
d) Total pore space **Ans : a**

355. Incorporating crop residues into the soil

a) Improves soil structure b) Improves soil texture
c) Improves both of the above d) None **Ans : a**

356. Soil permeability is affected mainly by

a) Soil texture b) Soil structure
c) Soil texture and structure d) Soil colour **Ans : c**

357. Bulk density is

a) Weight of oven dry soil/Volume
b) Volume/ Weight of oven dry soil
c) Weight of oven dry soil X Volume
d) None of the above **Ans : a**

358. If 100 cm^3 of soil has an oven dry weight of 135 grams, what is its bulk density?

a) 13,500 g/cm^3 b) 0.75 cm^3/g
c) 1.35 g/cm^3 d) 1.53 g/cm^3 **Ans : c**

359. Soil bulk density is greatest for

a) Sand b) Clay
c) Peat d) Muck Ans : a

360. Soil compaction results in (an) _______________ in soil bulk density.

a) Increase b) Decrease
c) No change d) None Ans : a

361. Soil colour is used as an indicator of

a) Soil texture b) Structure
c) Consistence d) Drainage Ans : d

362. Sandy loam is an example of

a) Soil texture b) Soil structure
c) Soil textural class d) Soil Ans : c

363. An example of a soil separate is

a) Sandy b) Sand
c) Loam d) Silt loam Ans : b

364. The volume used to calculate soil bulk density includes

a) The solid material and the pore space
b) Density of the solid particles only
c) Only the pore space
d) Moisture percentage Ans : a

365. Particle density is the

a) Solid material and the pore space
b) Density of the solid particles only
c) Total pore space
d) None of the above Ans : b

366. If a 200 cm^3 soil sample is saturated (all the pore space is filled with water) contains 80 grams of water, what is the percent pore space?

a) 60% b) 40%
c) 30% d) 50% Ans : b

367. When primary tillage is completely avoided and secondary tillage is restricted to seedbed preparation in the row zone only it is ....

a) Minimum tillage b) Zero tillage
c) Row zone tillage d) Advance Ans : b

368. The primary objective of conservation tillage is ____
   a) Removal of plant residues
   b) Partial retention of plant residues
   c) Complete retention of plant residues
   d) None of the above — Ans : b

369. In no tillage systems, the surface soil layer have
   a) High infiltration of water   b) High bulk density
   c) Low bulk density   d) None of the these — Ans : b

370. Conservation tillage tends to encourage
   a) Higher microbial population
   b) Low microbial population
   c) Reduced number of earthworms
   d) None of these — Ans : a

371. Zero tillage system was first used successfully in 1950 in pasture renovation
   a) USA   b) India
   c) Japan   d) France — Ans : a

372. Which of the following developments has helped in the success of "Zero tillage" ______
   a) Insecticides   b) Herbicides
   c) Fungicides   d) Soil amendments — Ans : b

373. Radioactive tracer technique is used for determining
   a) Translocation of minerals in plants
   b) Cations exchange capacity of soil
   c) Both these
   d) None of these — Ans : a

374. Montmorillonite Illite and Kaolinite are
   a) Kinds of soil microbes
   b) Clay minerals
   c) Kings of protozoa present in soil
   d) Kinds of soils in U.S.A. — Ans : b

375. The arrangement of soil particle is called ___________ and the percentage of sand, silt and clay is called ___________
   a) Structure and texture   b) Texture and structure
   c) Aggregate and particles   d) None — Ans : a

376. The sandy soil has more ___________ pores than ___________pores
   a) Macro and micro   b) Micro and macro
   c) Macro and water filled   d) None — Ans : a

377. .... is the most tolerant crop in respect of alkalinity and salts

a) Alfalfa b) Barely
c) Cotton d) Rice

378. In acidic soils the plant growth is affected due to ....ions

a) Fe and Cl b) Al and Cu
c) Fe and Al d) Cl and Cu

379. Plants tolerant to high soil salinity conditions

a) Heliophytes b) Hylophytes
c) Heliotropic d) None of these **Ans : a**

380. Saline soils occur mostly in

a) Utter Pradesh b) Haryana
c) Punjab d) Madhya Pradesh **Ans : b**

381. Addition of the following material makes it possible to take good crop in Sadic soil

a) FYM b) Green manure
c) Gypsum d) All of these **Ans : c**

382. Hydrogen sulphide is formed due to sulphate reduction and anaerobic decomposition of organic matter in

a) Flooded soils b) Saturated soils
c) Well drained soils d) None **Ans : a**

383. Akiochi soils are formed due to formation of

a) Hydrogen sulphide b) Sulphuric acid
c) Hydrochloric acid d) None **Ans : a**

384. The flooded soils where hydrogen sulphide is formed due to sulphate reduction and anaerobic decomposition of organic matter are called

a) Akiochi soils b) Acid sulphate soils
c) Hydrogen sulphide soils d) None **Ans : a**

385. A soil with a pH reaction of less than 7.0 is called

a) Acidic soil b) Basic soil
c) Neutral soil d) None **Ans : a**

386. An acid soil has a preponderance of

a) Hydrogen ions over hydroxyl ions
b) Hydroxyl ions over hydrogen ions
c) Both
d) None **Ans : a**

387. Blue litmus paper turns red in contact with moist.

a) Acidic soil b) Basic soil
c) Neutral soil d) None **Ans : a**

388. Very acid soil (pH<4) in which sulphuric acid is formed by the oxidation of S-bearing, ferrous pyrites minerals is called

a) Acid sulphate soil b) Strongly acidic soil
c) Acidic soil d) None **Ans : a**

389. Acid sulphate soil are found primarily in

a) Costal, deltaic and estuarine areas of the humid tropics
b) Clay soils of humid tropics
c) Hilly areas of semi arid tropics
d) Wet temperate regions **Ans : a**

390. Acid sulphate soil are also called as

a) Cat clay soils b) Rat clay soils
c) Sat clay soils d) None **Ans : a**

391. The activity of hydrogen ions in the aqueous phase of soil is called

a) Reserve acidity b) Active acidity
c) Passive acidity d) None **Ans : b**

392. .... activity is measured and expressed as pH value.

a) Reserve acidity b) Active acidity
c) Passive acidity d) None **Ans : b**

393. The amount of potassium hydroxide required in milligram to neutralize the acetic acid liberated by the hydrolysis of one g of the acetylated substance is called as

a) Acetyl value b) Acetylene value
c) Neutralizing value d) None **Ans : a**

394. Soluble salt content of soil is determined by measuring its

a) Electrical conductivity
b) pH value
c) Organic matter content carbon
d) None of these **Ans : a**

395. Degree of soil salinity is indicated by its

a) Total soluble salt content b) Organic matter content
c) Nitrogen content d) Phosphorus content **Ans : a**

396. Unit of measuring electrical conductivity is

a) Micro - gram/sq.cm. b) Angstron/cm.
c) Millimhos/cm d) None of these **Ans : c**

397. The negative logarithm of H+ ion concentration (activity) in the soil solution

a) Acidity b) Salinity
c) Alkalinity d) pH **Ans : d**

398. On the basis of pH measurements, the nearly neutral

a) 6.0 to 6.5 b) 6.6 to 7.5
c) 7.5 to 8.5 d) 5.0 to 6.0 **Ans : b**

399. pH value of 7.0 denotes that the soil reaction is

a) Acidic b) Alkaline
c) Neutral d) Highly alkaline **Ans : c**

❑❑❑

4 Chapter

Soil & Water Conservation

Multiple Choice Questions (MCQ's)

1. The most serious sheet erosion occurs in
   a) Red and Black soils
   b) Red and Alluvial soils
   c) Black and Alluvial soils
   d) Red and laterite soils
   Ans : a

2. Which of the following mechanical soil conservation measures is more suitable for fruit trees or other plantation crops on steep slopes?
   a) Contour bunds
   b) Half moon terraces
   c) Graded bunds
   d) Bench terraces
   Ans : b

3. Land degradation first starts with the
   a) Reduction in vegetative cover
   b) Exposing the soil surface to accelerated erosion
   c) Reduction in soil organic matter and nutrient content
   d) All the above
   Ans : d

4. The primary objectives of tillage is
   a) Seedbed preparation
   b) Provision of a good medium for plant roots
   c) Water infiltration and retention
   d) Erosion and weed control
   Ans : a

5. Conservation tillage can be defined as
   a) A crop planting system that allows minimum disturbance of the soil to allow seeds to be sown while ensuring maintenance of crop residues on the surface
   b) A tillage operation is mainly done to prevent the weed population
   c) A tillage operation is mainly done to prepare a fine seed bed
   d) All the above **Ans : a**

6. What is important feature of soil erosion?
   a) Erosion removes topsoil
   b) Reduces levels of soil organic matter
   c) It contributes to the breakdown of soil structure
   d) All the above **Ans : d**

7. Long-term soil erosion results in
   a) Persistent and large gullies
   b) Exposure of lighter colored subsoil at the surface
   c) Poorer plant growth
   d) All the above **Ans : d**

8. Wind erosion is calculated by
   a) Ramser's formula b) Chepil equation
   c) Chepil and Woodruff equation d) All **Ans : c**

9. In north-eastern hill regions of India, shifting cultivation is also called as
   a) Jhum cultivation b) Jhuming cultivation
   c) Both a and b d) None **Ans : c**

10. Total number of land capability classes in the land capability classification scheme
    a) Six b) Seven
    c) Eight d) Nine **Ans : c**

11. Total number of classes in the group, land suitable for cultivation is
    a) Eight b) Four
    c) Six d) Two **Ans : b**

12. In land capability classification scheme, the classes V, VI and VII are suitable for
    a) Cultivation
    b) Pasture and grazing
    c) Wildlife and water shed management
    d) All **Ans : b**

13. Rills with more than ______ cm depth are generally called as gullies.
    a) 15 b) 20
    c) 25 d) 30 **Ans : d**

14. Of the following, which one is important with regards to soil erosion by water?
    a) Intensity of rainfall b) Duration of rainfall
    c) Amount of rainfall d) Frequency of rainfall **Ans : a**

15. If the velocity of water is doubled, its erosion power increases _____________ times, and carrying capacity increases _____________ times.
    a) 2 and 16 b) 4 and 16
    c) 4 and 32 d) 4 and 64 **Ans : d**

16. An agronomist is primarily concerned with what use of soil?
    a) Construction
    b) Acting as a filter for the hydrologic cycle
    c) Supporting crop growth
    d) Natural beauty **Ans : c**

17. What is no-till crop production?
    a) Preparing a good seedbed for crops
    b) Ploughing ground, but not discing
    c) Planting the seeds in the previous year's crop residue
    d) Discing ground prior to planting **Ans : c**

18. CSWCRTI was established in which year?
    a) 1974 b) 1949
    c) 1958 d) 1985 **Ans : a**

19. First recorded soil conservation research in India started in which year?
    a) 1923 b) 1935
    c) 1958 d) 1974 **Ans : a**

20. First soil conservation scheme called Dry farming Scheme started at which place in India?
    a) Manjiri near Pune b) Dehradun
    c) Barapani at Shillong d) None of the above **Ans : a**

21. Which of the following is a type of terrace?
    a) Diversion b) Retention
    c) Bench d) All the above **Ans : d**

22. Soil erosion is a
    a) Three-stage process b) Two stage process
    c) Five stage process d) None of the above **Ans : a**

23. The uniform removal of thin layers of soil from a more or less smooth slope, carried by the distributed (rather than concentrated) flow of runoff water over the soil surface is called
    a) Sheet erosion b) Gully erosion
    c) Rill erosion d) None of the above **Ans : a**

24. Splash and sheet erosion are also known as
    a) S inter-rill erosion b) Gully erosion
    c) Both a and b d) None of the above **Ans : a**

25. The scouring and transport of soil by a concentrated flow of water is called
    a) Rill erosion b) Sheet erosion
    c) Both a and b d) None of the above **Ans : a**

26. Saltation occurs when particle size is
    a) 0.1–0.5 mm b) 0.5-1mm
    c) 2-5mm d) 10-20mm **Ans : a**

27. Different types of models includes
    a) Mathematical models b) Statistical models
    c) Simulation models d) All the above **Ans : d**

28. The erosive power of rainfall, running water, or wind is called
    a) Erosivity b) Erodibility
    c) Susceptibility d) Run-off **Ans : a**

29. The erosive power of rainfall depends upon
    a) Rainfall intensity b) Duration
    c) Both d) None **Ans : c**

30. The erosive power in the case of running water depends on
    a) The velocity b) Turbulence
    c) Both a and b d) None of the above **Ans : c**

31. An equation relating the height of the water-table between drains to depth of the drains and the distance between them, the hydraulic conductivity of the soil, and the rate of recharge by percolating water is called
    a) Hooghoudt equation b) Darcy equation
    c) Renould's equation d) None of the above **Ans : a**

32. The volume of water that runs off during an irrigation that exceeds the soil's infiltrability is called
    a) Run-off b) Erosion
    c) Splash erosion d) Sheet erosion
    Ans : a

33. A section of the land surface that drains its surface-water excess to a specified point along a stream is called
    a) Watershed b) Catchment
    c) Both a and b d) None of the above
    Ans : c

34. Vegetative windbreaks are also called as
    a) Shelterbelts b) Tree belts
    c) Watershed d) None of the above
    Ans : a

35. Water erosion is a
    a) 2 way process b) 1 way process
    c) 3 way process d) None of the above
    Ans : c

36. Which of the following soil properties influences erodability?
    a) Infiltration capacity b) Structural stability
    c) Both a and b d) None of the above
    Ans : c

37. Agroforestry interventions with adequate soil and moisture conservation methods can reduce the soil loss by
    a) 30% b) 40%
    c) 50% d) 70%
    Ans : d

38. Which of the following erosion can be easily removed by normal tillage operations?
    a) Rill erosion b) Gully erosion
    c) Sheet erosion d) Splash erosion
    Ans : a

39. Most common size of raindrop is
    a) 2mm b) 3mm
    c) 4mm d) 7mm
    Ans : d

40. The maximum impact of the raindrop will be at what angle when it hits the ground?
    a) 30° b) 45°
    c) 90° d) 180°
    Ans : c

41. Which of the following soil separates has more erodability?
    a) Sand b) Silt
    c) Clay d) Rocks
    Ans : b

42. Which of the following types of clay minerals will provide more stable aggregates?
    a) Nonexpanding type of clay minerals
    b) Expanding type of clay minerals
    c) Both a and b
    d) None of the above **Ans : a**

43. Which is the most important index of soil erodibility?
    a) Silica content b) Sesquioxide content
    c) Silica-sesquioxide ratio d) All the above **Ans : d**

44. The major role of inter-cultivation tillage is
    a) Improvement of soil texture b) Enhancement of soil aeration
    c) Weed control d) Improvement of soil structure **Ans : d**

45. According to "Universal soils loss equation- the factors on which ratio of erosion depends are:
    a) Rainfall, soil erodibility and slope length
    b) Rainfall, soil erodibility, slope length, slope gradient, crop management, erosion control practice
    c) Rainfall, wind velocity, crop management
    d) Soil erodibility, soil type, crop management. **Ans : b**

46. "Water harvesting" refers to
    a) Collection of runoff water into micro or macro reservoirs and recycling the store water for irrigation.
    b) Conservation of rain water in soil by preventing the runoff so that rain water is better utilized in dryland
    c) A technique of better harvesting the ground water to supplement the canal irrigation
    d) A method of preventing percolation loss of water by suitably applying moisture barrier substances like asphalt. **Ans : a**

47. Pure and free water has greater potential than soil water.
    a) True b) False
    c) Don't know **Ans : a**

48. Infiltration of soils with hard pan can be increased by
    a) Basin listing b) Mulching
    c) Sub soiling d) None **Ans : c**

49. Soil crusting reduces
    a) Permeability b) Infiltration
    c) Run-off d) All **Ans : b**

50. Poor water holding capacity of red soils is mainly due to
a) High infiltration b) Shallow departments
c) Coarse texture d) None **Ans : a**

51. The downward flow of water from the surface into the soil is called
a) Percolation b) Infiltration
c) Permeability d) Hydraulic conductivity **Ans : b**

52. Vertical down-ward movement of water within the soil is called
a) Infiltration b) Percolation
c) Seepage d) Capillary **Ans : b**

53. Clay content is higher in __________ soil
a) Red b) Desert
c) Black d) All **Ans : c**

54. Soil erosion can be avoided by
a) Maintaining a protective cover on the soil
b) Creating a barrier to the erosive agent
c) Modifying the landscape to control runoff amounts and rates
d) All the above **Ans : d**

55. Which type of erosion is the most highly visible?
a) Gully erosion b) Wind erosion
c) Ephemeral gullies d) Splash erosion **Ans : a**

56. The largest of these soil separates or particles is
a) Gravel b) Sand
c) Silt d) Clay **Ans : a**

57. Which of these is not a key ingredient of soil?
a) Organic matter b) Hemoglobin
c) Minerals d) Air **Ans : b**

58. What percentage of the average soil is organic matter?
a) 45% b) 5%
c) 25% d) 17% **Ans : b**

59. Erosion created by the activities of man and sometimes by animals is called as
a) Natural geological erosion b) Geologic erosion
c) Accelerated soil erosion d) All **Ans : c**

60. Scattering of detached soil particles by the impact of rain drops is called
a) Splash erosion b) Sheet erosion
c) Gully erosion d) Ravines Ans : a

61. Sheet erosion is observed on
a) Deeply sloping land b) Moderately sloping land
c) Slopy land d) Gently sloping land Ans : d

62. Muddy run-off from the field is an indication of
a) Splash erosion b) Sheet erosion
c) Rill erosion d) Ravines Ans : b

63. Formation of small channels a few centimeters deep al over the field along the watercourse is observed in
a) Rill erosion b) Slight gully erosion
c) Moderate gully erosion d) Ravines Ans : a

64. Erosion permitting crops is
a) Maize b) Wheat
c) Bajra d) All Ans : d

65. Erosion resisting crops is
a) Grasses b) Legumes
c) Cereals d) Both a and b Ans : d

66. Removal by rain of a very thin layer of soil from the entire surface of large area is called
a) Splash erosion b) Sheet erosion
c) Rill erosion d) Ravines Ans : b

67. The ratio of dispersion ratio to colloidal moisture equivalent ratio is
a) Erosion index b) Erosion ratio
c) Soil detachability d) Soil erosivity Ans : b

68. In Universal soil loss equation, A=RKLSCP, K denotes
a) Soil conservation practices b) Soil roughness
c) Soil erodability d) Rainfall factor Ans : c

69. In Universal soil loss equation, A=RKLSCP, P denotes
a) Soil conservation practices b) Soil roughness
c) Soil erodability d) Rainfall factor Ans : a

70. Growing of erosion resisting and erosion permitting crops on alternate strips of suitable width along the contour and across the slope is called as

a) Contour cropping b) Contour farming
c) Contour strip cropping d) Strip cropping **Ans : c**

71. Contour bunding is also called as

a) Level terraces b) Ridge-type terraces
c) Absorption type terraces d) All **Ans : d**

72. Soils more susceptible to water erosion are

a) Coarse textured soils b) Medium textured soils
c) Fine textured soils d) All **Ans : c**

73. Minimum wind speed required to initiate the movement of most erodable soil particles is

a) 10 km b) 12 km
c) 14 km d) 16 km **Ans : d**

74. Diameter of particles that is most susceptible to wind erosion is

a) 0.1 mm b) 0.2 mm
c) 0.02 mm d) 0.01 mm **Ans : a**

75. Dominant type of wind erosion in soil is

a) Saltation b) Surface creep
c) Suspension d) All **Ans : a**

76. Size range of soil particles in saltation type of wind erosion is

a) Less than 0.05 mm b) 0.5 - 1.0 mm
c) 0.5 - 2.0 mm d) 0.05 - 0.5 mm **Ans : d**

77. Size range of soil particles in suspension type of wind erosion is

a) Less than 0.05 mm b) 0.5 - 1.0 mm
c) 0.5 - 2.0 mm d) 0.05 - 0.5 mm **Ans : a**

78. Size range of soil particles in surface creep type of wind erosion is

a) Less than 0.05 mm b) 0.5 - 1.0 mm
c) 0.5 - 2.0 mm d) 0.05 - 0.5 mm **Ans : c**

79. In surface creep wind erosion, soil particles are moved primarily by the action of

a) Wind b) Saltation
c) Suspension d) All **Ans : b**

80. Most important factors that influence the erodability of soil by wind
a) Soil texture b) Soil structure
c) Soil water status d) All **Ans : d**

81. The crops which help to effectively check soil erosion and conserve soil moisture and nutrients are referred to as
a) Cover crops b) Buffer crops
c) Erosion crops d) Conservation crop **Ans : a**

82. Cover crops help to check soil erosion and conserve soil moisture and nutrients by
a) Checking movement of run-off water
b) Increasing absorption of rain water by soil
c) Preventing the rain and wind from beating the soil surface directly
d) All these ways **Ans : d**

83. Which of the following crops are most effective in protecting cultivated land from erosion
a) Legumes b) Millets
c) Cereals d) All these are equally effective **Ans : a**

84. The type of crop and cropping system that can be used to conserve soil against erosion varies from region to region depending on
a) Soil type b) Climate
c) Both these d) None of these **Ans : c**

85. Which of the following crops are known to provide effective cover against loss of soil nutrient and moisture
a) Cowpea, mung, Urad
b) Dhaincha, sunhemp, groundnut
c) Velvet bean, gaur, soyabean
d) All these **Ans : d**

86. Cowpea was found to provide best vegetative cover against soil erosion at
a) Dehra Dun b) Vasad
c) Rehmankhera d) All these places **Ans : d**

87. At Kota, the best vegetative cover was provided by
a) Velvet bean b) Cowpea
c) Mung bean d) Sunhemp **Ans : a**

88. At Kanpur, the best vegetative cover against soil and water loss was provided by
a) Velvet bean b) Cowpea
c) Mung d) Soyabean **Ans : c**

89. Water harvesting is defined as
    a) Application of water at the time of harvesting crops like colocasia, tapioca and yam
    b) Harvesting deep water from soils by pumping devices
    c) Unit of water applied to crops
    d) Collecting the excess run off from rain on the farm in ponds utilizing it later for agriculture

Ans : d

90. At Sholapur in Maharashtra the most effective check against soil and water loss was provided by
    a) North bean
    b) Horse gram
    c) Both these
    d) Cowpea

Ans : c

❑❑❑

5 Chapter

Mineral Nutrition & Soil Fertility

Multiple Choice Questions (MCQ's)

1. Available molybdenum in soils is extracted by
   a) Citric acid b) Ammonium acetate
   c) Hot water d) $CaCl_2$ **Ans : b**

2. What is the optimum pH for the availability of most of the plant nutrients?
   a) 5.0 – 6.0 b) 6.0 – 7.0
   c) 6.5 – 7.5 d) 6.0 – 8.0 **Ans : c**

3. In soil, the availability of phosphate ions to plants is considered to follow the order of
   a) $PO_4^{3-} > HPO_4^{2-} > H_2PO_4^-$ b) $H_2PO_4^- > HPO_4^{2-} > PO_4^{3-}$
   c) $H_2PO_4^- > HPO_4^{2-} > H_3PO_4$ d) $HPO_4^{2-} > H_2PO_4 > PO_4^{3-}$ **Ans : b**

4. Increase in soil water content and soil temperature, increases the availability of
   a) Nitrogen b) Phosphorous
   c) Potassium d) All **Ans : b**

5. The availability of calcium and magnesium in soil is low above the pH of
   a) 7.5 b) 8.5
   c) 9.5 d) 8.0 **Ans : b**

6. The availability of boron in soil is more in the pH range of
   a) 5.0-7.0 and > 8.5 b) 5.0 – 8.5
   c) 6.5 – 7.0 and > 8.5 d) 6.5 – 7.0 and > 9.0 **Ans : a**

7. The availability of molybdenum in soil is more in the pH of
   a) > 6.5 b) > 7.5
   c) > 8.5 d) > 8.0 **Ans : a**

8. The Law of Minimum was given by
   a) Juston von Liebig b) Spillman
   c) Mitcherlich d) Thomas Way **Ans : a**

9. The equation, log (A - y) = log A – Cx is called as
   a) Spillman's equation b) Mitcherlich's equation
   c) Liebig equation d) Gapon's equation **Ans : b**

10. The equation, $y = M (1-R^x)$ is called as
    a) Mitcherlich's equation b) Liebig equation
    c) Spillman's equation d) Gapon's equation **Ans : c**

11. Nutrient mobility concept was given by
    a) Cate and Nelson b) Arnon and Stout
    c) Larsen d) Bray **Ans : d**

12. Example for Bray's relatively immobile nutrients is
    a) Phosphorous b) Potassium
    c) Calcium d) All **Ans : d**

13. Examplefor Bray's mobile nutrients is
    a) Phosphorous b) NO_3 - nitrogen
    c) Potassium d) Calcium **Ans : c**

14. Nutrient diagnosis in soil by pot culture methods were first initiated by
    a) Bousingault b) Mitcherlich
    c) Neubauer d) Mehlich **Ans : b**

15. Nutrient diagnosis in soil by *Aspergillus niger* methods were first used by
    a) Bousingault b) Mitcherlich
    c) Neubauer d) Mehlich **Ans : d**

16. Nutrient diagnosis in soil by soil plaque method was first used by
    a) Bousingault b) Mitcherlich
    c) Sackett and Stewart d) Mehlich **Ans : c**

17. The range of concentration at which growth of plants is restricted in comparison with that of plant at higher nutrient level is called as
    a) Hidden hunger b) Critical nutrient level
    c) Limiting factor d) All **Ans : b**

18. In A - value [A/B = (1- y)/y], B denotes
    a) Available phosphorous
    b) Fertilizer phosphorous
    c) Phosphorous derived from soil
    d) Both b and c **Ans : b**

19. Common method(s) used for the determination of gypsum requirement of sodic soils
    a) Schofield's method b) Schoonover's method
    c) Sokonov's method d) All **Ans : b**

20. What is the value of low nutrient index (NI)?
    a) < 0.5 b) < 1.0
    c) < 1.5 d) < 2.0 **Ans : c**

21. What is the value of high nutrient index (NI)?
    a) > 4.5 b) > 3.5
    c) > 2.5 d) > 3.0 **Ans : c**

22. What is the percentage of nitrogen in cattle dung?
    a) 0.1 b) 0.2
    c) 0.3 d) 0.5 **Ans : c**

23. Highly hygroscopic fertilizer is
    a) Ammonium chloride b) Ammonium nitrate
    c) Urea d) CAN **Ans : b**

24. Urea is hydrolyzed by which of the following enzymes?
    a) Nitrogenase b) Urease
    c) Hydrogenase d) Both a and c **Ans : b**

25. The first unstable compound produced by the hydrolysis of urea is
    a) Ammonium carbamate b) Ammonium carbonate
    c) Ammonium d) Nitrate **Ans : a**

26. The ideal nitrogenous fertilizer suitable for foliar application is
    a) Ammonium nitrate b) Ammonium chloride
    c) Ammonium sulphate d) Urea **Ans : d**

27. ____________ is also called as Nitro-lime or Nitro-chalk.
a) Ammonium nitrate b) Calcium ammoinium nitrate
c) Ammonium sulphate d) Ammonium chloride
Ans : b

28. Nitrogen percentage in calcium ammonium nitrate (CAN) is ____________.
a) 20.6 b) 46.0
c) 25.0 d) 26.0
Ans : c

29. Fertilizer that supplies nitrogen in both ammonical and nitrate forms is ____________.
a) Calcium ammonium nitrate (CAN)
b) Urea
c) Ammonium sulphate
d) Ammonium chloride
Ans : a

30. ______________ is an example for explosive fertilizer.
a) Ammonium chloride b) Ammonium sulphate
c) Ammonium nitrate d) CAN
Ans : c

31. Fertilizer that is a by-product of soda ash manufacture is ________________.
a) Ammonium chloride b) Ammonium sulphate
c) Ammonium nitrate d) CAN
Ans : a

32. What is the formula of dicalcium phosphate?
a) $Ca (H_2PO_4)_2$ b) $2 CaHPO_4$
c) $Ca_2 HPO_4$ d) $Ca_3 (PO_4)_2$
Ans : b

33. Example for straight, water soluble phosphatic fertilizer is
a) Monocalcium phosphate b) Dicalcium phosphate
c) Tricalcium phosphate d) Diammonium phosphate
Ans : a

34. Example for citrate soluble phosphatic fertilizers is
a) Basic slag and pelophos b) Bone meal
c) Dicalcium phosphate d) All
Ans : d

35. What is the nitrogen use efficiency for crops other than rice?
a) 28 – 34 % b) 35 – 43 %
c) 40 – 60 % d) 42– 50 %
Ans : c

36. Which is a slow release nitrogenous fertilizer?
a) Oxamide b) N-serve
c) AM d) ST
Ans : a

37. Which of the following is a nitrification inhibitor?
    a) U-form b) IBDU
    c) SCU d) ST **Ans : d**

38. Which of the following is slow release nitrogenous fertilizers?
    a) Urea-form b) Isobutyledene diurea
    c) Sulphur coated urea d) All **Ans : d**

39. Which of the following is nitrification inhibitors?
    a) N-serve b) AM
    c) ST d) All **Ans : d**

40. Plant product used for the preparation of slow release fertilizer and also as nitrification inhibitors is
    a) Neem and karanji cake b) Neem and sal cake
    c) Karanji and sal cake d) Neem **Ans : d**

41. N-serve is also called as
    a) Sulphonylamide b) Nitrapyrin
    c) U-formaldehyde d) Both a and b **Ans : b**

42. Principal of soil acidity includes
    a) Active acidity b) Residual acidity
    c) Exchangeable acidity d) All **Ans : d**

43. Percent base saturation is also known as
    a) Acidity b) Acid saturation
    c) Non acid saturation d) None **Ans : c**

44. When ATP and ADP breakdown, the amount of energy released is
    a) 10 kcal/mol b) 12 kcal/mol
    c) 78 kcal/mol d) 786 kcal/mol **Ans : b**

45. Blossom end rot in tomato is caused by the deficiency of
    a) Ca b) Mg
    c) Mo d) B **Ans : a**

46. Bitter pit in apple is caused by the deficiency of
    a) Ca b) Mg
    c) Mo d) B **Ans : a**

47. The nutrient element that plays an important role in structural stability of proteins is
    a) P b) S
    c) Fe d) Mn **Ans : b**

48. Which of the following is a precursor of photorespiration in plant?
   a) Phosphoglycolate   b) Acetyl co-enzyme A
   c) Citric acid   d) Glycolate   **Ans : a**

49. What is the percentage of phosphorus use efficiency in soil?
   a) 10 – 30%   b) 20 – 30%
   c) 25 – 45 %   d) 20 – 25%   **Ans : a**

50. Lime requirement is calculated by
   a) Shoemaker *et al.*   b) Schoonover
   c) Chepil and Woodruff   d) White and Beckett   **Ans : a**

51. The law of minimum was given by Liebig in the year
   a) 1909   b) 1862
   c) 1961   d) 1940   **Ans : b**

52. Essentiality of chlorine for plants was given by
   a) Nicholas (1961)   b) Arnon and Stout
   c) Cate and Nelson   d) Broyer *et al.* (1954)   **Ans : d**

53. What is the ideal pH range for rice crop?
   a) 5.5-7.0   b) 5.0-6.5
   c) 6.5-8.0   d) 4-6.5   **Ans : d**

54. What is the ideal pH range for Maize, Cowpea and Groundnut?
   a) 5.5-7.0   b) 5.0-6.5
   c) 6.5-8.0   d) 4-6.5   **Ans : b**

55. What is the ideal pH range for soybean and peas?
   a) 5.5-7.0   b) 5.0-6.5
   c) 6.5-8.0   d) 4-6.5   **Ans : a**

56. What is the ideal pH range for wheat, barley, sugarcane, sunflower and sorghum?
   a) 6 – 7.5   b) 5.0 – 5.5
   c) 4 – 6.5   d) 6.5 – 8.0   **Ans : a**

57. What is the ideal pH range for sugar beet?
   a) 6 – 7.5   b) 5.0 – 5.5
   c) 4 – 6.5   d) 6.5 – 8.0   **Ans : d**

58. What is the ideal pH range for cotton and potato?
   a) 6 – 7.5   b) 5.0 – 5.5
   c) 4 – 6.5   d) 6.5 – 8.0   **Ans : a**

59. One Baule unit for nitrogen is

a) 250 lbs b) 241 lbs
c) 148 lbs d) 223 lbs **Ans : d**

60. One Baule unit for phosphorus is

a) 35 lbs b) 45 lbs
c) 48 lbs d) 38 lbs **Ans : b**

61. One Baule unit for potassium is

a) 66 lbs b) 76 lbs
c) 86 lbs d) 93 lbs **Ans : b**

62. Examples of natural chelates are

a) Citric acid and oxalic
b) EDTA and DTPA
c) Citric acid and oxalic and HEDTA
d) EDTA, DTPA and HEDTA **Ans : a**

63. Examples of artificial chelates are

a) Citric acid and oxalic
b) EDTA and DTPA
c) Citric acid and oxalic and HEDTA
d) EDTA, DTPA and HEDTA **Ans : d**

64. How many number of ATP molecules are required for the reduction of one mole of NO_3 to NH_3?

a) 12 b) 15
c) 18 d) 21 **Ans : a**

65. How many number of ATP molecules are required for assimilation of one mole NH_3?

a) 5 b) 10
c) 15 d) 18 **Ans : a**

66. The pH at which the concentration of both $H_2PO_4^-$ and HPO_4^{2-} becomes equal is

a) 7.2 b) 7.0
c) 6.5 d) 7.5 **Ans : a**

67. In biological methods of soil fertility evaluation, the *Aspergillus Niger* (Mulder) method is used to measure the availability of

a) Available K_2O b) Available P_2O_5
c) Cu and Mg d) N, P and K **Ans : c**

68. In biological methods of soil fertility evaluation, the pot culture (Mitcherlich) method is used to measure the availability of

a) N, P and K b) P, K, Ca and Micronutrients
c) Available P_2O_5 d) P and K **Ans : d**

69. Classification of nutrients based on their relative requirements was given by

a) Cate and Nelson b) Arnon
c) Nicholas d) Engelbert **Ans : b**

70. Classification of nutrients based on their biochemical behaviour and physiological function was given by

a) Cate and Nelson b) Emil Troug and Engel Bert
c) Nicholas d) Mengel and Kirk by **Ans : d**

71. Classification of nutrients based on their functions and content in plant tissues was given by

a) Cate and Nelson b) Emil Troug and Engel Bert
c) Nicholas d) Mengel and Kirk by **Ans : b**

72. Which country is top most in fertilizer consumption?

a) India b) China
c) USA d) All **Ans : b**

73. Fertilizer consumption in India is higher in

a) Irrigated areas b) Rainfed areas
c) Both a and b d) None **Ans : a**

74. Which of the following states has nutrient use of < 60 kg/ha?

a) Madhya Pradesh b) Karnataka
c) West Bengal d) Uttar Pradesh **Ans : a**

75. Which of the following states has nutrient use of < 100 kg/ha?

a) Madhya Pradesh b) Karnataka
c) West Bengal d) Uttar Pradesh **Ans : b**

76. Which of the following states has nutrient use of 100 – 150 kg/ha?

a) Madhya Pradesh b) Karnataka
c) West Bengal d) Gujarat **Ans : c**

77. Which of the following states has nutrient use of > 150 kg/ha?

a) Haryana b) Karnataka
c) West Bengal d) Punjab **Ans : d**

78. Low content of soil organic matter is due to

a) Continuous cropping b) Erosion
c) Both d) None **Ans : c**

79. Most important source of organic matter is
    a) FYM b) Compost
    c) Urban and industrial area d) All **Ans : d**

80. Fe chlorosis in plants is caused mainly due to
    a) High Mn/Fe ratio b) High pH
    c) Excess phosphate d) All the above
    e) None of the above **Ans : d**

81. High boron requiring crops is
    a) Cauliflower b) Radish
    c) Beet root d) All **Ans : d**

82. Low boron requiring crops are
    a) Wheat and Oats b) Corn and Barley
    c) Peas and Beans d) All **Ans : d**

83. Total phosphorus reserve in soils consists of
    a) Organic phosphorus b) Soluble phosphorus
    c) Adsorbed phosphorus d) Insoluble phosphorus
    e) All the above **Ans : e**

84. How many pounds of 5-10-10 fertilizer would be needed to supply 100 kg of N?
    a) 2000 b) 200
    c) 4444 d) 900 **Ans : c**

85. If you applied 500 kg of 10-20-20 fertilizer, how many pounds of nitrogen would be supplied?
    a) 500 b) 50
    c) 111 d) 250 **Ans : c**

86. The form of nitrogen that may volatilize from hog waste lagoons is
    a) Ammonia b) Nitrate
    c) Organic N d) Nitrite **Ans : a**

87. The rate of decomposition of organic residue depends on
    a) Environmental conditions b) The CN ratio of the material
    c) Neither of the above d) Both a and b **Ans : d**

88. Which organic residue has the greatest CN ratio?
    a) Pine straw b) Cow manure
    c) Red clover d) Rice straw **Ans : a**

89. Which plant residue would decompose more rapidly and release plant available N if incorporated into the soil?
a) Pine straw b) Oak leaves
c) Red clover d) All **Ans : c**

90. Fixation of N by organisms that live in the nodules on the roots of legumes is
a) Non-symbiotic nitrogen fixation
b) Symbiotic nitrogen fixation
c) Anaerobic nitrogen fixation
d) None **Ans : b**

91. Symbiotic nitrogen fixation can produce as much as
a) 100-200 lbs/ac/yr b) 10-20 lbs/ac/yr
c) 1-2 lbs/ac/yr d) 50-100lbs/ac/yr **Ans : a**

92. Denitrification occurs only if
a) Ammonium is present b) Nitrate is present
c) Ammonium nitrate is present d) None **Ans : b**

93. Denitrification occurs only when soil conditions are
a) Saturated
b) Well drained
c) Aeration does not affect denitrification
d) Submerged **Ans : a**

94. Conversion of ammonium to nitrate (Nitrification)
a) Requires aerobic soil conditions
b) Results in more acid soil conditions
c) Both of the above
d) Requires submerged conditions **Ans : c**

95. Nitrification requires
a) The presence of oxygen
b) The oxygen supply has no effect
c) The absence of oxygen
d) The presence of nitrate **Ans : a**

96. The main sources of the plant nutrients C, H. and O is
a) Slow release fertilizers b) Air and water
c) Phosphate fertilizers d) Organic matter **Ans : b**

97. The form of most of the nitrogen taken up by plants growing on well drained soil is
a) N_2 b) NO_3^-
c) NH_4^+ d) All of the above **Ans : b**

98. Nitrogen fixation refers to
   a) Reaction with Fe to form insoluble compounds
   b) Conversion of NH_4^+ to NO_3^-
   c) Conversion of N_2 to forms that plants can utilize
   d) Conversion of NO_3^- to NH_4^+ **Ans : c**

99. Soil phosphorus is more available for plant uptake at pH
   a) 4.5 b) 5.5-6.5
   c) Above 6.5 d) below 4.5 **Ans : b**

100. The secondary nutrient that strengthens plant cell walls is
   a) P b) Ca
   c) K d) S **Ans : b**

101. The source of nitrogen for manufacturing fertilizer and fixation by rhizobia is
   a) Air b) Water
   c) CO_2 d) All **Ans : a**

102. Phosphorus fixation is more of a problem on
   a) Soils high in Fe and Al b) Organic soils
   c) Soils with more SO_4^{2-} ions d) Both a and b **Ans : a**

103. Sulfur is taken up by plants as
   a) Elemental sulfur b) SO_4^{2-}
   c) SO_3 d) S^{2-} **Ans : b**

104. Nitrogen fixation by legumes is often increased by inoculating the seed with
   a) Mychorrizae b) Nematodes
   c) Rhizobia d) All **Ans : c**

105. Atmospheric pollutants contribute a significant amount of this nutrient for plant uptake.
   a) Ca b) Mg
   c) S d) P **Ans : c**

106. Potassium fixation refers to
   a) Atmospheric deposition of K^+
   b) Trapping of K^+ ions in the interlayer space of illite
   c) Reaction with iron
   d) Fixed by the microorganisms **Ans : b**

107. The amount of plant available nitrogen in the soil could be decreased by
   a) Ammonification b) Nitrification
   c) Denitrification d) Mineralization **Ans : c**

108. Most soil phosphorus is in the
   a) Organic form   b) Inorganic mineral form
   c) Both a and b   d) None   **Ans : b**

109. How many pounds of NH_4NO_3 (34% N) would be required to supply 150 kg of N?
   a) 150   b) 44
   c) 441   d) 241   **Ans : c**

110. Zn deficiency can be caused by which of the following conditions?
   a) Small amount of available Zn in the soil
   b) Planted varieties are susceptible to Zn deficiency (i.e., Zn-inefficient cultivars)
   c) High pH (close to 7 or alkaline under anaerobic conditions)
   d) All the above   **Ans : d**

111. Zn deficiency is the most widespread micronutrient disorder in rice. Its occurrence has increased due to which of the following?
   a) The introduction of modern varieties
   b) Crop intensification
   c) Increased Zn removal
   d) All the above   **Ans : d**

112. Which of the following soils are prone to Zn deficiency?
   a) Neutral and calcareous soils containing a large amount of bicarbonate.
   b) Intensively cropped soils where large amounts of N, P, and K fertilizers
   c) Paddy soils under prolonged inundation
   d) All the above   **Ans : d**

113. Which of the following soils are prone to Zn Deficiency?
   a) Sodic and saline soils
   b) Peat soils
   c) Soils with high available P and Si status
   d) All the above   **Ans : d**

114. Zinc is essential for which biochemical processes in the rice plant?
   a) Cytochrome and nucleotide synthesis
   b) Auxin metabolism
   c) Chlorophyll production
   d) Enzyme activation
   e) All the above   **Ans : e**

115. Which of the following is a method of management of Zn deficiency?
  a) Growing Zn-efficient varieties that are tolerant of high HCO_3^- and low plant-available Zn content
  b) Broadcasting $ZnSO_4$ in nursery seedbed
  c) Dipping seedlings or presoak seeds in a 2-4% ZnO suspension (e.g., 20-40 g ZnO L^{-1} H_2O)
  d) All the above

Ans : d

116. Which of the following is a method of management of Zn deficiency?
  a) Use of fertilizers that generate acidity
  b) Applying organic manure
  c) Allowing permanently inundated fields
  d) All the above

Ans : d

117. Zinc sulphate is
  a) Soluble and quick acting
  b) Insoluble but quick acting
  c) Insoluble and slow acting
  d) Quick acting

Ans : a

118. Zinc chelate is
  a) Soluble and quick acting
  b) Insoluble but quick acting
  c) Insoluble and slow acting
  d) Quick acting

Ans : a

119. Which of the following is method of application of fertilizer?
  a) Topdressing
  b) Broadcast
  c) Point injector
  d) Deep banding
  e) All the above

Ans : e

120. Which of the following is a method of application of fertilizer?
  a) Split application
  b) Side-dressing
  c) Fertigation
  d) Foliar sprays
  e) All the above

Ans : e

121. Utilization Efficiency of fertilizer is defined as
  a) The percentage of added fertilizer that is actually used by the plants.
  b) The change of economic return (often in term of rupees amount) after fertilization, or economic margins
  c) Both a and b
  d) None of the above

Ans : a

122. Economic Efficiency of fertilizer is defined as
  a) The percentage of added fertilizer that is actually used by the plants.
  b) The change of economic return after fertilization, or economic margins
  c) Both a and b
  d) None of the above

Ans : b

123. A mineral element is considered as essential, when
 a) Plants cannot complete reproductive stage of life cycle due to its deficiency
 b) Deficiency must be corrected only by supplying the element in question
 c) When the element is directly involved in the metabolism of the plant
 d) All the above
Ans : d

124. Mineral deficiency symptoms can be confused with which of the following events?
 a) Damage caused by insect-pest and disease
 b) Salt stress and water stress
 c) Light and temperature injury
 d) Herbicide damage.
 e) All the above
Ans : e

125. Greenish red in sweet potato, bluish green in chili, brown in birdsfoot trefoil or purplish tinge in sugar maple, blueberry and sugarcane are caused by which of the following nutrient deficiency?
 a) Se b) P
 c) N d) Cl toxicity
Ans : b

126. Brown spots on leaves, reduced expansion and premature leaf senescence in soybean crop is caused due to which of the following nutrient deficiency?
 a) Se b) P
 c) N d) Ca
Ans : d

127. Bitter pit in apple, leaf tip burn in cabbage and lettuce, black heart of celery, cavity spot of carrots vitrescence in melons are the typical disorders of which nutrient deficiency?
 a) B b) P
 c) Mg d) Ca
Ans : d

128. Mg deficiency may be induced in tomatoes by high levels of which of the following nutrient solution?
 a) N b) P
 c) Mg d) Ca
Ans : a

129. The leaves become narrow and small in chilies due to deficiency of which of the following nutrient deficiency?
 a) N b) P
 c) Zn d) Ca
Ans : c

130. 'Khaira' disease in rice results due to the deficiency of
a) N b) P
c) Zn d) Ca **Ans : c**

131. The symptoms of "little leaf" or "rosette" occur in deficiency of which of the following nutrient deficiency?
a) N b) P
c) Zn d) Ca **Ans : c**

132. 'Brown heart' in radish and crown choking in coconut are caused by deficiency of which of the following nutrient?
a) B b) P
c) Zn d) Ca **Ans : a**

133. 'Whiptail' in cauliflower is caused by deficiency of which of the following nutrients?
a) B b) Mo
c) Zn d) Ca **Ans : b**

134. Wilting and twisting of leaves in wheat are the symptoms of which nutrient deficiency?
a) B b) Mo
c) Zn d) Cl **Ans : d**

135. Which of the following crop is having high tolerance to Cl toxicity?
a) Wheat b) Tomato
c) Cotton d) All **Ans : d**

136. Which of the following is the most important deficiency symptom of nitrogen deficiency in cereals?
a) Severe restriction of tillering with stems drying and bronzing
b) Foliage restricted
c) No head formation
d) Number and size of roots severely restricted **Ans : a**

137. Which of the following is the most important deficiency symptom of nitrogen deficiency in brassica?
a) Severe restriction of tillering with stems drying and bronzing
b) Foliage restricted
c) No head formation
d) Number and size of roots severely restricted **Ans : b**

138. Which of the following is the most common symptom of P deficiency?
   a) Severe restriction in the growth of the plants in both tops and roots
   b) Development of lateral buds suppressed, narrow leaves
   c) Leaves may shed prematurely, flowering/fruiting may be delayed
   d) In cereals, foliage turns bluish green and then strong reddish-purple tints develop on the nodes
   e) All the above

Ans : e

139. Which of the following is the most common symptom of K deficiency?
   a) Chlorotic mottling between the veins and the marginal scorching of the older leaves
   b) Leaf tips and margins scorched and necrotic
   c) Both a and b
   d) None of the above

Ans : c

140. Which of the following is the most common symptom of Ca deficiency?
   a) Retarded growth of both tops and roots
   b) Chlorosis of young leaves
   c) Distortion of the growing points of the stem
   d) Development of die -back in fruit trees
   e) All the above

Ans : e

141. Which of the following is the most common symptom of Mg deficiency?
   a) Yellowing and chlorosis of older leaves leading to chlorotic stripping in the parallel -veined leaves
   b) Varied pattern of chlorosis/necrosis
   c) Pigmentation of older leaves in fruit trees
   d) All the above

Ans : d

142. Which of the following is the most common symptom of S deficiency?
   a) Brassica species highly susceptible to S deficiency
   b) Leaves showing curling
   c) Puckered inwardly between the veins
   d) All the above

Ans : d

143. Which of the following is the most common symptom of Zn deficiency?
   a) Chlorosis between veins
   b) Reduction in size of young leaves
   c) Bronzing/brown colouration of foliage
   d) All the above

Ans : d

144. Which of the following is/are the most important Zn deficiency symptoms in maize?
   a) Yellowish- green chlorotic mottling of young leaves, reduction in size of leaf, shoots die-back, restricted root system
   b) At seedling stage, leaves turning light yellow and developing necrotic spots
   c) Young leaves at 3-4 week old seedlings develop reddish brown pigmentation, tissue becomes papery/necrotic, plant growth arrested
   d) Symptoms appearing in third leaf in the form of brownish–grey necrotic lessons, leaf turning dry / papery
   e) Shortening of internodes, curvature of petioles downward curling of leaves **Ans : b**

145. Which of the following is/are the most important Zn deficiency symptoms in rice?
   a) Yellowish- green chlorotic mottling of young leaves, reduction in size of leaf, shoots die-back, restricted root system
   b) At seedling stage, leaves turning light yellow and developing necrotic spots
   c) Young leaves at 3-4 week old seedlings develop reddish brown pigmentation, tissue becomes papery/necrotic, plant growth arrested
   d) Symptoms appearing in third leaf in the form of brownish -grey necrotic lesons, leaf turning dry / papery
   e) Shortening of internodes, curvature of petioles downward curling of leaves **Ans : c**

146. Which of the following is the most important Zn deficiency symptomin wheat?
   a) Yellowish- green chlorotic mottling of young leaves, reduction in size of leaf, shoots die-back, restricted root system
   b) At seedling stage, leaves turning light yellow and developing necrotic spots
   c) Young leaves at 3-4 week old seedlings develop reddish brown pigmentation, tissue becomes papery/necrotic, plant growth arrested
   d) Symptoms appearing in third leaf in the form of brownish -grey necrotic lessons, leaf turning dry / papery
   e) Shortening of internodes, curvature of petioles downward curling of leaves **Ans : d**

147. Which of the following is/are the critical symptom(s) of boron deficiency in root crops?
   a) Development of roots severely retarded
   b) Cracking and distorted
   c) Browning and curling of young terminal leaves
   d) All the above **Ans : d**

148. Which of the following is the critical symptom of boron deficiency in cauliflower?
   a) Young curds developing small water-soaked areas
   b) Turning brown, necrotic and decay
   c) Leaf margins becoming chlorotic and curled
   d) Petioles developing small blisters
   e) All the above **Ans : e**

149. Which of the following is/are the critical symptom(s) of boron deficiency in maize?
   a) Young leaves developing white irregular areas spreading along the veins
   b) Apical margins of older leaves turning dry
   c) No cob formation under severe deficiency
   d) All the above **Ans : d**

150. Which of the following is the critical symptom of molybdenum deficiency in cauliflower?
   a) Leaves twisted and elongated with lamina showing various degree of narrowness irregularity known as whiptail
   b) Yellowing of young leaves
   c) White streaks on the lamina
   d) None of the above **Ans : a**

151. Which of the following is the critical symptom of molybdenum deficiency in cabbage?
   a) Young leaves becoming brown,
   b) Necrotic and malformed at the margins
   c) Both a and b
   d) None of the above Ans : c

152. Symptoms of tip burn in plants can be induced by
   a) Pathogens b) Air pollution
   c) Salinity d) Ca deficiency **Ans : c**

153. When a nutrient is suddenly no longer available to a rapidly growing plant it is called
   a) Acute deficiency b) Chronic deficiency
   c) Both a and b d) None **Ans : a**

154. Blossom-end rot of tomato (burning of the end part of tomato fruits), tip burn of lettuce, blackheart of celery and death of the growing regions in many plants are the characteristic deficiency symptoms of
   a) Ca b) Mg
   c) N d) Cu **Ans : a**

155. N deficiency is one of the most common problems in rice. It can be caused by which of the following?

a) Low soil N-supplying power
b) Insufficient application of mineral N fertilizer
c) Low N fertilizer use efficiency (losses from volatilization, denitrification, incorrect timing and placement, leaching, and runoff)
d) All the above

Ans : d

156. N is an essential constituent of

Amino acids
Nucleic acids
Nucleotides
Chlorophyll

Ans :

157. Excessive N may reduce yield because of which of the following?

a) Mutual leaf shading caused by excessive vegetative growth
b) Lodging caused by the production of long, weak stems
c) Increased number of unfilled grains
d) All the above

Ans : d

158. Excessive N may reduce yield because of which of the following?

a) Reduced milling recovery and poor grain quality
b) Increased incidence of diseases such as bacterial leaf blight (caused by *Xanthomonas oryzae*)
c) Increased incidence of insect pests, particularly leaf folder, *Cnaphalocrocis medinalis*
d) All the above

Ans : d

159. The highest recovery efficiency of applied N is achieved during

a) Late tillering to heading stages.
b) Fruiting stage
c) Transplanting stage
d) None of the above

Ans : a

160. Increasing N use efficiency involves which of the following measures?

a) N fertilizer placement in the reduced soil layer about 8-10 cm below the soil surface
b) Slow-release N fertilizer (S-coated urea) or poly-olefin coated urea incorporated as a basal dressing before planting
c) Both a and b
d) None of the above

Ans : c

161. Ammonia volatilization from different N fertilizer sources increases in which of the following order?
  a) Ammonium sulphate < Urea < Ammonium bicarbonate
  b) Urea < Ammonium sulphate < Ammonium bicarbonate
  c) Ammonium bicarbonate < Urea <Ammonium sulphate
  d) None of the above

Ans : a

162. The total crop N uptake per ton grain produced is approximately
  a) N - 15-20 kg
  b) N - 2-3 kg
  c) N - 30-40 kg
  d) None of the above

Ans : a

163. The total crop P uptake per ton grain produced is approximately
  a) P - 15-20 kg
  b) P - 2-3 kg
  c) P - 15-20 kg
  d) None of the above

Ans : b

164. The total crop K uptake per ton grain produced is approximately
  a) K - 30-40 kg
  b) K- 2-3 kg
  c) K - 15-20 kg
  d) None of the above

Ans : c

165. Nutrient requirements depend upon
  a) Target yield
  b) Soil nutrient supply
  c) Both a and b
  d) None of the above

Ans : c

166. To estimate fertilizer requirements what is/are needed to know?
  a) Target yield
  b) Crop yield with no fertilizer
  c) Fertilizer recovery
  d) Timing
  e) All the above

Ans : e

167. The factor to obtain P_2O_5 from P is
  a) 2.29
  b) 1.20
  c) 1.39
  d) 1.66

Ans : a

168. The factor to obtain K_2O from K is
  a) 2.29
  b) 1.20
  c) 1.39
  d) 1.66

Ans : b

169. The factor to obtain CaO from Ca is
  a) 2.29
  b) 1.20
  c) 1.39
  d) 1.66

Ans : c

170. The factor to obtain MgO from Mg is

a) 2.29
b) 1.20
c) 1.39
d) 1.66

Ans : d

171. The benefits of straw incorporation include which of the following?

a) Recycling of nutrients from straw without the trouble involved in common late incorporation and decomposition under anaerobic conditions
b) Improved re-oxidation of the topsoil layer during the fallow period
c) Substantial water savings
d) All the above

Ans : d

172. Lack of iron in a plant results in

a) Chlorotic (yellow to white) leaves
b) The veins may remain dark green
c) Petioles of tomato plants become purple
d) Stunted growth with a drop (up to 70%) in crop yield
e) All the above

Ans : e

173. A lack of manganese in a plant results in

a) Chlorosis (yellowing) of the leaves
b) Leaves becoming spotted (marbled effect)
c) Reduced growth and flowering
d) Cucumber, beans and many ornamentals are especially susceptible to manganese deficiency
e) All the above

Ans : e

174. A lack of molybdenum in a plant results in

a) Yellow/orange spots on the leaves between the veins
b) A reduced number of flowers which are small and sometimes do not open
c) Both a and b
d) None of the above

Ans : c

175. A lack of molybdenum in a plant results in

a) Deformed reproductive organs
b) "Whiptail leaf symptom" in cabbages
c) "Yellow spot" disease in citrus
d) All the above

Ans : d

176. Lack of magnesium in a plant results in

a) Chlorosis (yellowing) between the leaf veins
b) In some species, the lower leaves develop a purple colour preceding death
c) Both a and b
d) None of the above

Ans : c

177. Possible causes of nutrient deficiencies is
   a) Insufficient amount of available nutrient in soil
   b) The chemical characteristics of soil may reduce availability of nutrients
   c) The nutrient is positionally unavailable
   d) All the above **Ans : d**

178. Chlorosis is
   a) Yellowing of leaves
   b) Leaf tissue between veins turns yellow while the veins remain green
   c) Complete drying and death of plant tissue
   d) Shortened internodes
   e) Red, purple, brown colors caused by pigments **Ans : a**

179. Interveinal chlorosis - striping is
   a) Yellowing of leaves
   b) Leaf tissue between veins turns yellow while the veins remain green
   c) Complete drying and death of plant tissue
   d) Shortened internodes
   e) Red, purple, brown colors caused by pigments **Ans : b**

180. Necrosis is?
   a) Yellowing of leaves
   b) Leaf tissue between veins turns yellow while the veins remain green
   c) Complete drying and death of plant tissue
   d) Shortened internodes
   e) Red, purple, brown colors caused by pigments **Ans : c**

181. Stunting is?
   a) Yellowing of leaves
   b) Leaf tissue between veins turns yellow while the veins remain green
   c) Complete drying and death of plant tissue
   d) Shortened internodes
   e) Red, purple, brown colors caused by pigments **Ans : d**

182. Yellowing of leaves
   a) Leaf tissue between veins turns yellow while the veins remain green
   b) Complete drying and death of plant tissue
   c) Shortened internodes
   d) Red, purple, brown colors caused by pigments **Ans : a**

183. Internal cork of apple is due to the deficiency of

a) Ca b) B
c) Fe d) Mn Ans : b

184. Die back symptom is due to the deficiency of

a) Fe b) Mn
c) Zn d) Ca Ans : d

185. Bronzing in rice due to the toxicity of

a) Fe b) Mn
c) Zn d) Cu Ans : b

186. Brown heart or black heart of root crops is due to the deficiency of

a) Ca b) Mg
c) B d) Mo Ans : c

187. Crinkle leaf of cotton is due to the

a) Deficiency of Cu b) Toxicity of Cu
c) Deficiency of Mn d) Toxicity of Mn Ans : d

188. Nutrient element involved in the synthesis of tryptophane is

a) Cu b) Zn
c) Mn d) Mo Ans : b

189. Fern leaf in potato is due to the

a) Deficiency of Zn b) Toxicity of Zn
c) Deficiency of Cu d) Toxicity of Cu Ans : a

190. Little leaf of cotton is due to the

a) Deficiency of Zn b) Toxicity of Zn
c) Deficiency of Cu d) Toxicity of Cu Ans : a

191. Speckled yellow of sugar beet is due to the

a) Deficiency of Cu b) Toxicity of Cu
c) Deficiency of Mn d) Toxicity of Mn Ans : c

192. What is the N, P_2O_5 and K_2O content (%) in cow dung slurry from biogas plant?

a) 2.5: 1.5: 1.5 b) 1.8: 1.0: 1.0
c) 5.0: 3.0:2.0 d) 0.5: 0.5: 1.0 Ans : b

193. What is the percentage of nitrogen content in green manures?

a) 0.7 b) 0.5
c) 1.2 d) 1.5 Ans : d

194. Which of the following is an example for edible oil cake?

a) Castor cake
b) Karanji cake
c) Mahua cake
d) Mustard cake

Ans : d

195. Which of the following is an example for non- edible oil cake?

a) Groundnut oil cake
b) Niger cake
c) Neem cake
d) Sesame cake

Ans : c

196. Fertilizer that supplies both nitrogen and sulphur is

a) Urea
b) Ammonium sulphate
c) Ammonium chloride
d) CAN

Ans : b

197. Equivalent acidity of ammonium sulphate is

a) 80
b) 60
c) 100
d) 110

Ans : d

198. Equivalent acidity of calcium ammonium nitrate (CAN) is

a) 80
b) 60
c) 100
d) 110

Ans : b

199. Equivalent acidity of urea is

a) 80
b) 60
c) 100
d) 110

Ans : a

200. Gray speck of oats and marsh spot of peas is due to the deficiency of

a) B
b) Mo
c) Zn
d) Mn

Ans : d

201. Indicator plants used for the deficiency of N, Ca and Fe are

a) Rapseed and Sugarbeet
b) Oat and potato
c) Cauliflower and Cabbage
d) Potato and Sugarbeet

Ans : c

202. Indicator plant used for the phosphorus of deficiency of N, Ca and Fe is

a) Potato
b) Rapseed
c) Sugarbeet
d) Oat

Ans : b

203. Indicator plant used for the deficiency of potassium is

a) Potato
b) Raped
c) Sugarbeet
d) Oat

Ans : a

204. Indicator plant used for the deficiency of boron is
a) Cauliflower b) Cabbage
c) Sugarbeet d) Potato **Ans : c**

205. Indicator plant used for the deficiency of sodium is
a) Cauliflower b) Cabbage
c) Sugarbeet d) Potato **Ans : c**

206. Which of the following nutrient is highly mobile in plants?
a) N b) Mg
c) S d) B **Ans : a**

207. Which of the following nutrient is highly mobile in plants?
a) P b) Mg
c) S d) B **Ans : a**

208. Which of the following nutrient is highly mobile in plants?
a) K b) Mg
c) S d) B **Ans : a**

209. Which of the following nutrient is moderately mobile in plants?
a) P b) Mg
c) N d) B **Ans : b**

210. Which of the following nutrient is moderately mobile in plants?
a) N b) P
c) S d) B **Ans : c**

211. Which of the following nutrient is moderately mobile in plants?
a) N b) P
c) K d) Mo **Ans : d**

212. Which of the following nutrient is immobile in plants?
a) Ca b) P
c) S d) K **Ans : a**

213. Which of the following nutrient is immobile in plants?
a) B b) P
c) S d) K **Ans : a**

214. Which of the following nutrient is immobile in plants?
a) B b) Ca
c) S d) Both a & b **Ans : d**

215. Which of the following nutrient is less mobile in plants?

a) Cu b) P
c) N d) K

Ans : a

216. Leaves (first older ones) turn yellow/ brown, plants are spindly, lack vigour and may be dwarfed) This is a sign of which nutrient deficiency?

a) N b) P
c) S d) K

Ans : a

217. Interveinal chlorosis occurs first on older leaves. This is a sign of which nutrient deficiency?

a) Mg b) P
c) S d) K

Ans : a

218. Yellow/brown spots appear on older leaves and/or necrosis of edges. This is a sign of which nutrient deficiency?

a) Mg b) Ca
c) S d) K

Ans : c

219. Chlorosis of the tips of the youngest leaves and die-back of growing points. This is a sign of which nutrient deficiency?

a) Mg b) Ca
c) K d) Zn

Ans : b

220. In crops, other than cereals, the apical growing point on the main stem dies and lateral buds fail to develop shoots. This is a sign of which nutrient deficiency?

a) Mg b) Cu
c) B d) Zn

Ans : c

221. Marginal scorching and cupping of leaves. Wilting is common in brassica. This is a sign of which nutrient deficiency?

a) Mo b) Cu
c) B d) Zn

Ans : a

222. Interveinal chlorosis, first on older leaves. This is a sign of which nutrient deficiency?

a) Mo b) Cu
c) Mg d) Zn

Ans : c

223. For wetland the best nitrogenous fertilizer is

a) Urea b) Calcium nitrate
c) Ammonium sulphate d) Calcicum ammonia nitrate.

Ans : c

224. Earth worms present in the soil

a) Reduce soil fertility
b) increase soil fertility
c) Harm plants -
d) Cause plant diseases

Ans : b

225. Soils containing high proportions of organic matter have following colour

a) Light colour
b) Dark colour
c) Yellow colour
d) Red colour

Ans : b

226. Tarai soils are generally deficient in

a) Zinc
b) Nitrogen
c) Phosphorus
d) Potash

Ans : a

227. The inherent capacity of a soil to supply nutrients to plants in available and balanced proportion can be represented by -

a) Soil productivity
b) Soil taxonomy
c) Edaphology
d) Soil fertility

Ans : d

228. Humous celluloid are -composed basically of

a) Nitrogen
b) Hydrogen
c) Carbon
d) Helium

Ans : c

229. Non-humic substances are less complex and are to microbial attack than those belonging to humic group

a) More susceptible
b) Less susceptible
c) More resistant
d) Less resistant

Ans : d

230. The C:N. ratio in the organic matter of the, furrow slice (upper 15 cm) or arable soils

a) 2:1 to 5:1
b) 6:1 to 7:1
c) 8:1 to 15:1
d) 18:1 to '20:1

Ans : c

231. The carbon nitrogen ratio in cultivated soil is stabilized at-

a) 20: 1- to 30: 1
b) 4:1 to 9:1
c) 10:1 to 12:1
d) 20:1 to 30:1

Ans : c

232. Elements which are absorbed from the soil are known as

a) Plant nutrients
b) Plant components
c) Plant feeders
d) Plant supplements

Ans : a

233. Which of the following nutrient elements are regarded as energy exchangers?

a) Hydrogen and oxygen
b) Carbon and nitrogen
c) Potassium and magnesium
d) Iron manganese and molybdenum

Ans : a

234. Which of the following elements regulate oxidation reduction reactions in plant
a) Hydrogen, oxygen and carbon
b) Nitrogen, phosphorus and sulphur
c) Potassium calcium sodium, and magnesium
d) Iron, manganese, molybdenum, copper, boron and zinc **Ans : d**

235. Absorption of ions in plants occurring with the aid of metabolic energy is termed
a) Passive absorption b) Active absorption
c) Metabolic absorption d) Mass flow absorption **Ans : b**

236. Plants meet their oxygen requirement by
a) Breakdown of water in the plant body
b) Absorbing oxygen from air
c) Absorbing oxygen from soil
d) All the above methods **Ans : d**

237. Oxygen is required by the plants for
a) Respiration b) Transpiration
c) Photosynthesis d) Chlorophyll formation **Ans : a**

238. Carbon requirement of the plant is met by absorbing
a) Carbon from soil minerals
b) Carbon from soil organic matter
c) Carbon monoxide from atmosphere
d) Carbon dioxide from atmosphere **Ans : d**

239. Hydrogen requirement of plants is met by
a) Absorption of hydrogen from air
b) Absorption of hydrogen from water
c) Absorption of hydrogen from soil
d) Breakdown of water within the plant **Ans : d**

240. Organic carbon content in the soil is a measure of
a) Available potassium b) Available phosphorus
c) Available nitrogen d) Available magnesium **Ans : c**

241. Available nitrogen of less than 280 kg/hectare is con
a) Low b) Medium
c) High d) Trace **Ans : a**

242. A complex mixture of brown amorphous and colloidal substances synthesized by various soil organisms is referred to as
a) Compost b) Humus
c) FYM d) Peat super compost **Ans : b**

243. In acid soil rock phosphate should be applied

a) At the time of sowing
b) 5 days before sowing
c) 30 days after sowing
d) 30 days before sowing

Ans : d

244. Khaira disease of rice is diagnosed by

a) Necrotic lesions
b) Dark reddish brown pigmentation
c) Yellowish spots on leaves
d) Black pin head points on leaf

Ans : b

245. Upward translocation of fluid in xylem takes place due to

a) Pull of transpiration stream
b) Pumping action of guard cells
c) Suction by sun-rays
d) All above

Ans : a

246. Nitrogen, phosphorus, potassium, calcium, magnesium and sulphur, which are used by plants in relatively large quantities, are referred to as

a) Macronutrients
b) Micronutrients
c) Trace-nutrients
d) Bulk nutrients

Ans : a

247. Chlorosis is observed in upland rice due to deficiency of

a) S
b) Fe
c) Zn
d) Mn

Ans : b

248. Nutrients absorbed by plants are carried upward through the

a) Xylem
b) Phloem
c) Both these
d) None of these

Ans : a

249. Downward movement of food synthesized in leaves takes place through

a) Xylem
b) Phloem
c) Both these
d) None of these.

Ans : b

250. Which of the following is wrong?

a) Iron, manganese, copper, zinc, boron, molybdenum, chlorine and cobalt are called micronutrients.
b) Nitrogen phosphorus and potassium are primary nutrients.
c) Calcium, magnesium and sulphur are secondary nutrients.
d) Nitrogen is converted into nitrate by earthworms

Ans : d

251. Farmyard manure

a) Balanced food for plants
b) Recommended for salinity reclamation
c) Tends to make soil acidic
d) Tends to make soil alkaline

Ans : a

252. Movement of nutrient ions along with irrigation water or rainwater is called as

a) Diffusion b) Mass flow
c) Contact exchange d) All **Ans : b**

253. Movement of nutrient ions through irrigation water or rainwater is called as

a) Diffusion b) Mass flow
c) Contact exchange d) All **Ans : a**

254. What is the mineralization rate of organic nitrogen per year in Indian soils?

a) 1.5 % b) 2.0 %
c) 2.5 % d) 1.0 % **Ans : c**

255. Alkaline permanganate method is used for the estimation of

a) Total nitrogen b) Available nitrogen
c) Nitrate nitrogen d) Ammonical nitrogen **Ans : b**

256. The analytical method most suitable for the estimation of available phosphorous in acid soils is

a) Olsen's method b) Mehlich's method
c) Bray and Kurtz's method d) All **Ans : c**

257. The analytical method most suitable for the estimation of available phosphorous in all the soils except acid soils is

a) Olsen's method b) Bray and Kurtz's I method
c) Bray and Kurtz's II method d) Both b and c **Ans : a**

258. Phosphorus is essential for

a) Cell division
b) Development of meristematic tissues
c) Both of these
d) None of these **Ans : c**

259. Potassium is considered vital in

a) Synthesis proteins and amino acids
b) Photosynthetic activity of leaves
c) Movement of manufactured food from leaves to roots
d) All the above processes. **Ans : d**

260. Calcium is essential for

a) Cell wall formation b) Meristemati activity at roots
c) Both of these d) None of these **Ans : c**

261. The capability of soil to produce a plant or plant parts is called as
a) Soil fertility b) Soil quality
c) Soil productivity d) All **Ans : c**

262. The inherent capacity of soil to provide nutrients is called as
a) Soil fertility b) Soil quality
c) Soil productivity d) All **Ans : a**

263. The total number of essential nutrient elements required for plant growth is
a) 16 b) 17
c) 20 d) 21 **Ans : b**

264. The 'Functional or metabolism nutrients' was given by
a) Arnon and Stout b) Nicholas
c) Mayor and Brown d) Juston von Liebig **Ans : b**

265. N, P and K are called as
a) Macro nutrients b) Micronutrients
c) Secondary nutrients d) Beneficial nutrients **Ans : a**

266. Ca, Mg and S are called as
a) Macro nutrients b) Micronutrients
c) Secondary nutrients d) Beneficial nutrients **Ans : c**

267. Iron, zinc, boron and molybdenum are called as
a) Macro nutrients b) Micronutrients
c) Secondary nutrients d) Beneficial nutrients **Ans : b**

268. Na, Ve, Ni, Si, Co, Se and I are the examples of
a) Macro nutrients b) Micronutrients
c) Secondary nutrients d) Beneficial nutrients **Ans : d**

269. The form(s) of phosphorous absorbed by plants is
a) HPO_4^{3-} b) $H_2PO_4^-$
c) HPO_4^{2-} d) Both a and b
e) All the above **Ans : d**

270. The major portion of available boron is absorbed by plants as
a) BO_3^{3-} b) $HB_4O_7^-$
c) $H_2B_4O_7^-$ d) Both a and b **Ans : b**

271. Of the following, which is the available form of molybdenum to plant?

a) $HMoO_4^-$ b) MoO_4^-

c) $HMoO_4^{2-}$ d) MoO_4^{2-} Ans : a

272. Of the following, which nutrient element(s) is/are essential for protein synthesis?

a) Nitrogen b) Phosphorous

c) Potassium d) Ca and Mg Ans : a

273. ____________ is a constituent of sugar phosphates, viz. ADP, ATP?

a) Nitrogen b) Phosphorous

c) Sulphur d) Molybdenum Ans : b

274. Which nutrient element is involved in energy transformation?

a) Molybdenum b) Zinc

c) Sulphate d) Phosphorous Ans : d

275. The nutrient element involved in stomatal regulation of cell is

a) Ca and Mg b) Nitrogen

c) Potassium d) Both a and c Ans : c

276. The nutrient element essential for maintenance of the stability of cell wall is

a) Phosphorous b) Calcium

c) Magnesium d) Potassium Ans : b

277. Oil content in oil-bearing plants is increased by

a) Sulphur b) Molybdenum

c) Nitrogen d) Phosphorous Ans : a

278. Of the following, which nutrient element is responsible for the translocation of sugars across the membrane?

a) Boron b) Phosphorus and Zinc

c) Molybdenum and Iron d) All Ans : a

279. Nutrient element(s) essential for photosynthesis is/are

a) Manganese b) Copper and Zinc

c) Phosphorus d) All Ans : a

280. Nutrient element(s) essential for the functioning of sulphydryl compounds such as cysteine is/are

a) Sulphur b) Zinc

c) Phosphorous d) Both a and b Ans : b

281. Which nutrient is a constituent of nitrate reductase and nitrogenase enzymes?
a) Molybdenum b) Zinc
c) Copper d) Phosphorous Ans : a

282. Nutrients that are relatively immobile in the soil, move from soil particles to the root surface by the process of
a) Diffusion
b) Contact exchange
c) Mass flow
d) Bi-modal reciprocal transpiration inversion flow Ans : b

283. Nutrient elements whose soil requirement is directly related to yield goal, yield potential, or growth rate include
a) N, P and K b) N, S, B and Cl
c) Ca, Mg, K and P d) Fe, Mn, Zn and Cu Ans : a

284. Which nutrient is a constituent of chlorophyll and chromosomes?
a) Magnesium b) Iron
c) Phosphorous d) Both a and b Ans : a

285. Movement of nutrient ions form soil to plant roots by
a) Diffusion b) Mass flow
c) Contact exchange d) All Ans : d

286. Available potassium of more than, 280 kg/ha in soil is considered
a) Low b) Medium
c) High d) Trace Ans : c

287. Nitrogen is taken by the plants in the form of
a) Elemental nitrogen b) Nitrate
c) Nitrite d) None of these Ans : b

288. Available form of phosphorus ions is
a) Phosphoric acid b) Phosphorus chloride
c) Phosphorus sulfite d) Phosphorus nitrate Ans : a

289. Available phosphorous of more than 25 kg/ha of soil is considered as
a) Low b) Medium
c) High d) Trace Ans : c

290. Available phosphorus of less than 110 kg/ha of soil is considered as

a) Low b) Medium
c) High d) Trace **Ans : a**

291. When emerging leaves of maize show yellow to white bleached bands in lower parts, it shows the deficiency of

a) Zinc b) Iron
c) Copper d) Boron **Ans : a**

292. White, irregular spots between veins of upper leaves of maize plant shows deficiency of

a) Copper b) Molybdenum
c) Manganese d) Boron **Ans : d**

293. Pale green to yellow inveinal discoloration in young maize leaves shows deficiency of

a) Manganese b) Copper
c) Zinc d) Sulphur **Ans : a**

294. Which of the following is a nitrification inhibitor

a) A.M. (2, amino- 4 6 methyl pyridine)
b) Thio-urea
c) Sulfur coated urea
d) CBD (β-chloro. Ethyl, trimethyl ammonium chloride) **Ans : a**

295. "Hidden Hunger" meAns :

a) Deficiency symptoms are seen when the nutrient is deficient
b) Severe yield reduction may occur without appearance of deficiency symptoms
c) The nutrient is not deficient but apparently seems to be deficient.
d) Visual deficiency symptoms are suppressed by other elements. **Ans : b**

296. Which of the following statements is more correct as regards to "Luxury consumption" of K found in minute

a) Maize has mechanism to increase yields by taking excess K above the optimum requirement.
b) Maize takes excess K above the optimum requirement even though growth and yield are not promoted
c) Maize takes excess k above the optimum requirement the concentration which becomes toxic to plants
d) None of the above **Ans : b**

297. The quantity of single superphosphate needed for applying 80 kg P_2O_5lha to sugarcane will be
a) 250 kg b) 500 kg
c) 750 kg d) 1000 kg **Ans : b**

298. Single superphosphate is preferred for phosphorus application in groundnut because it also provides
a) Sulphur b) Calcium
c) Sulphur and calcium both d) None of these **Ans : c**

299. The tendency of the young upper leaves to remain as the lower leaves yellow or die is an indication of nitrogen in the plant
a) Immobility b) Mobility
c) Stability d) None of the above **Ans : b**

300. The sufficiency limit of iron is normally between ppm.
a) 10-45 b) 50-250
c) 350-400 d) 500-750 **Ans : b**

301. The zinc deficiency in rice causes disease
a) Khaira b) Bacterial leaf streak
c) Bacterial leaf blight d) Sheath blight **Ans : a**

302. The quantity of solid phase phosphorus that acts as reserve is often referred to as
a) Intensity factor b) Quality factor
c) Quantity factor d) None of these **Ans : b**

303. Which of the following methods is most useful to increase the availability of soil applied phosphorus
a) Broadcast
b) Band placement
c) Broadcast and soil incorporation.
d) None of these **Ans : b**

304. Rock phosphate can be directly applied as fertilizer in
a) Acidic b) Alkaline
c) Neutral d) All of these **Ans : a**

305. Phosphatase enzyme playa major role in ___________of organic phosphates in soil
a) Adsorption b) Mineralization
c) Immobilization d) None of these **Ans : b**

306. The mineralization of organic phosphorus has been studied in relation to-

a) C/N/P b) C/N/S
c) C/N/K d) C/N/Ca Ans : a

307. The maximum phosphorus availability in most of the soils in the pH range.

a) 4.0-4.5 b) 6.0-6.5
b) 7.0-7.5 d) None of these Ans : b

308. The highest nitrogen content among the commonly used nitrogenous fertilizers is embodied by

a) Anhydrous ammonia b) Urea
c) Ammonium nitrate d) Calcium ammonium nitrate Ans : b

309. Application of nitrogen in pulses at the time of planting is known as

a) Additional dose b) Starter dose
c) Synergistic dose d) Basic dose Ans : b

310. Ammonium sulphate is preferred for nitrogen application in 'groundnut because it also provides

a) Calcium b) Magnesium
c) Sulphur d) Copper Ans : c

311. The technique used for minimizing the loss of nitrogen in paddy fields comprises-

a) Use of neem cake b) Use of pre incubated urea
c) Use of mud ball d) All of these Ans : d

312. The quantity of urea needed for applying 115 kg nitrogen to maize will be (approx.)

a) 100 kg b) 250 kg
c) 200 kg d) 250 kg Ans : d

313. The oxidation of ammonia has been termed as nitrification is a _________ process.

a) Physical b) Chemical
c) Biological d) Physico-chemical Ans : c

314. The bacteria considered most important to bring about conversion of NH^+_4 to NO^-_2

a) Nitrosomonas b) Nitrobacter
b) Both a & b d) Azotobacter Ans : c

315. The losses of ammonia usually become greater with ______ soil

a) Increase in soil pH b) Decrease in soil pH
c) Increase in soil temperature d) Increase in soil calcium Ans : a

316. Anaerobic environment in paddy soil is responsible for gaseous loss of fertilizer nitrogen by

a) Ammonification b) Nitrification
c) Denitrification' d) Volatilization Ans : c

317. To check nitrification, nitrification inhibitors are most useful in

a) Paddy b) Cotton
c) Maize d) None of these Ans : a

318. To prevent denitrification losses of nitrogen in paddy field, nitrogen' should be applied in the

a) Reduced zone b) Oxidized zone
c) Hydronized zone d) Calcified zone Ans : a

319. In case of rice when urea is placed in the reduced zone it reduces .... losses

a) Leaching and denitrification
b) Leaching and volatilization
c) Runoff and leaching
d) Denitrification and volatilization Ans : b

320. What is the percentage of P_2O_5 in bone meal?

a) 10 b) 20
c) 15 d) 5 Ans : b

321. Fish meal contains more ____________.

a) Nitrogen b) Phosphorous
c) Potassium d) Ca and Mg Ans : a

322. Which organism is responsible for the production of methane from the biogas plant?

a) *Bacillus* b) *Pseudomonas*
c) *Arthrobacter* d) *Methanobacteria* Ans : d

323. Microbes belonging to the family methanobacteria are

a) Aerobes b) Anaerobes
c) Facultative aerobes d) Facultative anaerobes Ans : b

324. Which fertilizers enhance the manuring properties of legumes?
a) Nitrogenous fertilizers b) Phosphatic fertilizers
c) Potassic fertilizers d) All
Ans : b

325. Of the following, which one is concentrated organic manures?
a) FYM b) Compost
c) Bone meal d) Poultry litters
Ans : c

326. Of the following, which one is bulky organic manure?
a) Composts b) Oil cakes
c) Bone meals d) Fishmeals
Ans : a

327. When a fertilizer contains and is used for supplying single nutrient, it is called as
a) Straight fertilizers b) Mixed fertilizers
c) Complex fertilizers d) Compound fertilizers
Ans : a

328. The presence of two or more nutrients in one compound or mixture is called as
a) Complex fertilizers b) Compound fertilizers
c) Mixed fertilizers d) Both a and b
Ans : d

329. The physical mixture of two or more straight fertilizers or compound fertilizers is called as
a) Straight fertilizers b) Complex fertilizers
c) Mixed fertilizers d) Compound fertilizers
Ans : c

330. What is the percentage of nitrogen in urea?
a) 48 b) 46
c) 25 d) 42
Ans : b

331. What is the percentage of nitrogen in ammonium sulphate?
a) 20.6 b) 26.0
c) 25.0 d) 46.0
Ans : a

332. What is the percentage of nitrogen in ammonium chloride?
a) 20.6 b) 26.0
c) 25.0 d) 46.0
Ans : c

333. What is the percentage of P_2O_5 in single super phosphate?
a) 16.0 b) 46.0
c) 34.0 d) 28.0
Ans : a

334. What is the percentage of P_2O_5 in diammonium phosphate (DAP)?
a) 16.0 b) 48.0
c) 34.0 d) 46.0 **Ans : d**

335. The percentage of K_2O in muriate of potash (MOP) is
a) 58.0 b) 48.0
c) 23.0 d) 15.0 **Ans : a**

336. The percentage of K_2O in sulphate of potash (MOP) is
a) 58.0 b) 48.0
c) 23.0 d) 15.0 **Ans : b**

337. Muriate of potash is a
a) Straight fertilizer b) Compound fertilizer
c) Mixed fertilizer d) Complex fertilizer **Ans : a**

338. Diammonium phosphate (DAP) is a
a) Compound or complex fertilizer
b) Mixed fertilizer
c) Straight fertilizer
d) Complete complex fertilizer **Ans : a**

339. Which form of nitrogen is present in urea?
a) Nitrate form b) Ammonical form
c) Amide form d) Both a and b **Ans : c**

340. The most deficient nutrient in Indian soils is
a) Nitrogen b) Zinc
c) Copper d) Boron **Ans : a**

341. The second most deficient nutrient in Indian soils after nitrogen is
a) Nitrogen b) Zinc
c) Copper d) Boron **Ans : b**

342. What is the temperature maintained in ammonia production by Claude-Haber-Bosch synthesis process?
a) 400-500 °C b) 500-600 °C
c) 550-600 °C d) 600-650 °C **Ans : a**

343. The most ground water polluting fertilizers are ......
a) Nitrate fertilizers b) Ammonical fertilizers
c) Potassic fertilizers d) Amide fertilizers **Ans : a**

344. Hydrolysis of urea results in the loss in nitrogen in the form of
a) NH_4 b) NH_3
c) NH_2 d) NH_4OH Ans : b

345. Person who established the relationship between plant growth factor
a) Liebig b) Van Helmont
c) Tisdale d) Mistscherlich Ans : d

346. Availability of H_2PO_4 is greatest when soil pH is
a) Low b) High
c) Neutral d) None of these Ans : c

347. Single superphosphate contains phosphorus in
a) Citrate soluble form b) Insoluble form
c) Water soluble form d) None of these Ans : a

348. Which of the following in considered as bio-fertilizer
a) Farm yard manure b) Compost.
c) Green manure d) Blue green algae Ans : d

349. Rhizobium culture is used increase biological nitrogen fixation in
a) Pulse crops b) Cereals crops
c) Oilseeds d) Cotton Ans : a

350. The presence of autotrophic mycorrhiza enhances the uptake of nutrients in general, but particularly of
a) Iron b) Molybdenum
c) Phosphorus d) Nitrogen Ans : c

351. Which of the following organism is responsible for decomposition of organic matter
a) Algal b) Bacteria
c) Fungi d) Microorganism Ans : b

352. Volume of soil under the influence of roots of growing plants is known as
a) Phyto-bio-sphere b) Phyto-bio-zone
c) Rhizosphere d) Phytozone Ans : c

353. Rhizospheres affect microbial population and nutrient availability by
a) Exerting physical pressure
b) Secreting nutritive metabolites and organic solvents
c) Increasing pore space
d) Changing soil structure Ans : b

354. Symbiotic association of fungal hyphae with plant roots is known as

a) Mycorrhiza
b) Mutual infection
c) Lichen
d) Co-parasitism

Ans : a

355. Which of the following statement is correct?

a) Blue-green algae causes plant diseases
b) Blue-green algae fixes atmospheric nitrogen in soil
c) Blue-green algae grows best in maize crop
d) Dry conditions are good for growth of blue-green algae

Ans : b

356. Mycorrhiza-infected plants show better ability of absorption of the nutrients, especially

a) Iron
b) Manganese
c) Potassium
d) Phosphorus

Ans : c

357. Mycorrihiza-infected plants show better growth than uninfected plants, especially on

a) Low fertility soils
b) High - fertility soils
c) Marshy soils
d) Desert soils

Ans : a

358. *Azotobacter* and *Rhizobium* bacteria are

a) Antibiotic producers
b) Nitrogen fixers
c) Plant pathogens
d) Animal pathogens

Ans : b

359. *Azotobacter* sp. fix atmospheric nitrogen

a) As symbiouts on sugarcane
b) As symbiouts on rice
c) As symbiouts on legumes
d) In free soil

Ans : d

360. *Rhizobium* sp. fix atmospheric nitrogen

a) As symbiouts on wheat
b) As Symbiouts on cucumber
c) As symbiouts on pulses
d) In free soil

Ans : c

361. Blue-green algae grows best in

a) Upland rice
b) Jute
c) Lowland rice
d) Sorghum

Ans : a

362. If soybean in being grown for the first time in a field, the seed should be treated with_____________ culture.

a) *Azospirillium*
b) *Azotobacter*
c) *Rhizobium*
d) *Phosphorbacterim*

Ans : c

363. Which of the following strain is used for biological nitrogen fixation in soybean

a) *Rhizobium leguminoseum*
b) *Rhizobium Japonicum*
c) *Rhizobium melilotic*
d) *Rhizobium 'Trifolii'*

Ans : b

364. The bacteria responsible for fixation of nitrogen in soybean is
a) *Rhizobium leguminoserum* b) *Rhizobium phaseoli*
c) *Rhizobium glycicum* d) *Rhizobium Japonicum* **Ans : d**

365. Soil productivity is measured in terms of
a) Cost in rupees of per acre field
b) Yield per unit area
c) Total production obtained for the total cropped area
d) None of these. **Ans : b**

366. Application of organic materials with C: N ratio wider than 20 to 30: 1 leads to .... of soil N
a) Nitrification b) Fixation
c) Mineralization d) Immobilization **Ans : d**

367. 'A' value concept was given by
a) Sorenson b) Beckett
c) Schofield d) Fried and Dean **Ans : d**

368. Which one of the following has organic form of sulphur
a) Purine b) Cysteine
c) RNA d) Phytin **Ans : b**

369. The process of formation of nitrogen and nitrous oxide gases from ammonical fertilizers in soil is known as
a) Ammonification b) Nitrification
c) Denitrification d) Mineralization **Ans : c**

370. Dolomite is
a) $CaCO_3$ b) $MgSO_4$
c) $Ca(OH)_2$ d) $MgCO_3CaCO_3$ **Ans : d**

371. The organic residues with wider C:N ratios as compared to narrow C:N ratios are decomposed at
a) Faster rate
b) Slow rate
c) Equal rate
d) Faster rate followed by slow rate **Ans : b**

372. Which form of nitrogen is available in urea
a) Ammonical b) Amide
c) Nitrite d) Nitrate **Ans : b**

373. At what pH value, phosphate availability is the highest in the soil
a) 5.5 b) 6.5
c) 7.5 d) 8.5 **Ans : b**

374. Application of potash increases
a) Disease resistance in plants b) Resistance for water logging
c) Frost resistance in plants d) None of these **Ans : a**

375. LEISA is related to
a) Organic farming b) Inorganic farming
c) Natural farming d) All of these **Ans : a**

376. Phosphatic fertilizer that supplies both phosphorous and sulphur is
a) Dicalcium phosphate b) Single super phosphate
c) Tricalcium phosphate d) Diammonium phosphate **Ans : b**

377. Most popular phosphatic fertilizer in India is
a) Single super phosphate b) Double super phosphate
c) Rock phosphate d) Diammonium phosphate **Ans : a**

378. Phosphatic fertilizersuitable for acid soils is
a) Rock phosphate b) Dicalcium phosphate
c) Basic slag d) All **Ans : d**

379. Phosphatic fertilizer obtained from steel industry as a by-product is
a) Dicalcium phosphate b) Schoenite
c) Basic slag d) Pelophos **Ans : c**

380. What is the percentage of P_2O_5 in Indian basic slag?
a) 3 – 8 % b) 20 – 25 %
c) 10 - 12 % d) 12 – 18 % **Ans : a**

381. Potassic fertilizer most suitable for potato and tobacco is
a) Muriate of potash b) Sulphate of potash
c) Schoenite d) All **Ans : b**

382. What is the nitrogen use efficiency for rice crop?
a) 40 – 60% b) 28 – 34%
c) 35 – 43% d) 42– 50% **Ans : b**

383. The principal way in which P and K ions move from the soil to the root of field crops
a) Root interception b) Mass flow
c) Diffusion d) None **Ans : c**

384. Immobilization of sulphur take place when the 'S' content of organic matter is less than
a) 0.45% b) 0.60%
c) 0:30% d) 0.15% **Ans : b**

385. Organic carbon is a measure of

a) Available N in soil b) Available P in soil
c) Available K in soil d) Available Mg in soil **Ans : a**

386. 'Tarai' soils are deficient in

a) N b) P
c) S d) Zn **Ans : b**

□□□

6 Chapter

Manures & Fertilizers

Multiple Choice Questions (MCQ's)

1. Which fertilizer contains nutrients other than primary nutrients?

   a) Diammonium phosphate b) Urea
   c) Ammonium sulphate d) Ammonium nitrate **Ans : c**

2. What is the acid equivalent of 100 kg of ammonium sulphate nitrate?

   a) 60 kg $CaCO_3$ b) 80 kg $CaCO_3$
   c) 94 kg $CaCO_3$ d) 110 kg $CaCO_3$ **Ans : c**

3. What is the acidifying capacity of ammonium nitrate expressed in kg of $CaCO_3$ per kg of N supplied?

   a) 5.41 b) 3.61
   c) 1.89 d) 1.81 **Ans : c**

4. Which fertilizer is physiologically neutral?

   a) Calcium ammonium nitrate b) Ammonium sulphate nitrate
   c) Ammonium nitrate d) Ammonium chloride **Ans : a**

5. Which is the chemical formula of calcium ammonium nitrate?

   a) $Ca(NO_3)_2 . NH_4NO_3$ b) $CaNH_4NO_3$
   c) $CaCO_3 . NH_4NO_3$ d) $CaCl_2 . NH_4NO_3$ **Ans : c**

6. Highly concentrated widely used solid nitrogenous fertilizer is
   a) Ammonia b) Ammonium nitrate
   c) Urea d) Ammonium sulphate **Ans : c**

7. Which fertilizer would you prefer for groundnut grown on light soil?
   a) Ammonium sulphate b) Urea
   c) Ammonium nitrate d) Ammonium sulphate nitrate **Ans : a**

8. Which fertilizer would you prefer for low land paddy?
   a) Urea b) Ammonium nitrate
   c) Ammonium sulphate nitrate d) Ammonium sulphate **Ans : a**

9. The fertilizer that may cause maximum phytotoxicity is
   a) Urea phosphate b) Urea
   c) Ammonium nitrate d) Ammonium chloride **Ans : b**

10. Which is ammonium forming fertilizer?
    a) Ammonium sulphate b) Ammonium chloride
    c) Calcium ammonium nitrate d) Urea **Ans : d**

11. Percentage of calcium in single super phosphate is
    a) 16-18% b) 12-14%
    c) 26-29% d) 18-21% **Ans : d**

12. Percentage of sulphur in single super phosphate is
    a) 16% b) 12%
    c) 26% d) 18% **Ans : b**

13. Percentage of P_2O_5 in single super phosphate is
    a) 48-52% b) 26%
    c) 16% d) 34% **Ans : c**

14. Percentage of calcium in triple super phosphate is
    a) 12-14% b) 18-21%
    c) 48-52% d) 36-38% **Ans : a**

15. Percentage of P_2O_5 in triple super phosphate is
    a) 12-14% b) 18-21%
    c) 48-52% d) 36-38% **Ans : c**

16. Percentage of calcium in $Ca(NO_3)_2$ is
    a) 16% b) 19%
    c) 26% d) 34% **Ans : b**

17. Nitrogen percentage in $Ca(NO_3)_2$ is
   a) 16% b) 14%
   c) 15% d) 20% **Ans : c**

18. Percentage of magnesium in magnesia (MgO) is
   a) 45% b) 55%
   c) 60% d) 65% **Ans : b**

19. Percentage of magnesium in basic slag is
   a) 3-4 b) 3-8%
   c) 20 d) 26% **Ans : a**

20. Percentage of P_2O_5 in basic slag is
   a) 3-4% b) 3-8%
   c) 20% d) 26% **Ans : b**

21. What is the percentage of calcium in basic slag?
   a) 18% b) 20%
   c) 34% d) 26% **Ans : b**

22. What is the percentage of magnesium in serpentine?
   a) 18% b) 20%
   c) 34% d) 26% **Ans : d**

23. What is the percentage of magnesium in $Mg(NO_3)_2$?
   a) 14% b) 16%
   c) 19% d) 21% **Ans : b**

24. Single super phosphate is best suitable for
   a) Neutral and alkaline soils b) Neutral and acid soils
   c) Acid and alkaline soils d) Neutral soils only **Ans : a**

25. A potential inhibitor of nitrification is
   a) CS_2 b) Nitrogin
   c) ST d) U-Form **Ans : a**

26. Percentage of nitrogen in ammonium thiosulphate $[(NH_4)_2S_2O_3]$ is
   a) 26% b) 18%
   c) 12% d) 8% **Ans : c**

27. Percentage of sulphur in ammonium thiosulphate $[(NH_4)_2S_2O_3]$ is
   a) 26% b) 18%
   c) 12% d) 8% **Ans : a**

28. In S reduction, SO_4 to S^o is carried out by
a) Desulphovibrio b) Desulphomaculatum
c) Desulphomonas d) Both a and b **Ans : d**

29. Percentage of nitrogen in $CaCN_2$ is
a) 24.0 b) 21.0
c) 12- 14 d) 30.0 **Ans : b**

30. What is the conversion factor for the conversion of Mg in to MgO?
a) 0.751 b) 0.399
c) 1.658 d) 0.608 **Ans : c**

31. Which nutrients element is the basic constituent of life?
a) Nitrogen b) Phosphorous
c) Sulphate d) Ca and Mg **Ans : a**

32. Which nutrient is essential for the increases in protein content in food fodder crops?
a) Nitrogen b) Phosphorous
c) Sulphate d) Ca and Mg **Ans : a**

33. ____________ establishes early root development and growth, thereby helping to establish seeding quickly.
a) Nitrogen b) Phosphorous
c) Sulphate d) Both a and b **Ans : b**

34. ____________ brings about the early maturity of crops, particularly the cereals, and counteracts the effects of excessive nitrogen.
a) Phosphorous b) Potassium
c) Ca and Mg d) Sulphate **Ans : a**

35. Which nutrients stimulates flowering and aids in seed form?
a) Phosphorous b) Potassium
c) Ca and Mg d) Sulphate **Ans : a**

36. Which of the following nutrients, when applied to legumes, enhances the activity of rhizobia and increases the formation of root nodules?
a) Phosphorous b) Nitrogen
c) Potassium d) Molybdenum **Ans : a**

37. Which of the following elements, when excess may cause deficiencies of trace elements, particularly iron and zinc?
a) Phosphorous b) Nitrogen
c) Potassium d) Molybdenum **Ans : a**

38. ____________ increases the vigour and disease resistance to plants.

a) N b) K
c) S d) Ca and Mg **Ans : b**

39. Of the following nutrients, which one increases the winter hardiness, reduces lodging of crops.

a) N b) K
c) S d) Ca and Mg **Ans : b**

40. _____________ is essential is the formation and transfer of starches and sugars

a) N b) K
c) S d) Ca and Mg **Ans : b**

41. Which nutrient is essential for stomatal regulations in plants?

a) Ca b) Mg
c) K d) Mo **Ans : c**

42. What is the purpose of nitrification inhibitors?

a) To stop conversion nitrite to nitrate and hold nitrite by anion exchange to reduce its chances of leaching to ground water
b) To stop rapid mineralization of NH_4^+ ions to reduce leaching losses of N to ground water
c) To convert nitrate to NH_4^+ to reduce the potential losses of N as nitrate to ground water
d) Inhibit conversion of NH_4^+ to nitrite to keep NH_4^+ on exchange sites, reducing losses of nitrate to ground water **Ans : b**

43. ______________ is not a constituent of chlorophyll but, it helps in chlorophyll formation and encourages vegetative plant growths.

a) Phosphorous b) Magnesium
c) Sulphur d) Nitrogen **Ans : c**

44. ______________ is an essential constituent of certain volatile compounds and helps in the reduction-oxidation system in respiration.

a) Phosphorous b) Magnesium
c) Sulphur d) Nitrogen **Ans : c**

45. Which is the element essential for the calcium metabolism, both with its uptake by roots and its efficient use in plants?

a) Mg b) N
c) S d) B **Ans : d**

46. ______________ helps in absorption of nitrogen.
a) Mn b) B
c) Mo d) Zn
Ans : b

47. Element which is a constituent of cell membrane and in essential for cell division is
a) Mn b) B
c) Mo d) Zn
Ans : b

48. Of the micronutrients, which element is essential for translocation of sugars in plant?
a) B b) Mo
c) Zn d) Mn
Ans : a

49. Of the micronutrients which element is essential for flowering and fruiting process and pollen germination?
a) B b) Mo
c) Zn d) Mn
Ans : a

50. ______________ is not a constituent of chlorophyll but, it helps in its formation.
a) Fe b) Mn
c) Zn d) Cu
Ans : a

51. Which nutrient element influences the formation of some growth hormones in the plant?
a) Fe b) Mn
c) Zn d) Cu
Ans : c

52. Which nutrient element brings about the reduction of nitrates to ammonia prior to amino acid and protein synthesis in the cells of the plant?
a) Boron b) Molybdenum
c) Zinc d) Iron
Ans : b

53. ______________ is essential for nitrogen fixation, both (symbiotic and non-symbiotic)
a) Boron b) Molybdenum
c) Zinc d) Iron
Ans : b

54. ______________ acts as "electron carrier" in enzymes which brings about oxidation-reduction reactions in plants.
a) Fe b) Mn
c) Zn d) Cu
Ans : d

55. ______________ helps in utilization of iron in chlorophyll synthesis.
a) Fe b) Mn
c) Zn d) Cu **Ans : d**

56. Based on the functional basis, which of the following elements are classified as accessory structural elements?
a) C and H b) C, H and O
c) N and P d) All **Ans : c**

57. Based on the functional basis, which of the following elements is classified as accessory structural elements?
a) C b) H
c) O d) S **Ans : d**

58. Based on the functional basis, which of the following elements are classified as regulators and carriers?
a) C, H, O b) N, P ,S
c) K, Ca, Mg d) Ca, Mg and Fe **Ans : c**

59. Based on the functional basis, which of the following elements are classified as main structure elements?
a) C, H, O b) N, P ,S
c) K, Ca, Mg d) Ca, Mg and Fe **Ans : a**

60. Based on the functional basis, which of the following elements are classified as catalysts and activators?
a) Fe, Mn, Zn, Cu b) B, Mo, Cl
c) Fe, Mn, Cu, Mo d) All **Ans : d**

61. Which nutrient element is essential for the synthesis and translocation of carbohydrates in plant?
a) K b) Ca
c) Mg d) S **Ans : a**

62. Of the following nutrients, which nutrients loss is most by leaching?
a) N b) P
c) K d) Zn **Ans : a**

63. Example(s) for nitrification inhibitors
a) Nitropyrin and ST b) ST and AM
c) KN_3 d) Both b and c
e) All the above **Ans : e**

64. Oxamide, CDU and IBDU are
   a) Low water soluble compounds
   b) Sparingly water soluble compounds
   c) Soluble or relatively soluble compounds
   d) Nitrification inhibitors
   e) Urease inhibitors
   Ans : a

65. What is the percentage of nitrogen in urea-formaldehyde?
   a) 30 b) 35
   c) 38 d) 41
   Ans : c

66. What is the percentage of nitrogen in sulphur coated urea (SCU)?
   a) 30 b) 35
   c) 38 d) 41
   Ans : c

67. What is the percentage of nitrogen oxamide?
   a) 36.8 b) 31.8
   c) 32 d) 33 – 35
   Ans : b

68. What is the percentage of nitrogen in thiourea?
   a) 33 – 35 b) 33 – 38
   c) 31. 8 d) 36.8
   Ans : d

69. Nitrogen percentage in raw bone meal is
   a) 4 – 8 b) 2 – 5
   c) 3 – 5 d) 5 – 8
   Ans : c

70. Percentage of citrate soluble P_2O_5 in rock phosphate is
   a) 25 – 35 % b) 5 – 17 %
   c) 22 – 26 % d) 12 – 18 %
   Ans : b

71. What is the product of the reaction $NH_3 + H_4P_2O_7$?
   a) Monoammonium phosphate b) Diammonium
   c) Triammonium phosphate d) Dicalcium phosphate
   Ans : c

72. Dicalcium phosphate is the product of the reaction of
   a) Rock phosphate + HCl b) Rock phosphate + H_3PO_4
   c) Rock phosphate + $H_4P_2O_7$ d) Rock phosphate + H_2SO_4
   Ans : a

73. Who first discovered that plants need mineral nutrients for their growth and developments?
   a) Nicholas Theodore de Saussure b) Jean Baptiste Bousingault
   c) Justin von Liebig d) Both a and b
   e) All the above
   Ans : d

74. Who discovered the essentiality of iron to plants?
a) Gris b) Sprengel
c) L. Sommer and P. Lipman d) Theodore de Saussure **Ans : a**

75. Who discovered the essentiality of nitrogen to plants?
a) Gris b) Sprengel
c) L. Sommer and P. Lipman d) Theodore de Saussure **Ans : d**

76. Who discovered the essentiality of boron to plants?
a) Sprengel b) L. Sommer and P. Lipman
c) K. Warrington d) I. Arnon and P.R. Stout **Ans : c**

77. Who discovered the essentiality of molybdenum to plants?
a) Sprengel b) L. Sommer and P. Lipman
c) K. Warrington d) I. Arnon and P.R. Stout **Ans : d**

78. Who discovered the essentiality of zinc to plants?
a) Gris b) Sprengel
c) L. Sommer and P. Lipman d) Theodore de Saussure **Ans : c**

79. Which is a primary plant nutrient
a) Sulphur b) Cobalt
c) Sodium d) Potassium **Ans : d**

80. Which is an essential nutrient element?
a) Aluminium b) Iodine
c) Boron d) Barium **Ans : c**

81. Which of the following is an organic chemical fertilizer?
a) Calcium ammonium nitrate b) Urea
c) Ammonium phosphate d) Muriate of potash **Ans : b**

82. Which of the following is a complex fertilizer?
a) Ammonium sulphate nitrate b) Calcium ammonium nitrate
c) Single superphosphate d) Ammonium phosphate **Ans : b**

83. Which of the following is an ultra micronutrient?
a) Boron b) Molybdenum
c) Zinc d) Iron **Ans : b**

84. Of the followings, which is a micronutrient fertilizer?
a) Calcium sulphate b) Magnesium sulphate
c) Potassium sulphate d) Zinc sulphate **Ans : d**

85. The number of essential elements required for the growth of most higher plant is

a) 17 b) 6
c) 8 d) 16 Ans : a

86. Which fertilizer is basic fertilizer?

a) Urea b) Calcium ammonium nitrate
c) Sodium nitrate d) Ammonium sulphate Ans : c

87. Which fertilizer material is partly soluble in water?

a) Urea b) Calcium ammonium nitrate
c) Sodium nitrate d) Ammonium sulphate Ans : b

88. Which fertilizer is highly hygroscopic?

a) Urea b) Calcium ammonium nitrate
c) Sodium nitrate d) Ammonium sulphate Ans : a

89. Which fertilizer contains sulphur?

a) Urea b) Calcium ammonium nitrate
c) Sodium nitrate d) Ammonium sulphate Ans : d

90. Percent nitrogen in urea is

a) 20.5 - 21 b) 33.5 - 34
c) 26 d) 44-46 Ans : d

91. The permissible maximum biuret content of urea produced in India is

a) 0.5% b) 1.0%
c) 1.5% d) 2.0% Ans : c

92. Single superphosphate is produced by the reaction of rock phosphate with

a) HNO_3 b) H_3PO_4
c) H_2SO_4 d) HCl Ans : c

93. Concentrated super phosphate is produced by the reaction of rock phosphate with

a) HNO_3 b) H_3PO_4
c) H_2SO_4 d) HCl Ans : b

94. The chief component of single superphosphate that supplies phosphorous to plant is

a) Gypsum b) Free phosphoric acid
c) Monocalcium phosphate d) Rock phosphate Ans : c

95. Which is the acid soluble phosphatic fertilizer?

a) Super phosphate b) Dicalcium phosphate

c) Rock phosphate d) Monocalcium phosphate **Ans : c**

96. The per cent content of N and P in DAP is

a) 20-20 b) 18-46

c) 46-18 d) 22-48 **Ans : b**

97. Which is the water soluble phosphate?

a) $Ca(H_2PO_4)_2$ b) $CaHPO_4$

c) $Ca_3(PO_4)_2$ d) CaH_2PO_4 **Ans : a**

98. The conversion factor to convert % P to % P_2O_5 is

a) 1.29 b) 2.90

c) 2.29 d) 0.4 **Ans : c**

99. The fertilizer that contains maximum sulphur is

a) Nitric phosphate b) Concentrated super phosphat

c) Single super phosphate d) Enriched super phosphate **Ans : c**

100. The super phosphate is largely produced, now-a-days from

a) Phosphate guanos b) Rock phosphate

c) Bone meal d) Basic slag **Ans : b**

101. The per cent K_2O in KCl is

a) 48 b) 64

c) 127 d) 80 **Ans : b**

102. The formula for potassium meta phosphate is

a) $K_4P_2O_7$ b) KH_2PO_4

c) KPO_3 d) K_2HPO_4 **Ans : c**

103. Which form of nitrogen absorbed by plant is toxic?

a) NH_4^+ b) NO_3^-

c) NO_2^- d) Amide **Ans : c**

104. Ammonia volatilization is highest from

a) NH_4Cl b) NH_4NO_3

c) $CaCO_3.NH_4NO_3$ d) $(NH_4)_2SO_4$ **Ans : d**

105. The process in which nitrate is reduced is called

a) Chemo-denitrification b) Indirect denitrification

c) Non-biological nitrification d) Denitirification **Ans : d**

106. By denitrification, which gas largely escapes to the atmosphere?
a) N_2O b) NO
c) NO_2 d) N_2 Ans : a

107. Which organism fixes atmospheric nitrogen?
a) Azotobacter b) Nitrobacter
c) Nitrosomonas d) Nitrosococcus Ans : a

108. Which microorganism is responsible for nitrification or nitrite to nitrate?
a) Nitrosomonas b) Nitrobacter
c) Clostridium d) Nitrosococcus Ans : b

109. Conversion of ammonical nitrogen to nitrate nitrogen is known as
a) Ammonification b) Nitrification
c) Denitrification d) Nitrate formation Ans : b

110. Volatilization loss of nitrogen is comparatively more in
a) Acid soils b) Alkaline soils
c) Neutral soils d) Both a and b Ans : b

111. Ammonia volatilization is highest with
a) NH_4NO_3 b) $(NH_4)_2.SO_4$
c) NH_4Cl d) $CO(NH_2)_2$ Ans : d

112. The predominant phosphate in single superphosphate is
a) Tricalcium phosphate b) Dicalcium phosphate
c) Monocalcium phosphate d) Apatite Ans : c

113. Root development is stimulated due to supply of
a) N b) P
c) K d) Ca Ans : b

114. Fixation of phosphate is more in
a) Acid soils b) Neutral soils
c) Submerged soils d) None Ans : a

115. Trench method of preparing FYM has been recommendation by
a) Schmidt Eggersgluss b) Wager and Krosender
c) S.V. Desai d) N. Acharya Ans : d

116. What is the C:N ratio of organic matter?

a) 8:1 b) 10:1
c) 20:1 d) 15: 1 **Ans : b**

117. What is the chemical formula of anhydrous ammonia?

a) NH_4 b) $NH_3 . nH_2O$
c) NH_3 d) $NH_4 . nH_2O$ **Ans : c**

118. When anhydrous ammonia is applied to the soil, what will happen to the soil pH?

a) pH will increase to 9.0 and above
b) pH will decrease to 3.0 and below
c) pH will increase to 8.0 and above
d) pH will decrease to 4.0 and below **Ans : a**

119. Which is the enzyme involved in the hydrolysis of urea?

a) Urease b) Nitrogenase
c) Nitrate reductase d) All **Ans : a**

120. Triple superphosphate is produced by the reaction of

a) Rock phosphate with sulphuric acid
b) Rock phosphate with hydrochloric acid
c) Rock phosphate with phosphoric acid
d) Phosphoric acid with sulphuric acid **Ans : d**

121. Single superphosphate is produced by the reaction of

a) Rock phosphate with sulphuric acid
b) Rock phosphate with hydrochloric acid
c) Rock phosphate with phosphoric acid
d) Phosphoric acid with sulphuric acid **Ans : a**

122. Treating rock phosphate with sulphuric acid for the production of phosphoric acid is called as

a) Green process b) Wet process
c) Yellow process d) Both a and b **Ans : d**

123. Phosphoric acid made by burning is called as

a) White acid b) Furnace acid
c) Green acid d) Both a and b **Ans : d**

124. Single super phosphate is

a) A neutral fertilizer b) An acid producing fertilizer
c) A alkali producing fertilizer d) None **Ans : a**

125. Which of the following phosphatic fertilizers is called as concentrated superphosphate?
a) Single superphosphate b) Double superphosphate
c) Triple superphosphate d) All
Ans : c

126. Monoammonium phosphate is produced by the reaction of
a) Single superphosphate with NH_3
b) Triple superphosphate with NH_3
c) Double superphosphate with NH_3
d) Both a and b
e) All the above
Ans : d

127. What is the conversion factor for the conversion of MgO in to Mg?
a) 0.751 b) 0.399
c) 1.658 d) 0.608
Ans : d

128. What is the conversion factor for the conversion of Ca in to CaO?
a) 0.751 b) 0.399
c) 1.129 d) 0.347
Ans : b

129. What is the conversion factor for the conversion of CaO in to Ca?
a) 0.751 b) 0.399
c) 1.129 d) 0.347
Ans : a

130. Percentage of sulphur in potassium sulphate (K_2SO_4) is
a) 21.0 b) 18.5
c) 22.0 d) 12-14
Ans : b

131. What is the percentage of sulphur in gypsum?
a) 21.0% b) 23.5%
c) 22.0% d) 18.5%
Ans : a

132. Percentage of sulphur in ammonium sulphate [$NH_4 \cdot (SO_4)_2$] is
a) 21.0% b) 18.5%
c) 22.0% d) 12-14%
Ans : a

133. Percentage of calcium in calcium ammonium nitrate (CAN) is
a) 30.0% b) 18 - 21%
c) 8.0% d) 26.0%
Ans : c

134. What is the percentage of calcium in gypsum?
a) 30.0% b) 18 - 21%
c) 8.0% d) 26.0%
Ans : a

135. Percentage of calcium in single super phosphate is
a) 40.0% b) 33.0%
c) 23.0% d) 18-21% **Ans : d**

136. What is the percentage of sulphur in triple superphosphate?
a) 12 - 14% b) 4 - 8%
c) 2 - 2.5% d) 1- 1.5 % **Ans : d**

137. What is the percentage of P_2O_5 in mono ammonium phosphate?
a) 46 - 53% b) 48 - 62%
c) 43 - 44% d) 13 - 42% **Ans : b**

138. What is the percentage of P_2O_5 in urea ammonium phosphate?
a) 46 - 53% b) 48 - 62%
c) 43 - 44% d) 13 - 42% **Ans : d**

139. What is the percentage of sulphate in urea sulphate?
a) 46 - 53% b) 43 - 44%
c) 6 - 11 % d) 24.0% **Ans : c**

140. Which is the first fertilizer produced in India?
a) Single superphosphate b) Triple superphosphate
c) Urea d) Diammonium phosphate **Ans : a**

141. Which is the high fertilizer consuming state in India?
a) Punjab b) Delhi
c) Haryana d) Bihar **Ans : a**

142. What is the desired ratio of NPK India?
a) 2 : 1 : 1 b) 3 : 2 : 1
c) 4 : 2 : 1 d) 4 : 2 : 2 **Ans : c**

143. Commonly used NPK ratio in India is
a) 8.9:2.7:1 b) 8:3:1
c) 9:3:1 d) 4:2:1 **Ans : a**

144. In 1954, who discovered the essentiality of Cl^- to plants?
a) T.C. Broyer b) A.B. Carlton
c) C.M. Johnson d) P.R. Stout
e) All the above **Ans : e**

145. Of the following, which is an example of fertilizer elements?
a) N, P and K b) Ca, Mg and S
c) Both a and b d) C, H, O, N, P and K **Ans : a**

146. Examples of secondary nutrients are
a) N, P and K
b) Ca, Mg and S
c) Fe, Mn, Cu, Cl, Mo, and B
d) All
Ans : b

147. Examples of major nutrients are
a) N, P and K
b) Ca, Mg and S
c) Fe, Mn, Cu, Cl, Mo, and B
d) All
Ans : a

148. Examples of macro nutrients are
a) N, P and K
b) Ca, Mg and S
c) Fe, Mn, Cu, Cl, Mo, and B
d) Both a and b
Ans : d

149. Fertilizers containing Cl^- ions are not suitable for
a) Sugarcane
b) Potato
c) Wheat
d) Pigeonpea
Ans : b

150. Stem nodulating green manure plant is
a) Sunnhemp
b) Sesbania aculiata
c) Sesbania rostrata
d) Tephrosea sps.
Ans : c

151. A fodder and pasture legume
a) Glyricidia
b) Stylosanthus
c) Cowpea
d) Sesbania
Ans : b

152. How much of muriate of potash is required to supply 50 kg K_2O/ha
a) 50 kg
b) 73.3kg
c) 83.3kg
d) 125.3 kg
Ans : c

153. Amongst oil cakes, the highest nitrogen content is in
a) Castor cake
b) Neem cake
c) Coconut cake
d) Groundnut cake
Ans : d

154. Among the following the set of essential nutrient elements, which are absorbed as anions by plants is
a) Boron, chlorine, copper, iron, manganese
b) Copper, iron, phosphorus, sulphur, zinc
c) Potassium, manganese, molybdenum, phosphorus, zinc
d) Boron, chloride, molybdenum, phosphorus, sulphur
Ans : d

155. A fertilizer which supplies three essential plant nutrients is
a) DAP
b) MOP
c) SSP
d) SOP
Ans : c

156. The Fe-chelate suitable for application to crops grown on acid soils is

a) Fe-EDTA b) Fe-EDDHA
c) Fe-DTPA d) Fe-HEDTA **Ans : a**

157. Which one of the following fertilizers contain water soluble P

a) SSP b) DCP
c) MAP d) Both a) and c) **Ans : d**

158. Stem nodulation occurs in green manure crop of

a) Sesbania aculeate b) Sesbania rostrata
c) Crotolaria juncea d) Aeschynomene afraspera **Ans : b**

159. Which one of the following is a green house gas

a) Oxygen b) Ammonia
c) Methane d) Chlorine **Ans : c**

❑❑❑

7 Chapter

Irrigation Water Management

Multiple Choice Questions (MCQ's)

1. The largest quantity of water is in

   a) Ocens b) Polar ice
   c) Ground water d) Atmospheric water **Ans : a**

2. The quantity of fresh water in the earth is

   a) 97% b) 2.60%
   c) 0.02% d) 77.23% **Ans : b**

3. The quantity of water in the oceans is

   a) 97% b) 2.60%
   c) 0.02% d) 77.23% **Ans : a**

4. The average rainfall of India has been estimated to be

   a) 1194mm b) 1294mm
   c) 1394mm d) 1094mm **Ans : a**

5. The total geographical area of India is

   a) 328m ha b) 308m ha
   c) 318m ha d) 338m ha **Ans : a**

6. The total amount of precipitation in India accounts for

   a) 400m ha m b) 328m ha m
   c) 450m ha m d) 300m ha m **Ans : a**

7. The total amount of precipitation received between June-September accounts for

a) 400m ha m b) 328m ha m
c) 450m ha m d) 300m ha m **Ans : d**

8. The total amount of runoff from outside India accounts for

a) 400m ha m b) 20m ha m
c) 450m ha m d) 300m ha m **Ans : b**

9. The ultimate irrigation potential of India has been estimated to be

a) 100m ha b) 200m ha
c) 150m ha d) 139m ha **Ans : d**

10. ....river basin has highest ground water potential

a) Ganga b) Godavari
c) Narmada d) Brhamputra **Ans : a**

11. The diameter of water molecule is

a) $3x10^{-8}$cm b) $3x10^{-2}$cm
c) $3x10^{-4}$cm d) $3x10^{-6}$cm **Ans : a**

12. In water molecule the two hydrogen atoms link to the oxygen atom at an angle of

a) 104.5^{o} b) 109°
c) 180° d) 90° **Ans : a**

13. The latent heat of vaporization of water is ....calories/g of water

a) 540 b) 100
c) 180 d) 90 **Ans : a**

14. Water passes from liquid to gaseous state by absorbing ....calories/g of water

a) 540 b) 100
c) 180 d) 90 **Ans : a**

15. The attraction of water molecules for each other is termed as

a) Adhesion b) Cohesion
c) Both d) None **Ans : b**

16. The attraction of water molecules for solid surface is termed as

a) Adhesion b) Cohesion
c) Both d) None **Ans : a**

17. Capillarity is due to
a) Adhesion b) Cohesion
c) Both d) None
Ans : c

18. There is ....relationship between height of rise and size of pore through which water rises
a) Positive b) Inverse
c) Both d) None
Ans : a

19. When all the pores are filled with water and the soil is said to be saturated this condition is termed as
a) Maximum water holding capacity
b) Field capacity
c) Water holding capacity
d) None
Ans : a

20. When all macropores are filled with air, micropores filled with water and water movement is through capillary this water is termed as
a) Capillary water b) Hygroscopic water
c) Available water d) None
Ans : a

21. At field capacity
a) All macropores are filled with water, micropores filled with air
b) All macropores are filled with air, micropores filled with water
c) All macropores and micropores are filled with air
d) All macropores and micropores are filled with water
Ans : b

22. At field capacity the soil water potential ranges from
a) -0.1to -0.3 bars b) -0.1to -0.3 M Pa
c) -0.01to -0.03 bars d) -1to -3 bars
Ans : a

23. Field capacity is the........... limit of soil water availability to plants
a) Upper b) Lower
c) Middle d) All
Ans : a

24. Movement of soil water takes in the direction of
a) Increasing tension b) Decreasing tension
c) Both d) None
Ans : a

25. At PWP the soil water potential ranges from
a) -15 bars b) -15 M Pa
c) -0. 1to -0. 3 bars d) -1to -3 bars
Ans : a

26. At soil water potential of ....water can move only in vapour phase
a) -0.1to -0.3 bars b) -3 M Pa
c) -0.01to -0.03 bars d) -1to -3 bars
Ans : b

27. At Ultimate wilting point the soil water potential ranges from
   a) -15 to -30 bars b) –30 to -45
   c) -45 to -60 bars d) -60 bars **Ans : d**

28. When water is held tightly as thin film around soil particles and can move only in vapour phase this condition is termed as
   a) Ultimate wilting point b) Field capacity
   c) Hygroscopic water d) PWP **Ans : c**

29. When plant is not dead but remains in wilted condition this condition is termed as
   a) Ultimate wilting point b) Field capacity
   c) Hygroscopic water d) PWP **Ans : d**

30. When plants can not absorb any water and die eventually this condition is termed as
   a) Ultimate wilting point b) Field capacity
   c) Hygroscopic water d) PWP **Ans : a**

31. Water movement in the soil comprises of
   a) Infiltration b) Redistribution
   c) Withdrawal d) All **Ans : d**

32. The water available to the plants is the capillary water between
   a) PWP and ultimate wilting point
   b) Field capacity and PWP
   c) Field capacity and hygroscopic water
   d) PWP and hygroscopic water **Ans : b**

33. The available water holding capacity can be determined by
   a) FC - ultimate wilting point
   b) FC- PWP
   c) Field capacity -hygroscopic water
   d) PWP -hygroscopic water **Ans : b**

34. ....soil has lowest quantity of available water
   a) Sandy loam b) Silty loam
   c) Clay loam d) Clay **Ans : a**

35. ....soil has highest quantity of available water
   a) Sandy loam b) Silty loam
   c) Clay loam d) Clay **Ans : d**

36. ....soil texture has greater quantity of available water
   a) Fine b) Coarse
   c) Medium d) All **Ans : a**

37. The soil water in terms of reducing availability is in the sequence of
   a) FC, ultimate wilting point, hygroscopic water, PWP
   b) FC, PWP, hygroscopic water, ultimate wilting point
   c) Field capacity, hygroscopic water, PWP, ultimate wilting point
   d) PWP, hygroscopic water, ultimate wilting point, PWP **Ans : b**

38. Water potential is expressed as
   a) Volume b) Mass
   c) Weight d) All **Ans : d**

39. 1bar equals to
   a) 100 kPa b) 0.99 atm
   c) 14.5 psi d) All **Ans : d**

40. 1psi equals to
   a) 6.9 kPa b) 2.31 feet head
   c) Both d) None **Ans : c**

41. 1kPa equals to
   a) 1 J kg^{-1} b) 0.99 bar
   c) 14.5 psi d) All **Ans : a**

42. 1M Pa equals to
   a) 1 J kg^{-1} b) 10 bar
   c) 100 psi d) All **Ans : b**

43. Osmotic potential is due to presence of
   a) Solutes b) Solvents
   c) Both d) All **Ans : a**

44. When soil water is at a hydrostatic pressure greater than atmosphere , itspressure potential is considered to be
   a) +ve b) -ve
   c) Zero d) None **Ans : a**

45. When soil water is at a hydrostatic pressure lesser than atmosphere , its pressure potential is considered to be
   a) +ve b) -ve
   c) Zero d) None **Ans : b**

46. When soil water is at a hydrostatic pressure equal to atmosphere, its pressure potential is considered to be
   a) +ve b) -ve
   c) Zero d) None **Ans : c**

47. Water risen in a capillary tube is characterized by a
a) +ve pressure potential b) -ve pressure potential
c) Zero pressure potential d) None **Ans : b**

48. Free water is characterized by a
a) +ve pressure potential b) -ve pressure potential
c) Zero pressure potential d) None **Ans : c**

49. The -ve pressure potential is often termed as
a) Matric potential b) Suction potential
c) Pressure potential d) None **Ans : a**

50. The pressure due to solutes is termed as
a) Osmotic pressure b) Imbibitional pressure
c) Suction pressure d) None **Ans : a**

51. Diffusion pressure deficit is expressed as
a) DPD= OP-TP b) DPD= OP+TP
c) DPD= OP-TP +IP d) DPD= OP-TP-IP **Ans : c**

52. The potential is expressed in terms of
a) Energy/unit mass b) Energy/unit volume
c) Energy/unit weight d) All **Ans : d**

53. Soil moisture tension is expressed in terms of
a) Atmospheres b) Bars
c) Psi d) None **Ans : a**

54. The soil moisture characteristic curve is strongly affected by soil
a) Texture b) Structure
c) Both d) None **Ans : a**

55. The process of gradually drying a saturated soil sample by applying increasing suction and recording successive measurement of water content is termed as
a) Sorption b) Desorption
c) Hysteresis d) None **Ans : b**

56. The process of gradually wetting a dry soil sample while reducing the suction is termed as
a) Sorption b) Desorption
c) Hysteresis d) None **Ans : a**

57. The difference in soil moisture content at a given suction pressure during the process of desorption and sorption is termed as
a) Sorption b) Desorption
c) Hysteresis d) None **Ans : c**

58. The soil moisture content at a given suction pressure during the process of hysteresis is greater for
a) Sorption b) Desorption
c) Equal in both d) None **Ans : b**

59. Neutron moisture meter is suitable for ....moisture content in the field
a) Surface b) Profile
c) Both d) None **Ans : b**

60. Neutron moisture meter is not suitable for ....moisture content in the field
a) Surface b) Profile
c) Both d) None **Ans : a**

61. Hydraulic conductivity is expressed as
a) V=Kf b) f=VK
c) K=Vf d) None **Ans : a**

62. The air entry value to tensiometer cup is as high as ....
a) 0.75 bar b) 0.80 bar
c) 0.85 bar d) 0.90 bar **Ans : c**

63. The tensiometer are suitable for...........soils
a) Clayey b) Sandy
c) Silty d) All **Ans : b**

64. The tensiometer are more accurate in............range of soil moisture
a) Dry b) Wet
c) Both d) None **Ans : b**

65. The tensiometer indicates
a) Soil moisture content b) Soil moisture suction
c) Soil moisture potential d) All **Ans : b**

66. The nylon blocks are sensitive for tensions less then............
a) -2 bars
b) -1 to -15 bars
c) Entire range of available soil moisture
d) All **Ans : a**

67. The plaster of paris blocks are sensitive for tensions between............

a) -2 bars
b) -1 to -15 bars
c) Entire range of available soil moisture
d) All

Ans : b

68. The fiberglass blocks are sensitive for tensions ....

a) -2 bars
b) -1 to -15 bars
c) Entire range of available soil moisture
d) All

Ans : c

69. If a crop is to be irrigated at 0.75 IW/CPE ratio with 60 mm depth, it should be irrigated after ....CPE

a) 60 mm b) 75 mm
c) 80 mm d) 70 mm

Ans : b

70. If a crop is to be irrigated at 0.6 IW/CPE ratio with 6 cm depth, it should be irrigated after ....CPE

a) 60 mm b) 10 mm
c) 10 cm d) 70 mm

Ans : c

71. Under saline soil conditions ....irrigation could help in obtaining better crop stand and yield

a) Sprinkler b) Subsurface
c) Check basin d) Furrow

Ans : b

72. At soil water potential of ....water can move only in vapour phase

a) -0.1to -0.3 bars b) -30 bars
c) -0.01to -0.03 bars d) -1to -3 bars

Ans : b

73. Use of brackish water for agriculture has become more relevant due to .... irrigation:

a) Sprinkler b) Furrow
c) Check basin d) Micro

Ans : d

74. Drainage lowers underground water table so as to facilitate increased ....

a) Water holding capacity b) Root zone depth
c) Microbial activity d) Porosity

Ans : b

75. If the field capacity is 18% and permanent wilting points is 6%, 50% of the available soil moisture will be:

a) 8 % b) 10 %
c) 6 % d) 14 %

Ans : c

76. Better drainage helps crop growth most due to ....
   a) Better aeration of the soil
   b) Better mineralisation of manures and fertilizers
   c) Better root growth
   d) Better availability of nutrients
   Ans : c

77. Tillering in rice is encouraged by ....
   a) Optimum moisture content of the soil
   b) Saturation of the soil
   c) Shallow depth of standing water
   d) Higher depth of standing water
   Ans : b

78. One acre inch of water is equivalent to.......... litres of water:
   a) 102800 b) 126800
   c) 140000 d) 180000
   Ans : a

79. One cubic foot of water is equivalent to............. litres of water:
   a) 102800 b) 28.31
   c) 10^3 d) 102.79
   Ans : b

80. One cubic metre of water is equivalent to............. litres of water:
   a) 102800 b) 28.31
   c) 10^3 d) 102.79
   Ans : c

81. One acre inch of water is equivalent to............. cubic metres of water:
   a) 102800 b) 28.31
   c) 10^3 d) 102.79
   Ans : d

82. One cusec of water is equivalent to............. litres of water:
   a) 102800 b) 28.31
   c) 10^3 d) 102.79
   Ans : b

83. One cusec of water is equivalent to............. litres/min of water:
   a) 16.66 b) 28.31
   c) 1698.96 d) 101.94
   Ans : c

84. One cusec of water is equivalent to............. cubic metres/ hour of water:
   a) 16.66 b) 28.31
   c) 1698.96 d) 101.94
   Ans : d

85. One cumec of water is equivalent to............. litres/sec of water:

a) 10^3 b) 28.31
c) 1698.96 d) 101.94 **Ans : a**

86. One cumec of water is equivalent to............. cubic metres/ hour

a) 10^3 b) 28.31
c) 3600 d) 101.94 **Ans : c**

87. One cumec of water is equivalent to............. cubic feet/sec of water:

a) 10^3 b) 28.31
c) 35.31 d) 101.94 **Ans : c**

88. The generalized expression of Darcy's law is

a) $q = -k x\Delta H$ b) $q = k x\Delta H$
c) $q = k + \Delta H$ d) None **Ans : a**

89. In Darcy's law equation the negative sign represents

a) The flow is in the direction of the gradient
b) The direction of flow is towards decreasing head
c) Both
d) None **Ans : c**

90. In Darcy's law the proportionality factor k is designated as

a) Hyadraulic conductivity
b) Pressure gradient
c) Hyadraulic head drop between two heads
d) None **Ans : a**

91. In Darcy's law ΔH is designated as

a) Hyadraulic conductivity
b) Pressure gradient
c) Hyadraulic head drop between two heads
d) None **Ans : c**

92. Darcy's law is valid for

a) Saturated flow b) Unsaturated flow
c) Both d) None **Ans : c**

93. Unsaturated hyadraulic conductivity is also known as

a) Capillary conductivity b) Pressure gradient
c) Hyadraulic head drop d) None **Ans : a**

94. Unsaturated hyadraulic conductivity/ capillary conductivity is variable and depends upon

a) Soil moisture content b) Pressure gradient
c) Hyadraulic head drop d) None **Ans : a**

95. Watershed Management helps to increase productivity in

a) Irrigated areas b) Rainfed areas
c) Humid areas d) None of these areas **Ans : b**

96. Which one of the following is considered as rain water recycling

a) Farm pond b) Wells
c) Percolation pond d) Compartmental bunding **Ans : a**

97. At permanent wilting point the approximate value of PF would be

a) 1.6 b) 5.8
c) 4.2 d) 1.0 **Ans : c**

98. Moisture moves through soil pores at a rate proportional to the

a) Hydraulic gradient b) Hydraulic head
c) Soil porosity d) None of the above **Ans : a**

99. Water that is available to the plants and readily absorbed is:

a) Gravitational water b) Hygroscopic water
c) Rain water d) Capillary water **Ans : d**

100. Water that occurs as thin film (4 to 5 mili micron) and held with a tension of 31 atmosphere or more and not available. to plants is:

a) Hygroscopic water b) Capillary water
c) Gravitational water d) Surface water **Ans : a**

101. The most common means of expressing soil water content is the............................... of water associated with their specified quantity of soil solids.

a) Volume b) Weight
c) Vapour d) None of the above **Ans : a**

102. Indirect methods of measuring soil water content

a) Gravimetric b) Volumetric
c) Neutron scattering d) None of the above **Ans : c**

103. .... method is direct method of measuring soil water content

a) Gravimetric
b) Tensiometer
c) Neutron scattering
d) None of the above

Ans : a

104. The movement of water after rain or irrigation after 48-72 hours, the soil is said to be at -

a) Incipient wilting point
b) Field capacity
c) Permanent writing point
d) None of the above

Ans : b

105. Which of the following criteria of irrigation scheduling is more acceptable to farmers:

a) Soil moisture tension
b) Pan evaporation
c) Water potential of plant
d) Critical growth stages

Ans : d

106. Water available to plants to the maximum extent is the one held ______

a) 1/3 atmosphere
b) Pan evaporation
c) Between field capacity and wilting point
d) At 20 atmospheres

Ans : c

107. Hygroscopic water, which occurs as thin film held tenaciously around soil particles with a tension of 31 atmosphere or more, is

a) Readily available to plants
b) Not available to plants
c) Harmful to plants
d) None of these

Ans : b

108. The moisture, which is present in the form of a continuous film outside hygroscopic water around soil particles and micropore spaces is called

a) Available moisture
b) Interspace water
c) Capillary water
d) None of these

Ans : c

109. Water requirement of a crop is related to its

a) Evapotranspiration
b) Potential absorption coefficient
c) Dry matter content
d) Soil mineralogy

Ans : a

110. In most of the cases field capacity represents the water held at a tension of ____

a) 5 atmospheres
b) 1/3 atmospheres
c) 2 atmospheres
d) 10 atmospheres

Ans : b

111. Major source of water used by the plants is

a) Capillary water
b) Hygroscopic water
c) Gravitational water
d) Inter space water

Ans : a

112. Capillary water will not be available to the plants if it is held at a tension of

a) Less than 15 atmospheres b) 20 atmospheres
c) More than 15 atmospheres d) Less than 5 atmospheres **Ans : c**

113. The moisture content at which the wilting is complete and the plant die is called the

a) Permanent wilting b) Wilting co-efficient
c) Wilting range d) Ultimate wilting point **Ans : d**

114. Transpiration is

a) Evaporation b) Water loss from soil
c) Water loss from plants d) Changing status of water **Ans : c**

115. The soil moisture content at which the plants remain wilted day and night or in a humid chamber, the moisture content is called

a) Permanent wilting point b) Incipient wilting point
c) Temporary wilting point d) None of these **Ans : a**

116. The process of water entry into the soil through soil surface and vertically downward is called

a) Infiltration b) Percolation
c) Seepage d) None of these **Ans : a**

117. The flux passing through soil surface and flowing into the profile is called

a) Infiltration rate b) Percolation rate
c) Seepage rate d) None of these **Ans : a**

118. The ratio between evapotranspiration of crop and potential evapotranspiration is called

a) Water requirement b) Irrigation coefficient
c) Crop coefficient d) All **Ans : c**

120. The soil infiltration capacity depends upon

a) Texture b) Structure
c) Uniformity of soil profile d) All **Ans : d**

121. The Infiltration rate can be measured using

a) Infiltrometer b) Tensiometer
c) Both d) None of these **Ans : a**

122. Infiltration rate can be obtained from cumulative infiltration verses

a) Time b) Depth
c) Amount of water d) None of these

Ans : a

123. Permeability is often used synonymous to

a) Hydraulic head b) Hydraulic conductivity
c) Infiltrability d) None of these

Ans : b

124. The readiness with with which a porous medium transmit water is termed as

a) Percolation b) Seepage
c) Permiability d) Infiltration

Ans : c

125. The downward movement of water through the soil profile is termed as

a) Percolation b) Seepage
c) Permiability d) Infiltration

Ans : a

126. The percolation rate is governed by

a) Hydraulic conductivity b) Seepage
c) Permiability d) Infiltration rate

Ans : a

127. The first phase of percolation is

a) Hydraulic conductivity b) Seepage
c) Permiability d) Infiltration

Ans : d

128. The infiltration can not exceed

a) Hydraulic conductivity b) Seepage
c) Permiability d) Percolation

Ans : d

129. .... can not exceed the rate at which water enters the soil

a) Hydraulic conductivity b) Seepage rate
c) Permiability d) Percolation rate

Ans : d

130. Percolation occurs when the suction is less than.............atm

a) 1 b) 1/10
c) 1/2 d) 1/4

Ans : b

131. The downward and lateral movement of water inthe soil is termed as

a) Percolation b) Seepage
c) Permiability d) Infiltration

Ans : b

132. Crop coefficient is the ratio between is called
   a) Water requirement and Irrigation requirement
   b) Irrigation requirement and potential evapotranspiration
   c) Evapotranspiration of crop and potential evapotranspiration
   d) All **Ans : c**

133. Crop coefficient depends on
   a) Soil cover b) Soil moisture
   c) Crop height d) All **Ans : d**

134. ....usually contributes to water table
   a) Percolation b) Seepage
   c) Permiability d) Infiltration **Ans : a**

135. The water movement as vapour through soil profile occurs by
   a) Cohesion b) Seepage
   c) Diffusion d) Infiltration **Ans : c**

136. Vapour movement takes from warm to ....part of the soil
   a) Wet b) Dry
   c) Cold d) None **Ans : c**

137. Vapour movement tends to ....during day and ....during night
   a) Upward, downward b) Downward, upward
   c) Upward, lateral d) Lateral, downward **Ans : b**

138. ....% of water absorbed by the plant is used for photosynthesis
   a) 1 b) 99
   c) 111 d) 89 **Ans : a**

139. ...% of water absorbed by the plant escapes as vapour in the process of transpiration
   a) 1 b) 99
   c) 111 d) 89 **Ans : b**

140. ....is the major constituent of vegetative biomass
   a) Water b) Carbon
   c) Nitrogen d) All **Ans : a**

141. Water constitutes more than ....% of vegetative biomass
   a) 10 b) 60
   c) 90 d) 80 **Ans : c**

142. A soil having wet weight of 225 grams and dry weight of 200 grams will have .... % moisture by weight

a) 10 b) 25
c) 12.5 d) 15 **Ans : c**

143. A soil having 15 % moisture by weight and bulk density of 1.5 will have ....cm/m water content on volumetric basis

a) 10 b) 25
c) 22.5 d) 13.5 **Ans : c**

144. A soil having 30 cm/m water content on volumetric basis with a sampling depth of 40 cm will ....cm depth of water

a) 10 b) 12
c) 22 d) 1.33 **Ans : b**

145. A soil having 1000 cm^3 volume and dry weight of 1.3 kg will have bulk density of ....g/cm^3

a) 1300 b) 130
c) 1.3 d) 13 **Ans : c**

146. A soil auger having height of 20 cm and inside diameter of 10 cm will have ....cm^3 volume of soil core

a) 1070 b) 1570
c) 200 d) 2.00 **Ans : b**

147. Tensiometer indicates soil moisture

a) Pressure b) Suction
c) Tension d) None **Ans : b**

148. The ratio of actual water content to water content at saturation is termed as

a) Actual water content b) Relative water content
c) Water content at Saturation d) None **Ans : b**

149. Actual water content can be obtained by

a) (FW-DW / FW) x 100 b) (TW-FW / TW- DW) x 100
c) (FW-DW / TW- DW) x 100 d) None **Ans : a**

150. Relative water content can be obtained by

a) (FW-DW / FW) x 100 b) (TW-FW / TW- DW) x 100
c) (FW-DW / TW- DW) x 100 d) None **Ans : b**

151. Water content at Saturation can be obtained by

a) (FW-DW / FW) x 100 b) (TW-FW / TW- DW) x 100
c) (FW-DW / TW- DW) x 100 d) None **Ans : c**

152. The soil depth from which the crop extracts most of the water needed to meet its evapotranspiration requirements is known as

a) Effective root zone depth b) Soil moisture depth
c) Rhizosphere d) None **Ans : a**

153. ....is used to determine irrigation water requirements for design

a) Effective root zone depth b) Soil moisture depth
c) Rhizosphere d) None **Ans : a**

154. Evaporation measurement can be obtained from weight changes of

a) Lysimeter b) Tensiometer
c) Pan evaporimeter d) None **Ans : a**

155. Evaporation measurements can be obtained by using water balance equation

a) ET=Weight change + water added - percolation
b) ET=Weight change + water added + percolation
c) ET=Weight change - water added - percolation
d) None **Ans : a**

156. Consumptive use can be obtained by

a) CU= E
b) CU= E+ T
c) CU= E+ T- plant metabolic needs
d) CU= E+ T+ plant metabolic needs **Ans : d**

157. Irrigation requirement can be obtained by

a) IR = WR b) IR = WR +ER
c) IR = ER – (WR +S) d) IR =WR - (ER+S) **Ans : d**

158. Water requirement can be obtained by

a) WR= IR b) WR= IR+ER
c) WR= IR + ER+S d) WR=IR-ER+S **Ans : c**

159. Flow of water in pipe is measured by

a) Pipe orifices b) Water meters
c) Co-ordinate method d) All **Ans : d**

160. The method of irrigation exclusive for low land rice is

a) Flooding b) Check basin
c) Basin d) All **Ans : a**

161. Flow of water in channel is measured by

a) Parshall flume b) V notch
c) Co-ordinate method d) Both a & b **Ans : d**

162. ....is an excellent device for measuring small flow of water in channel

a) Parshall flume b) V notch
c) Co-ordinate method d) Both a & b
Ans : b

163. The method of irrigation having least labour requirement is

a) Flooding b) Check basin
c) Basin d) All
Ans : a

164. Flow of water in channel using V notch is computed by formula

a) $Q=0.0138H^{25}$ b) $Q=0.0138H^{5}$
c) $Q=0.0138H^{2}$ d) None
Ans : a

165. .... is an automated method of surface irrigation

a) Flooding b) Check basin
c) Basin d) Cablegation
Ans : d

166. The term trickle irrigation is used synonymously with

a) Sprinkler irrigation b) Drip irrigation
c) Both d) None
Ans : b

167. Drip irrigation has been originated in

a) USA b) Israel
c) India d) None
Ans : b

168. The first Drip irrigation was developed by

a) Symcha Blase b) Throntwait
c) Panmann d) None
Ans : a

169. The most common method among surface methods of irrigation

a) Flooding b) Check basin
c) Basin d) All
Ans : b

170. The method of irrigation suitable for fruit crops is

a) Flooding b) Check basin
c) Basin d) All
Ans : c

171. Corrugation is a type of ....irrigation

a) Surface b) Sub surface
c) Furrow d) None
Ans : c

172. The method of irrigation in which intermittent application of water to field surface under gravity flow is practiced

a) Flooding b) Check basin
c) Basin d) Surge
Ans : d

173. .... is a form of gated pipe system

a) Flooding b) Check basin
c) Basin d) Cablegation
Ans : d

174. The major problem of drip irrigation system is..................

a) Flooding b) Clodding
c) Clogging d) All
Ans : c

175. .... is defined as yield of marketable crop produced per unit of water used in evapotranspiration

a) WUE b) WCE
c) Both d) None
Ans : a

176. Salt content of irrigation water is measured as.................

a) EC b) CEC
c) PBS d) None
Ans : a

177. Water containing total dissolved salts more than 1.5 m mhos are called

a) Saline b) Alkaline
c) Normal water d) None
Ans : a

178. Water containing total dissolved salts more than 1.5 m mhos are called

a) Saline b) Alkaline
c) Normal water d) None
Ans : a

179. Gross irrigation requirement is

a) NIR
b) NIR x Efficiency of irrigation system
c) NIR / Efficiency of irrigation system
d) None
Ans : c

180. Saline waters contains............. as the predominant salt

a) Sodium chloride b) Sodium carbonate
c) Ammonium chloride d) None
Ans : a

181. Based on EC irrigation water is classified into ......categories

a) 2 b) 3
c) 4 d) 5
Ans : d

182. EC is expressed as

a) DS/m b) Kilograms
c) % d) All **Ans : a**

183. Irrigation water containing more than........ppm of boron is harmful to crops

a) 2 b) 3
c) 4 d) 5 **Ans : b**

184. Boron waters contains ....ppm of boron

a) 2 b) 3 - 4
c) 4 - 5 d) 5 - 10 **Ans : d**

185. The difference in soil moisture content between field capacity and the soil moisture content in the root zone before application of irrigation water is called

a) NIR b) GIR
c) WR d) None **Ans : a**

186. The amount of irrigation water that must be applied to the soil surface to meet the net irrigation requirement for each irrigation is called

a) IR b) GIR
c) WR d) None **Ans : b**

187. Irrigation interval is a function of

a) Crop b) Soil
c) Both d) None **Ans : c**

188. Design interval is

a) Net depth of application
b) Net depth of application x Peak period use rate
c) Net depth of application / Peak period use rate
d) None **Ans : c**

189. The number of days that can be allowed for applying one irrigation to a given design area during the peak period consumptive use of the crop is called

a) Irrigation requirement b) Irrigation period
c) Water requirement d) None **Ans : b**

190. The method of irrigation where water is allowed to run over the soil surface and allowed to infiltrate is called as

a) Drip irrigation b) Surface irrigation
c) Sub surface irrigation d) Sprinkler irrigation **Ans : b**

191. The method of irrigation where water is allowed to fall on the plants and soil as simulated rainfall is called as

a) Drip irrigation b) Surface irrigation
c) Sub surface irrigation d) Sprinkler irrigation **Ans : d**

192. The method of irrigation where water is applied directly to the root zone is called as

a) Drip irrigation b) Surface irrigation
c) Sub surface irrigation d) Sprinkler irrigation **Ans : a**

193. The amount of irrigation water inclusive of evapotranspiration, consumptive use and that necessary to regulate the salt concentration in the soil solution to the total volume of water delivered is called as

a) Irrigation efficiency b) Water conveyance efficiency
c) Water application efficiency d) Water distribution efficiency **Ans : a**

194. The ratio between the amount of water delivered to irrigated plot to the total volume of water delivered at the source is called as

a) Irrigation efficiency b) Water conveyance efficiency
c) Water application efficiency d) Water distribution efficiency **Ans : b**

195. The ratio between the water stored in the root zone to the amount of water delivered to irrigated plot is called as

a) Irrigation efficiency b) Water conveyance efficiency
c) Water application efficiency d) Water storage efficiency **Ans : c**

196. The ratio between water stored in the root zone to the amount of water needed in the root zone prior to the irrigation is called as

a) Irrigation efficiency b) Water conveyance efficiency
c) Water application efficiency d) Water storage efficiency **Ans : d**

197. The ratio between the consumptive use of water by the crop to the water depleted from the root zone is called as

a) Consumptive use efficiency b) Water use efficiency
c) Field water use efficiency d) None **Ans : a**

198. The depth of water applied in the first watering to crop is known as

a) Delta b) Kor depth
c) Base depth d) Kor watering **Ans : b**

199. The ratio between the marketable crop yield and the water used in evapotranspiration is called as

a) Consumptive use efficiency b) Water use efficiency
c) Field water use efficiency d) None **Ans : b**

200. EC is expressed as

a) DS/m b) M mhos/cm
c) ppm d) All **Ans : d**

201. 1 m mhos/cm equals to

a) 1dS/m b) 1ppm
c) 1% d) All **Ans : a**

202. 1 ds/m equals to

a) 10 m mhos/cm b) 10ppm
c) 690 mg/l d) All **Ans : c**

203. Fresh water will have salt content

a) <500 mg/l b) 1500 -5000 mg/l
c) 3,50,000 mg/l d) All **Ans : a**

204. Bittern water will have salt content

a) <500 mg/l b) 1500 -5000 mg/l
c) 3,50,000 mg/l d) All **Ans : c**

205. Brackish water will have salt content

a) <500 mg/l b) 1500 -5000 mg/l
c) 3,50,000 mg/l d) All **Ans : b**

206. Which of the following is most sensitive to boron

a) Apple b) Maize
c) Berseem d) All **Ans : a**

207. Which of the following is most tolerant to boron

a) Apple b) Maize
c) Berseem d) All **Ans : c**

208. Which of the following is semi tolerant to boron

a) Grape b) Berseem
c) Potato d) All **Ans : c**

209. The total area irrigated by a project including the area covered by roads, buildings, villages, unculturable area etc within the boundaries of cultivated area is called

a) Gross command area b) Culturable command area
c) Net command area d) All **Ans : a**

210. The actual crop area under irrigation in the gross command area is called

a) Gross command area b) Culturable command area
c) Net command area d) All Ans : b

211. The depth of water required by a crop during its life cycle in the field is called

a) Duty b) Delta
c) Base period d) All Ans : b

212. The entire duration of crop from irrigation for preparatory tillage to last irrigation is called

a) Duty b) Delta
c) Base period d) All Ans : c

213. The ratio between the irrigated area and the quantity of water applied to the crop is called

a) Duty b) Delta
c) Base period d) All Ans : a

214. The first watering to the crop is known as

a) Duty b) Delta
c) Base irrigation d) Kor watering Ans : d

215. The part of the base period during which first watering is given to crop is known as

a) Kor period b) Kor depth
c) Base irrigation d) Kor watering Ans : a

216. One cumec per day equals to

a) 8.64 ha cm b) 8.64 ha m
c) 8.64 ha d) None Ans : b

217. The first irrigation before sowing the crop for seed germination and seedling establishment is known as

a) Palco b) Blind irrigation
c) Base irrigation d) None Ans : a

218. ....is also called as consumptive use of water

a) Absolute water requirement b) Relative water requirement
c) Gross water requirement d) None Ans : a

219. The quantity of water in ha-cm per crop season absorbed by the crop together with the evaporation from the crop producing land is called as

a) Absolute water requirement b) Relative water requirement
c) Gross water requirement d) None Ans : a

220. Consumptive use of water includes
   a) The water used by evaporatranspiration
   b) The water retained in plant body
   c) Both
   d) None **Ans : c**

221. The amount of water that would cover an acre of land to a depth of one foot assuming no seepage evaporation and run-off losses is called
   a) Acre foot of water b) Acre inch
   c) Acre yards d) None **Ans : a**

222. The quantity of flow of water equal to a flow which will cover one acre to a depth of one inch is called
   a) Acre foot of water b) Acre inch
   c) Acre yards d) None **Ans : b**

223. Absorption of water and other substances against the concentration gradient is called
   a) Active absorption b) Mass flow
   c) Passive absorption d) None **Ans : a**

224. .... involves expenditure of energy in contrast to imbibition and osmosis
   a) Active absorption b) Mass flow
   c) Passive absorption d) None **Ans : a**

225. The rate of evapo-transpiration equal to all smaller than potential evapo-transpiration as affected by the level of the available soil water, salinity, filed size, or other causes is termed as mm/day
   a) Actual crop evapo-transpiration b) PET
   c) ET d) None **Ans : a**

226. The rate of actual evapo-transpiration is expressed in terms of as
   a) Cm/hr b) Mm/day
   c) M/day d) None **Ans : a**

227. The ratio of number of hectare days required in mono culture to the number of hectare days used in intercropping to produce identical quantities of each of the component crop is called
   a) ATER b) NAR
   c) CGR d) None **Ans : a**

228. Sufficient soil moisture in __________stage is essential for optimum productivity of blackgram

a) Flowering and pod development
b) Branching and pod development
c) Branching and flowering
d) Pod development and maturity **Ans : a**

□□□

8 Chapter

Dryland Agriculture

Multiple Choice Questions (MCQ's)

1. The International Centre for Agriculture Research in Dry Areas (ICARDA) was established in
   a) 1987 b) 1977
   c) 1967 d) None of these **Ans : b**

2. The ICRISAT was established in
   a) 1987 b) 1972
   c) 1967 d) None of these **Ans : b**

3. The CRIDA was established in
   a) 1987 b) 1982
   c) 1967 d) None of these **Ans : b**

4. The dryland areas can be designated where
   a) Annual rainfall is equal to potential evapotranspiraiton
   b) Potential evapotranspiration is more than annual rainfall
   c) Potential evapotranspiration is equal to annual rainfall
   d) None to these **Ans : b**

5. Under stress condition, which amino acid accumulated in crop plants
   a) Methionine b) Tryptophan
   c) Proline d) Phenyl alanine **Ans : c**

6. The areas where drought has occurred in ____________ per cent of years during the period are considered drought areas
   a) 40 b) 30
   c) 10 d) 20 **Ans : c**

7. What sort of aberration in rainfall affects crop growth most:
   a) Lesser rainfall b) Higher rainfall
   c) Uncertain distribution d) No rain in reproductive stage **Ans : c**

8. Which of these influence most the water requirement of a crop:
   a) Weather b) Canopy development
   c) Water table d) Soil condition **Ans : a**

9. Which one can be called as Dryland?
   a) When rainfall received is less than 1000 mm
   b) When rainfall received is less than 500 mm
   c) When rainfall received is less than 750 mm
   d) Hilly terrain with low rainfall less than 500mm **Ans : a**

10. In undulating terrain one should go for ....
    a) Normal crop raising b) Strip cropping
    c) Pastoral cropping d) Mixed pastoral and tree cropping **Ans : b**

11. Which one can be called as Dryfarming?
    a) When rainfall received is less than 1000 mm
    b) When rainfall received is less than 500 mm
    c) When rainfall received is less than 750 mm
    d) Hilly terrain with low rainfall less than 500mm **Ans : c**

12. Which one can be called as Rainfed farming area?
    a) When rainfall received is more than 1150 mm
    b) When rainfall received is less than 500 mm
    c) When rainfall received is less than 800 mm
    d) Hilly terrain with low rainfall less than 500mm **Ans : c**

13. Dry farming areas are characterized by a growing season of
    a) <75 days b) 75-120 days
    c) >120 days d) All **Ans : a**

14. Dryland farming areas are characterized by a growing season of
    a) <75 days b) 75-120 days
    c) >120 days d) All **Ans : b**

15. Rainfed farming areas are characterized by a growing season of
    a) <75 days   b) 75-120 days
    c) >120 days   d) All   Ans : c

16. Dryland constitutes ....% of total cultivated area in India
    a) 72   b) 82
    c) 62   d) 42   Ans : a

17. Dryland contributes ....% of total food production in India
    a) 72   b) 82
    c) 62   d) 42   Ans : d

18. In dry farming areas emphasis is on
    a) Moisture conservation practices   b) Drainage
    c) Disposal of excess water   d) All   Ans : a

19. In dryland farming areas emphasis is on
    a) Moisture conservation practices   b) Drainage
    c) Disposal of excess water   d) All   Ans : b

20. In rainfed farming areas emphasis is on
    a) Moisture conservation practices   b) Drainage
    c) Disposal of excess water   d) All   Ans : c

21. Higher yield of rice could be ensured by adequate supply of ....fertilizer at the tillering stages
    a) Nitrogen   b) Potash
    c) Both N and K   d) Phosphorus   Ans : a

22. By delayed planting of rice in the second season the yield decreases most due to ....
    a) Poor tillering   b) Higher temperature at flowering
    c) Heavy insect infestation   d) For want of water   Ans : b

23. The first dry farming research station started in the year 1923 at
    a) Bijapur   b) Sholapur
    c) Rohtak   d) Manjri   Ans : d

24. The dry farming research station started in the year 1935 at
    a) Bijapur   b) Sholapur
    c) Manjri   d) Rohtak   Ans : d

25. The dry farming research station was started in the year 1934 at

a) Hagari b) Raichur
c) Manjri d) Both a & b Ans : d

26. The dry farming research station was started in the year 1933 at

a) Bijapur b) Hagari
c) Manjri d) Rohtak Ans : a

27. The first scientific approach to tackle the problem of dry farming areas was initiated by

a) Kanitkar b) Swaminathan
c) Tamhane d) None Ans : c

28. .... is a permanent climatic feature and is the culmination of long term process

a) Aridity b) Drought
c) Dryland d) All Ans : a

29. .... is a condition when the actual seasonal rainfall is deficient by more than twice the mean deviation

a) Aridity b) Drought
c) Dryland d) All Ans : b

30. Drought has been classified into .... kinds

a) 2 b) 3
c) 4 d) 5 Ans : c

31. Which of the following is not a kind of drought

a) Invisible b) Climatic
c) Seasonal d) Permanent Ans : b

32. Desert climate is the characteristic of .... drought

a) Invisible b) Climatic
c) Seasonal d) Permanent Ans : d

33. Sparse vegetation adopted to drought and crop raising only through irrigation during entire period is the characteristic of .... drought

a) Invisible b) Climatic
c) Seasonal d) Permanent Ans : d

34. The drought found in climates with well defined rainy and dry seasons is
a) Invisible b) Climatic
c) Seasonal d) Permanent **Ans : c**

35. The drought characterized by abnormal failure of rainfall is
a) Invisible b) Contingent
c) Seasonal d) Permanent **Ans : b**

36. When rainfall is inadequate to meet the ET losses and result in borderline deficiency of water in soil resulting in less than quantum yield is
a) Invisible b) Climatic
c) Seasonal d) Permanent **Ans : a**

37. When annual precipitation is less than the normal over an area for prolonged period is
a) Meteorological drought b) Atmospheric drought
c) Hydrological drought d) Agricultural drought **Ans : a**

38. Drought due to low air humidity and accompanied by hot dry winds is
a) Meteorological drought b) Atmospheric drought
c) Hydrological drought d) Agricultural drought **Ans : b**

39. Meteorological drought for prolonged period with depletion of surface water and consequent drying of reservoirs is
a) Meteorological drought b) Atmospheric drought
c) Hydrological drought d) Agricultural drought **Ans : c**

40. Drought due to imbalance between available soil moisture and ET of a crop is
a) Meteorological drought b) Atmospheric drought
c) Hydrological drought d) Agricultural drought **Ans : d**

41. Which of the following is not a plant adaptation to drought
a) Escaping drought b) Drought resistance
c) Drought tolerance d) Drought survival **Ans : d**

42. The concept of aridity index was proposed by
a) Pannman b) Thronthwaite
c) Kanitkar d) Tamhane **Ans : b**

43. The departure of aridity index from normal by ....represents moderate drought

a) <0.5 b) 0.5-1.0
c) 1.0-2.0 d) >2 **Ans : a**

44. The departure of aridity index from normal by ....represents disastrous drought

a) <0.5 b) 0.5-1.0
c) 1.0-2.0 d) >2 **Ans : d**

45. The departure of aridity index from normal by ....represents large drought

a) <0.5 b) 0.5-1.0
c) 1.0-2.0 d) >2 **Ans : b**

46. The departure of aridity index from normal by ....represents severe drought

a) <0.5 b) 0.5-1.0
c) 1.0-2.0 d) >2 **Ans : c**

47. The drought zones areas with .... % probability of rainfall departure from normal

a) 25 b) 35
c) 40 d) 50 **Ans : a**

48. The chronic drought zones areas with .... % probability of rainfall departure from normal

a) 25 b) 35
c) 40 d) 50 **Ans : c**

49. Which of the following is not a moisture conservation practice in drylands

a) Tillage b) Fallowing
c) Mulching d) None **Ans : d**

50. Vertical mulching has been adopted in

a) Heavy soils b) Light soils
c) Loamy soils d) All **Ans : a**

51. Trenches of 30 x60 cm are dug at 5-6 m interval and filled with stalk to keep the cracks open for better water intake in

a) Vertical mulching b) Dust mulch
c) Stubble mulch d) All **Ans : a**

52. The broad beds and furrow method is suitable for
   a) Black soils b) Sandy soils
   c) Loamy soils d) All **Ans : a**

53. Evaporation from soil surface can be restricted by
   a) Reducing external evaporativity
   b) Reducing energy supply to evaporating surface
   c) Using growth retardants
   d) All **Ans : d**

54. Moisture stress in soils and plants during crop growth due to imbalance between soil moisture and evapotranspiration of a crop is ....
   a) Agricultural drought b) Hydrological drought
   c) Meteorological drought d) Physiological drought **Ans : d**

55. Which one be called as Dryland
   a) When rainfall received is less than 1000 mm
   b) When rainfall received is less than 500 mm
   c) When rainfall received is less than 750 mm
   d) Hilly terrain with low rainfall less than 500mm **Ans : c**

56. Transpiration can be restricted by
   a) Use of antitranspirants b) Increasing leaf reflectance
   c) Use of growth retardants d) All **Ans : d**

57. Atrazine used as an antitranspirant
   a) Reduces the growth of the crop
   b) does not reflect light from plant leaf surface
   c) Affects the closure and opening of stomata
   d) Forms thin layer on leaf surface **Ans : a**

58. Phenylmercuric acetate (PMA) is a chemical used in agricultural crops in order to
   a) Increase CO_2 uptake b) Reduce respiration
   c) Reduce transpiration d) Increase transpiration **Ans : c**

59. Water requirement (WR) includes the losses due to
   a) Evapotranspiration and application of water
   b) Consumptive use of water and application of water
   c) Consumptive use, water required for special operations and other economically unavoidable losses of water.
   d) ET and water required for special operations. **Ans : d**

60. Dust mulch is very effective in
   a) Coarse textured soil b) Heavy textured soil
   c) Sandy soil d) None of these **Ans : b**

61. The value of albedo in most of the dry land soils ranges from
a) 0.15 to 0.35 b) 0.00 to 0.10
c) 0.50 to 0.75 d) 0.76 to 0.80 **Ans : ....**

62. Normally drought is assessed using
a) Aridity b) Dryness
c) Aridity anomaly index d) Deficient rainfall **Ans : a**

63. Materials which reduces water loss from leaf surface by reducing the size or number of stomatal openings are called
a) Growth retardants b) Growth promoters
c) Antitranspirants d) All **Ans : d**

64. .... is a stomata closure type of antitranspirants
a) PMA b) Kaolin
c) Mobileaf d) All **Ans : a**

65. .... is a film forming type of antitranspirants
a) PMA b) Kaolin
c) Mobileaf d) All **Ans : c**

66. .... is a leaf reflectance type of antitranspirants
a) PMA b) Kaolin
c) Mobileaf d) All **Ans : b**

67. Coating of the leaf surface with white material increases
a) Albedo b) Stomatal closure
c) Growth d) All **Ans : a**

68. .... type of antitranspirants reduce transpiration loss by reducing leaf temperature and vapour pressure gradient
a) Film forming b) Stomatal closure
c) Leaf reflectance d) All **Ans : c**

69. .... type of antitranspirants produce external physical barrier to reduce vapour loss from leaf surface
a) Film forming b) Stomatal closure
c) Leaf reflectance d) All **Ans : a**

70. .... At low concentrations also serves as stomata closing type of antitranspirants
a) 2,4-D b) MCPA
c) Atrazine d) All **Ans : c**

71. .... are stomata closure type of antitranspirants
a) Hexadecanol b) Silicon
c) Mobileaf d) All **Ans : d**

72. CCC is a .... type of antitranspirants
a) Film forming b) Stomatal closure
c) Growth retardant d) All **Ans : c**

73. The process of runoff collection during peak periods of rainfall in storage tanks and ponds is
a) Water harvesting b) Runoff collection
c) Water storage d) All **Ans : a**

74. Which of the following is a characteristic of dryland crop
a) Shorter duration b) Faster growth
c) Strong root system d) All **Ans : d**

75. Which of the following is not a characteristic of dryland crop
a) Potential for high yield b) Fertilizer responsive
c) Strong root system d) High water requirement **Ans : d**

76. .... is the most popular dryland crop
a) Sorghum b) Maize
c) Wheat d) All **Ans : a**

77. Soil mulch reduces
a) Deep cracking b) Reduces evaporation
c) Improves water storage d) All **Ans : d**

78. Hedgerow intercropping is the
a) Cultivation of food crops in alleys formed by hedge row of trees / shrubs in arable lands
b) Cultivation of food crops in alleys formed by hedge row of trees / shrubs in waste lands
c) Cultivation of food crops in alleys formed by hedge row of trees / shrubs in forest lands
d) All **Ans : a**

79. Which is a alley system
a) Forage-alley cropping b) Forage-cum-mulch system
c) Forage-cum-pole system d) All **Ans : d**

80. In agri-horticulture system for arable land the tree component is
a) Fruit b) Timber
c) Hedgerow d) All **Ans : a**

81. Horti/silvi-pastural system is suitable for .... lands
   a) Arable b) Non arable
   c) Hills d) All **Ans : b**

82. Horti/silvi-pastural system is suitable for .... lands
   a) Class IV and above b) Class IV and below
   c) Hills d) All **Ans : a**

83. Growing of pasture for grazing and conservation is called
   a) Alternate husbandary b) Mixed farming
   c) Both a & b d) None **Ans : c**

84. Area covered with pasture which are reseeded at regular intervalsis termed as
   a) Alternate husbandary b) Mixed farming
   c) Both a & b d) Ley farming **Ans : d**

85. Ley farming is the
   a) Cultivation of pasture in arable lands
   b) Cultivation of pasture in non arable lands
   c) Cultivation of food pasture in forest lands
   d) All **Ans : b**

86. TIMFIB represents
   a) Timber cum fiber system b) Timber cum food system
   c) Timber cum fodder system d) All **Ans : a**

87. Moisture stress in soils and plants during crop growth due to imbalance between soil moisture and evapotranspiration of a crop is ....
   a) Agricultural drought b) Hydrological drought
   c) Meteorological drought d) Physiological drought **Ans : a**

88. Which of the following antitranspirant belongs to stomatal closing group
   a) Cycocel b) PMA
   c) Silicone d) Paraquat **Ans : b**

89. Which one of the following chemical is used as antitranspirant in dryland crops
   a) 2, 4-D b) MCPA
   c) Cycocel d) Paraquat **Ans : c**

90. The most appropriate classification scheme which relates climate to vegetation is
   a) Penman classification b) Thomthwite classification
   c) Koppens classification d) None of the above **Ans : b**

91. The plastic mulch helps to

a) Reduce soil moisture b) Increase the evaporation
c) Increase soil temperature d) None of the above **Ans : c**

92. A year in which rainfall of a particular place is short by more than ....is known as drought year:

a) Twice the mean deviation b) Twice the standard deviation
c) Twice the median d) Twice standard error **Ans : a**

❑❑❑

9 Chapter

Weed Management

Multiple Choice Questions (MCQ's)

1. Weed is a

a) Unwanted plant	b) Undesirable plant	
c) Plant out of place	d) All	**Ans : d**

2. In India weeds account for losses due to pests as high as

a) 45%	b) 55%	
c) 25%	d) 35%	**Ans : a**

3. Globally weeds account for losses....... of the total food production

a) 11.5%	b) 10%	
c) 15.5%	d) 25%	**Ans : a**

4. Which of the following weed brought reduction in land value

a) *Cynodon dectylon*	b) *Cyperus rotundus*	
c) *Setaria glauca*	d) *Melilotus indica*	**Ans : b**

5. Which of the following weed results in reduction in human efficiency

a) Ambrosia	b) Parthenium	
c) Argemone mexicana	d) All	**Ans : d**

6. Maximum number of seeds is produced by which of the following weeds
   a) Echinochloa colonum b) Parthenium
   c) Utrica sp d) Orobanche cernua **Ans : d**

7. Maximum number of seeds is produced by which of the following weeds
   a) Echinochloa colonum b) Parthenium
   c) Utrica sp d) Argemone mexicana **Ans : a**

8. Maximum number of seeds is produced by which of the following weeds
   a) Amaranthus b) Parthenium
   c) Utrica sp d) Argemone mexicana **Ans : a**

9. Which one of the following has maximum viability
   a) *Convolvulus arvensis* b) *Cyperus rotundus*
   c) *Setaria glauca* d) *Melilotus indica* **Ans : a**

10. Which of the following type of dormancy is found in weed seeds
    a) Innate b) Induced
    c) Enforced d) All **Ans : d**

11. Enforced dormancy is due to
    a) Deeper placement of seeds b) Genetic factor
    c) Physiological change d) All **Ans : a**

12. Innate dormancy is due to
    a) Deeper placement of seeds b) Genetic factor
    c) Physiological change d) All **Ans : b**

13. Induced dormancy is due to
    a) Deeper placement of seeds b) Genetic factor
    c) Physiological change d) All **Ans : c**

14. Wild oat exhibits which of the following type of dormancy
    a) Innate b) Induced
    c) Enforced d) All **Ans : d**

15. Which of the following does not affect dormancy
    a) Temperature b) Light
    c) Soil moisture d) Soil fertility **Ans : d**

16. Dormancy may be due to
    a) Seed coat b) Immature embryo
    c) Both a & b d) None **Ans : c**

17. If the rate of application per hectare is 3.00 kg ai., the quantity of simazine WP (80%) a i. required to be sprayed in 0.20 hectare area would be
a) 0.50 kg b) 0.75 kg
c) 1.25 kg d) 1.87 kg **Ans : b**

18. Which one of the following pairs is not correctly matched

Herbicide group	*Common name*
a) Phenoxy acids	2,4,5-T
b) Urea	Isoproturon
c) Acetamide	Butachlor
d) Carbamates	Dinoseb

Ans : a

19. Weeds completing their life cycle in a season/year are known as
a) Annual weeds b) Biennial weeds
c) Perennial weeds d) None **Ans : a**

20. Weeds that grow in one season and completing their life cycle in the following season/year are known as
a) Annual weeds b) Biennial weeds
c) Perennial weeds d) None **Ans : b**

21. Weeds surviving for more then two years are known as
a) Annual weeds b) Biennial weeds
c) Perennial weeds d) None **Ans : c**

22. ....is an annual weed
a) Bidens pilosa b) Chenopodium album
c) Phalaris minor d) All **Ans : d**

23. ....is an biennial weed
a) Phalaris minor b) Cyperus rotundus
c) Alternenthera pungens d) Convolvulus arvensis **Ans : c**

25. ....is an perennial weed
a) Sonchus arvensis b) Cyperus rotundus
c) Convolvulus arvensis d) All **Ans : d**

26. ....is an simple perennial weed
a) Phalaris minor b) Cyperus rotundus
c) Sonchus arvensis d) Convolvulus arvensis **Ans : c**

27. ....is an bulbous perennial weed
a) Phalaris minor b) Cyperus rotundus
c) Allium sp d) Convolvulus arvensis **Ans : c**

28. ....is an corm perennial weed
a) Phalaris minor b) Cyperus rotundus
c) Timothy d) Convolvulus arvensis **Ans : c**

29. ....is an creeping perennial weed
a) Cynodon dactylon b) Cyperus rotundus
c) Convolvulus arvensis d) All **Ans : d**

30. ....is a dryland weed
a) Phalaris minor b) Cyperus rotundus
c) Cynotis cucullata d) Convolvulus arvensis **Ans : c**

31. ....is a wet land weed
a) Phalaris minor b) Cyperus rotundus
c) Eclipta prostrata d) Convolvulus arvensis **Ans : c**

32. ....is a acidophile weed
a) Phalaris minor b) Cyperus rotundus
c) Rumex acetosella d) Convolvulus arvensis **Ans : c**

33. ....is basophile weed
a) Phalaris minor b) Cyperus rotundus
c) Taraxacum stricta d) Convolvulus arvensis **Ans : c**

34. ....is neutrophile weed
a) Taraxacum stricta b) Rumex acetosella
c) Achylapha indica d) All **Ans : c**

35. Weeds grow in acid soils are
a) Neutrophile weed b) Acidophilel weed
c) Basophile weed d) All **Ans : b**

36. Weeds grow in neutral soils are
a) Neutrophile weed b) Acidophilel weed
c) Basophile weed d) All **Ans : a**

37. Weeds grow in saline and alkaline soils are
a) Neutrophile weed b) Acidophilel weed
c) Basophile weed d) All **Ans : c**

38...................is monocot weed
a) Commelina benghalensis b) Convolvulus arvensis
c) Crotolaria verucosa d) All **Ans : a**

39. ....is dicot weed
a) Commelina benghalensis b) Setaria gluca
c) Crotolaria verucosa d) All **Ans : c**

40. ....is grassy weed

a) Commelina benghalensis b) Setaria gluca
c) Crotolaria verucosa d) All **Ans : b**

41. ....is broadleaf weed

a) Commelina benghalensis b) Setaria gluca
c) Cyperus rotundus d) Phalaris minor **Ans : a**

42. ....is an example of sedges weed

a) Commelina benghalensis b) Cyperus rotundus
c) Fimbrystylis miliaceae d) Both b & c **Ans : d**

43. ....is an example of poisinous weed

a) Commelina benghalensis b) Cyperus rotundus
c) Fimbrystylis miliaceae d) Datura **Ans : d**

44. ....is an example of parasitic weed

a) Orobanche b) Striga
c) Cuscuta d) All **Ans : d**

45. ....is total root parasitic weed

a) Orobanche b) Striga
c) Cuscuta d) All **Ans : a**

46. ....is partial root parasitic weed

a) Orobanche b) Striga
c) Cuscuta d) All **Ans : b**

47. ....is total stem parasitic weed

a) Orobanche b) Striga
c) Cuscuta d) All **Ans : c**

48. ....is partial stem parasitic weed

a) Orobanche b) Striga
c) Cuscuta d) Cassytha filiformis **Ans : d**

49. Orobanche is

a) Total stem parasitic weed b) Partial root parasitic weed
c) Total root parasitic weed d) All **Ans : c**

50. Striga is

a) Total stem parasitic weed b) Partial root parasitic weed
c) Total root parasitic weed d) All **Ans : b**

51. Cuscuta is

a) Total stem parasitic weed b) Partial root parasitic weed
c) Total root parasitic weed d) All **Ans : a**

52. Loranthus longifolius is

a) Total stem parasitic weed
b) Partial root parasitic weed
c) Total root parasitic weed
d) Partial stem parasitic weed

Ans : d

53. Fimbrystylis miliaceae is

a) Total stem parasitic weed
b) Partial root parasitic weed
c) Total root parasitic weed
d) None

Ans : d

54. An annual weed commonly found in wheat field is

a) Allagalliis arvellsis
b) Echinochloa crusgalli
c) Amaranthus spinosus
d) Phalaris minor

Ans : d

55. The herbicides containing carbon and hydrogen in their molecule are called

a) Arsenic
b) Acid
c) Organic herbicide
d) Salt

Ans : c

56. Which one of the following is non selective herbicide

a) Alachlor
b) Butachlor
c) Paraquat
d) Atrazine

Ans : c

57. Which one of the following is not narrow leaved weeds

a) Cynodon dectylon
b) Cyperus rotundus
c) Setaria glauca
d) Melilotus indica

Ans : d

58. Which one of the following is not *kharif* season weed

a) Chenopodium album
b) Amarauthus viridis
c) Echinocloa crusgalli
d) Commelina benghalensis

Ans : a

59. Which one of the following is a aquatic weed

a) Salvinia
b) Amarauthus viridis
c) Echinocloa crusgalli
d) Commelina benghalensis

Ans : a

60. Which one of the following is a Submersed weed

a) Salvinia
b) Utricularia flexosus
c) Pistia stratiotes
d) Nymphea pubescens

Ans : b

61.Which one of the following is not a Floating weed

a) Salvinia
b) Utricularia flexosus
c) Pistia stratiotes
d) Nymphea pubescens

Ans : b

62. Which one of the following is a Floating weed

a) Salvinia
b) Nymphea pubescens
c) Pistia stratiotes
d) All

Ans : d

63. Which one of the following is a emersed weed
   a) Salvinia b) Nelumbium speciosum
   c) Pistia stratiotes d) Nymphea pubescens **Ans : b**

64. Absolute density represents the actual
   a) Number of plants /unit area b) Number of plants in the farm
   c) Number of plants in the crop c) None **Ans : a**

65. Relative density is given by
   a) Absolute density of a given species/Total absolute density of all species X100
   b) Total absolute density of all species/Absolute density of a given species X100
   a) Absolute density of a given species/Absolute frequency X100
   d) None **Ans : a**

66. The degree of uniformity of the occurrence of individuals of a species within an area is called
   a) Relative density b) Absolute density
   c) Relative frequency d) Absolute frequency **Ans : d**

67. The number of occurrence of individuals of a species per quadrat is called
   a) Relative dominance b) Dominance
   c) Relative frequency d) Absolute frequency **Ans: b**

68. The passage of toxicity from one plant to another is termed as
   a) Teletoxicity b) Phytotoxicity
   c) Relative toxicity d) All **Ans: a**

69. The alternate host for gram caterpillar is
   a) Amaranthus b) Crotolaria sp
   c) Echinocloa sp d) Agropyron repens **Ans : a**

70. .... is an objectionable weed in wheat
   a) Amaranthus b) Convolvulus arvensis
   c) Echinocloa sp d) Agropyron repens **Ans : b**

71. .... is an objectionable weed in Mustard
   a) Amaranthus b) Argemone maxicana
   c) Echinocloa sp d) Agropyron repens **Ans : b**

72. .... is an objectionable weed in Barseem
   a) Amaranthus b) Cichorium intybus
   c) Echinocloa sp d) Agropyron repens **Ans : b**

73. .... is an objectionable weed in lucern
a) Amaranthus b) Dodder
c) Echinocloa sp d) Agropyron repens **Ans : b**

74. .... is an objectionable weed in methi
a) Amaranthus b) Melilotus alba
c) Echinocloa sp d) Agropyron repens **Ans : b**

75. An ingredient which when added to a formulation enhances action of toxicant
a) Adjuvant b) Surfactant
c) Antidote d) None **Ans : a**

76. The amount of active ingredient expressed in terms of parent acid is
a) Active ingredient b) Acid equivalent
c) Phytochemical d) None **Ans : b**

77. The active toxic material present in the formulation is
a) Active ingredient b) Acid equivalent
c) Phytochemical d) None **Ans : a**

78. The alternate host for hairy caterpillar is
a) Amaranthus b) Crotolaria sp
c) Echinocloa sp d) Agropyron repens **Ans : b**

79. The alternate host for stem borer is
a) Amaranthus b) Crotolaria sp
c) Echinocloa sp d) Agropyron repens **Ans : c**

80. The alternate host for black rust is
a) Amaranthus b) Crotolaria sp
c) Echinocloa sp d) Agropyron repens **Ans : d**

81. The alternate host for ergot is
a) Amaranthus b) Cenchrus ciliaris
c) Sacharum spontaneum d) Agropyron repens **Ans : b**

82. The alternate host for downey mildew is
a) Amaranthus b) Cenchrus ciliaris
c) Sacharum spontaneum d) Agropyron repens **Ans : c**

83. Dermotitis is caused by
a) Ambrosia b) Parthenium.
c) Utrica sp d) Brush weed **Ans : b**

84. Hay fever is caused by

a) Ambrosia b) Parthenium
c) Utrica sp d) Brush weed **Ans : a**

85. Itching and inflammation is caused by

a) Ambrosia b) Parthenium
c) Utrica sp d) Brush weed **Ans : c**

86. African sleeping sickness is caused by

a) Ambrosia b) Parthenium
c) Utrica sp d) Brush weed **Ans : d**

87. Cyperus rotundus is an example of

a) Annual weed b) Perennial weed
c) Biennial weed d) None **Ans : b**

88. Cyperus rotundus is an example of

a) Grass b) Sedge
c) Broad leaf weed d) None **Ans : b**

89. The world's most problematic weed is

a) Ambrosia b) Parthenium
c) Utrica sp d) Cyperus rotundus **Ans : d**

90. The weed with maximum number of seeds per plant is

a) Ambrosia b) Parthenium
c) Lambs quarters d) Cyperus rotundus **Ans : c**

91. The world's most problematic weed is

a) Ambrosia b) Parthenium
c) Utrica sp d) Cyperus rotundus **Ans : d**

92. The effect of one plant on germination and growth of other plant in their vicinity is

a) Allelopathy b) Competition
c) Critical period d) None **Ans : a**

93. The shortest period during which weeding results in maximum economic return is called

a) Critical period of weed competition
b) Critical period of crop competition
c) Critical period of weed association
d) None **Ans : a**

94. The dormancy of weed seeds due to presence in deeper layers is called

a) Induced b) Enforced
c) Innate d) *In situ* **Ans : b**

95. The dormancy of weed seeds due to genetical factors is called

a) Induced b) Enforced
c) Innate d) *In situ* **Ans : c**

96. The induced dormancy of weed seeds is due to

a) High temperature b) High CO_2
c) Waterlogging d) All **Ans : d**

97. Which of the following is a type of dormancy

a) Induced b) Enforced
c) Innate d) All **Ans : d**

98. Which of the following is not a type of dormancy

a) Forced b) Enforced
c) Innate d) Induced **Ans : a**

99. Which of the following chemical is used for suicidal germination of striga seeds

a) Ethylene b) Bromine
c) Methylene d) All **Ans : a**

100. Which of the following is not a weed control measure

a) Chemical b) Cultural
c) Physical d) None **Ans : d**

101. Which of the following is a weed control measure

a) Biological b) Dormancy
c) Allelopathy d) None **Ans : a**

102. Which of the following is not a cultural weed control method

a) Tillage b) Irrigation
c) Herbicide application d) Fertilizer application **Ans : c**

103. Which of the following is not a physical method of weed control

a) Hand weeding b) Digging
c) Planting d) Cutting **Ans : c**

104. Weeds most suitable for biological control

a) Facultative weeds b) Obligate weeds
c) Introduced weeds d) Associated weeds **Ans : c**

105. Which of the following is not a physical method of weed control

a) Hand weeding b) Digging
c) Planting d) Cutting **Ans : c**

106. 2,4-D was first used as a herbicide for weed control in

a) 1934 b) 1954
c) 1944 d) 1964 **Ans : c**

107. Which of the following is a selective herbicide

a) Glyphosate b) Paraquat
c) Diallate d) Diquat **Ans : c**

108. Which of the following is a non selective herbicide

a) Glyphosate b) Paraquat
c) Diquat d) All **Ans : d**

109. Glyphosate belongs to herbicide group

a) Aliphatics b) Amide
c) Carbamates d) Thiocarbamates **Ans : a**

110. Metolachlor belongs to herbicide group

a) Aliphatics b) Amide
c) Carbamates d) Thiocarbamates **Ans : b**

111. Chlorpropham belongs to herbicide group

a) Aliphatics b) Amide
c) Carbamates d) Thiocarbamates **Ans : c**

112. Diallate belongs to herbicide group

a) Aliphatics b) Amide
c) Carbamates d) Thiocarbamates **Ans : d**

113. Isoproturon belongs to herbicide group

a) Substituted Unea b) Amide
c) Carbamates d) Thiocarbamates **Ans : a**

114. The weeds that belong to the family .... are known as sedges

a) Gramineae b) Convolvulaceae
c) Euphorbiaceae d) Cyperaceae **Ans : d**

115. The critical period of weed competition for transplanted rice is .... days after transplanting

a) 30-50 b) 15-45
c) 20-60 d) 25-60 **Ans : a**

116. Butachlor and propanil belong to .... group of herbicide

a) Ureas b) Carbamates
c) Aliphatic d) Amides Ans : d

117. The .... under the trade name Collego is normally applied to rice and soybean

a) Bioherbicide b) Chemical herbicide
c) Mycoherbicide d) Natural plant origin herbicide Ans : c

118. If the recommended dose of oxydiazon (25% a i) to rice is 0.5 kg ai/ha, how much amount of commercial product you will apply in the crop.

a) 1 kg b) 2 kg
c) 1.5 kg d) 2.5 kg Ans : b

119. The alternate host of gram caterpillar is one of the following:

a) Amaranthus b) Crotalaria sp.
c) Agropyron repens d) Cenchrus sp. Ans : a

120. Skin allergy is caused by one of the following:

a) Amaranthus b) Brush weeds
c) Cyperus d) Parthenium Ans : d

121. Which one of the following weed comes in the category of sedges:

a) Chloris barbata b) Eclipta alba
c) Cyperus rotundus d) Leuscus aspera Ans : c

122. Which one of the following weeds belong to the Leguminaecae family:

a) Elesusine indica b) Melilotus alba
c) Panicum repens d) Cyperus deformis Ans : b

123. Which of the following herbicides is used for chemical tillage:

a) 2, 4, D b) Grammaxone
c) MCPA d) MCPB Ans : b

124. Which one of the following weedicides is available in wettable powder form:

a) 2, 4-D b) Pendamethaline
c) Isoproturon d) MCPB Ans : c

125. Which one of the following weedicide is recommended for wheat + mustard:

a) 2, 4-D b) Pendamethaline
c) Isoproturon d) MCPB Ans : c

126. Which of the following statement is false
   a) 2, 4-D is specific to kill monocots
   b) MCPA is specific for killing dicots
   c) Aquatic weed "chara chinensis" can be controlled by 2, 4-D
   d) Glyphosate kills woody plants
Ans : a

127. Chemical name of herbicide 'Machete' is:
   a) Isproturon b) Fluchloralin
   c) Alachlor d) Butachlor
Ans : d

128. Glyphosate is a .... herbicide
   a) Selective b) Pre-emergence
   c) Non-selective d) Preplanting
Ans : c

129. Orobanche is a parasite on
   a) Sorghum b) Sugarcane
   c) Safflower d) Tobacco
Ans : d

130. Growth of Parthenium weed can be controlled by weed
   a) Cassia serecia b) Cyperus rotundus
   c) Cynodon doctylon d) Mimosa pudica
Ans : a

131. Herbicide is a chemical which used to
   a) Destroy weeds b) Destroy crops
   c) Destroy weeds and crops both d) None of the above
Ans : a

132. Weed competition in rice is more severe in ________
   a) Direct seeded crop b) Transplanted crop
   c) Late sown crop d) Flooded crop
Ans : a

133. The critical period of weed control in upland rice varies between ___________ days after sowing.
   a) 10-15 b) 15-20
   c) 20-25 d) 15–30
Ans : d

134. The herbicide which become active only when under go - 3- Oxidation is
   a) 2, 4-D b) Atrazine
   c) MCPA d) 2, 4-D B
Ans : d

135. Urea herbicides under go conjunction with
   a) Proteins b) Carbohydrates
   c) Fats d) All of these
Ans : a

136. Photodecomposition plays an important role' in loss of urea herbicides from the soil surface in

a) Moist b) Dry
c) Wet d) None of these **Ans : b**

137. .... Degradation plays a major role in the persistence and behavior of herbicides in the soil.

a) Microbial b) Physical
c) Thennal d) Chemical **Ans : a**

138. Triazine herbicides are said as notorious herbicides because of there

a) Long residual toxicity b) Short residual toxicity
c) Non-selectiveness d) None of these **Ans : a**

139. Propanil a postemergence herbicide is readily absorbed by the

a) Stem b) Root
c) Leaf d) None of these **Ans : c**

140. Phenoxy herbicide moves inside the plant

a) Only simplistically b) Only apoplectically
c) By both mechanisms d) None of these **Ans : c**

141. Phalaris minor is a crop associated weed with

a) Wheat b) Maize
c) Rice d) Cotton **Ans : a**

142. Most common herbicide used in wheat is

a) 2, 4-D b) Atrazine
c) Isoproturon d) Pendimethalin **Ans : a**

143. Which of the following physiological growth stages of wheat crop is most susceptible to 2,4-D

a) Milk stage b) Panicle initiation stage
c) Late tillering stage d) Dough stage **Ans : a**

144. Phalaris minor and Avenafatua weeds can be controlled effectively by using _______ herbicide in wheat

a) Isoproturon b) 2,4,-D
c) Fluchloralin d) Pendimethalin **Ans : a**

145. Use of .... herbicide provides effective control of weeds in groundnut

a) 2,4-D b) Atrazine
c) Isoproturn d) Pendimethalin **Ans : d**

146. One of the following herbicides is a selective herbicide
a) Atrazine b) Paraquat
c) Diquat d) Glyphosate **Ans : a**

147. 2, 4-D is a ________ herbicide
a) Selective b) Non selective
c) Contact d) Selective and systemic **Ans : c**

148. For more effective weed control in soybean fluchloralin should be used as
a) Pre-emergence b) Postemergence
c) Early postemergence d) Preplant soil incorporation **Ans : d**

149. Under dry land conditions most preferred herbicide for maize is
a) Atrazine b) Fluchloralin
c) Propanil d) Butachlor **Ans : a**

150. Striga is crop associated weed with
a) Pearl millet b) Mustard
c) Maize d) Gram **Ans : a**

151. .... incorporation of fluchloralin provides relatively more effective control of weeds in groundnut
a) Preemergence b) Pre plant
c) Postemergence d) None of above **Ans : b**

152. ___________ has been recommended for the control of weeds in blackgram
a) Atrazine 0.5 kg/kg b) 2,4-D 0.75 kg/ha
c) Fluchloralin 0.75.kg/ha d) Simazin 0.5 kg/ha **Ans : c**

153. The stages of a crop more prone to weed competition
a) Germination to seedling b) Vegetative
c) Reproductive d) Maturity **Ans : a**

154. Atrazine is a selective herbicide for controlling weeds in crop.
a) Maize b) Wheat
c) Rice d) All **Ans : a**

155. Urea derivative types of herbicides are normally............... applied
a) Foliage b) Soil
c) Soil incorporated d) None **Ans : a**

156. Simazine is selective herbicide in maize crop because of
a) Differential absorption of herbicide by crop and weeds
b) Protein synthesis blockage
c) Germination inhibition
d) All
Ans : d

157. Conversion of a non-phytotoxic compound into phytotoxic compound is called
a) Beta oxidation b) Oxidation
c) Reduction d) All
Ans : a

158. Gesatop is the commercial name of the herbicide
a) Simazine b) Atrazine
c) Fluchloralin d) Gramaxone
Ans : a

159. Couch grass is
a) Sorghum b) Sugarcane
c) Striga d) Agropyron repens
Ans : d

160. The first herbicide used is
a) Sulphuric acid b) Sodium chlorate
c) Ammonium sulphamate d) Copper sulphate
Ans : d

161. Dalapon belongs herbicide group
a) Haloalkanoic acid b) Phenoxyalkanoic acid
c) Aromatic acids d) Nitriles
Ans : a

162. MCPA belongs herbicide group
a) Haloalkanoic acid b) Phenoxybutyric
c) Aromatic acids d) Nitriles
Ans : b

163. Dicamba belongs herbicide group
a) Haloalkanoic acid b) Phenoxybutyric
c) Aromatic acids d) Nitriles
Ans : c

164. Bromoxynil belongs herbicide group
a) Haloalkanoic acid b) Phenoxybutyric
c) Aromatic acids d) Nitriles
Ans : d

165. .... is a post emergence grass killer used in dicotyledonous crops
a) Dicamba b) Dalapon
c) Fluazifop-butyl d) Bromoxynil
Ans : c

166. .... is a total weed killer used in non cropped areas
a) Chlorofenac b) Dalapon
c) Fluazifop-butyl d) Bromoxynil
Ans : a

167. .... is a herbicide used for control of wild oat

a) Di-allate b) Dalapon
c) Fluazifop-butyl d) Propanil **Ans : a**

168. .... is a post emergence contact herbicide used in rice

a) Dicamba b) Dalapon
c) Fluazifop-butyl d) Propanil **Ans : d**

169. .... is a alien weed

a) Cynodon b) Cyperus
c) Phalaris d) Maxican poppy **Ans : d**

170. .... is a weed introduced from some other part of the world

a) Cynodon b) Cyperus
c) Phalaris d) Water hyacinth **Ans : d**

171. Weed introduced from some other part of the world is called as

a) Alien weed b) Anthrophyte
c) Both a & b d) None **Ans : c**

172. The complementary use of resources by two or more plants species is called

a) Annidation b) Allelopathy
c) Both a & b d) Antagonism **Ans : a**

173. Metabolic process in which organic acids are shortened by two carbon increments is

a) Beta oxidation b) Reduction
c) Oxidation d) Both a & b **Ans : a**

174. Conversion of 2,4-DB to 2,4 -D in some plants is an example of

a) Beta oxidation b) Reduction
c) Oxidation d) Both a & b **Ans : a**

175. Enzyme responsible for conversion of 2,4-DB to 2,4 -D in some plants is

a) Beta oxidase b) Beta reductase
c) Nitrate reductase d) Both a & b **Ans : a**

176. 2,4-DB can be used in pulses but 2,4 -D can not be used because of

a) Absence of Beta oxidase b) Presence of Beta reductase
c) Presenceof Beta oxidase d) Both a & b **Ans : a**

177. Gesaprin and primatol is the commercial name of the herbicide

a) Simazine b) Atrazine
c) Fluchloralin d) Gramaxone
Ans : b

178. The herbicide group known to inhibit pantothemic acid synthesis in plants is

a) Thiocarbamate b) Sulfonylurea
c) Amide d) All
Ans : b

179. The phytotoxicity of monuron in cotton increases when applied with

a) Urea b) DAP
c) Phorate d) All
Ans : c

180. Synergistic phytoxicity is observed on oats with phorate and

a) Urea b) DAP
c) Monuron d) All
Ans : c

181. The most problemtic weed in deep water rice is ____________________.

a) Echinocloa crusgalli b) Water hycinth
c) Cyperus rotundus d) All
Ans : c

182. Plant growing where it is not desired is called ___________________.

a) Off types b) Weeds
c) Crop plants d) All
Ans : b

183. Plant whose economic value has not been discovered is called

a) Off types b) Weeds
c) Crop plants d) All
Ans : b

184. _______________ is a non-selective systemic herbicide.

a) Dicamba b) Dalapon
c) Glyphosate d) Bromoxynil
Ans : c

185. Addition of ____________________at 5% concentration to glyphosate enhances the control of Imperata spp.

a) DAP b) MOP
c) Urea d) All
Ans : c

186. Soil application of phorate and _____________ herbicide increases the cotton yield)

a) Urea b) DAP
c) Monuron d) All
Ans : c

187. Coupling of intact herbicide molecules with plant cell is called ________________.

a) Conjugation b) Addehision
c) Sticking d) All Ans : a

188. A substance that is added to herbicide to increase its efficiency is called

a) Adjuvant b) Surfactant
c) Sticker d) All Ans : a

189. The loss of herbicide after application to soil is due to:

a) Volatalization b) Absorption
c) Both a & b d) None Ans : c

190. Weeds in pearlmillet can be controlled by spraying _____________.

a) Isoproturon b) Pendimethalin
c) Atrazine d) None Ans : c

191. The control of monocot weeds in wheat crop can be done by spraying herbicide.

a) 2, 4-D b) Metsuffuron methyl
c) Isoproturon d) Both a & b Ans : c

192. In irrigated areas, the competition by weeds will be more for _____ than for water.

a) Light b) Space
c) Nutrients d) None Ans : c

193. The dispersion of finely divided solid particles in the liquid is called

a) Absorption b) Suspension
c) Conjugation d) None Ans : b

194. All the three types, namely, enforced, innate and induced dormancies are present in the weed plant

a) Wild oat b) Amaranthus
c) Pathenium d) All Ans : a

195. Paraquat is available as _____________.

a) Wettable Powder b) Soluble Concentrate
c) Emulsifiable Concentrate d) None Ans : b

196. The name of the weed was used by Jethro Tull in his famous writing

a) Horse hoeing husbandary b) Weed management
c) Crop and weeds d) None **Ans : a**

197. Reduction in yield of maize due to weeds is less in __________ period

a) Harvesting b) Seedling
c) Knee Height d) Flowering **Ans : a**

198. Typha is an example of ___________ type of aquatic weed

a) Free floating b) Submerged
c) Marginal d) Emersed **Ans : c**

199. ________ is an aquatic weed having aesthetic value for a city dweller

a) Typha b) Nymphaea
c) Nelumbium d) None **Ans : c**

200. ________________is the best suited physical method for weed control, in soils with uneven topography

a) Digging b) Cutting
c) Burning d) None **Ans : c**

201. Factor influencing the effectiveness of foliage applied herbicide are

a) Leaf age b) Surface
c) Concentration d) All **Ans : d**

202. Surfactants, based on chemical nature, are grouped into

a) 1 b) 2
c) 3 d) 4 **Ans : c**

203. Based on use, surfactants are classified into

a) Wetting agents b) Spreaders
c) Penetrants d) All **Ans : d**

204. The main effects of surfactants is

a) Emulsifier b) Spreading
c) Wetting d) All **Ans : d**

205. Differential deactivation of herbicides in the plants may be due to the following processes

a) Oxidation b) Hydrolysis
c) Conjugation d) All **Ans : d**

206. The loss of herbicide after application to soil is due to
a) Microbial degradation b) Leaching
c) Volatilization d) All Ans : d

207. The aquatic weeds can be classified into
a) Submersed b) Emersed
c) Marginal d) All Ans : d

208. Weeds in pearlmillet can be controlled by spraying ____________.
a) Atrazine b) 2,4-D
c) Pendimethalin d) Both a & b Ans : d

209. The control of monocot weeds in wheat crop can be done by spraying herbicide
a) Isoproturon b) Pendimethalin
c) 2,4-D d) Both a & b Ans : d

210. The control of dicot weeds in wheat crop can be done by spraying herbicide
a) Isoproturon b) Pendimethalin
c) 2,4-D d) Both a & c Ans : d

211. The control of dicot and monocot weeds in wheat crop can be done by spraying herbicide
a) Isoproturon b) Pendimethalin
c) 2,4-D d) Both a & b Ans : a

212. All the three types, namely, enforced, innate and induced dormancies are present in the weed plant
a) Phalaris minor b) Wild oat
c) Both d) None Ans : b

213. Paraquat is available as ____________
a) WP b) SC
c) SP d) GR Ans : b

214. The most common weed all over the world is__________ which is difficult to distinguish from rice seedlings in the nursery.
a) Echinocloa b) Phalaris minor
c) Avena fatua d) None Ans : d

215. The stage of crop at which the presences of weeds cause maximum loss to crops is called
a) Acute period b) Critical period
c) Both d) None Ans : b

216. A single parthenium plant is capable of producing more ___________ than seeds.

a) 10000 b) 500
c) 25000 d) 250000 **Ans : a**

217. The application of herbicides on the field before seeding or planting is termed as

a) Pre emergence application b) Pre plant application
c) Pre sowing application d) None **Ans : b**

218. Common snail is a bioagent to control............... aquatic weeds.

a) Typha b) Water Hycinth
c) Hydrilla d) All **Ans : c**

219. Dredging is a practice to control ____________ weeds.

a) Teresstrial b) Aquatic
c) Crop land d) Waste land **Ans : b**

220. If urea is mixed with Stam F-34 (Propanil) it increases ____________ in the plant.

a) Absorption b) Penetration
c) Movement d) None **Ans : a**

221. Stam F-34 (Propanil) is

a) Contact b) Systemic
c) Both a & b d) None **Ans : a**

222. Stam F-34 (Propanil) controls

a) Grasses b) Sedges
c) Broad leaved weeds d) All **Ans : d**

223. Parasitic plants form __________ near host plant roots to suck water and nutrients.

a) Root association b) Stem association
c) Haustoria d) None **Ans : c**

224. Balloon is a modified papery calyx that encloses the fruit loosely with entrapped air as seen in ___________ weed

a) Phyllanthus niruri b) Calotropis
c) Physalis minima d) None **Ans : c**

225. The pollen of _____________ weed reduces, human efficiency.

a) Ambrossia b) Amaranthus
c) Artemisia d) All **Ans : a**

226. A liquid dispersed in another liquid is called ____________.

a) Solution b) Emulsion
c) Concentrate d) Suspension **Ans : b**

227. Flooding kills the weeds by excluding the ______________ from their environment.

a) Air b) Water
c) Nutrients d) All **Ans : a**

228. The herbicide Nitrofen is grouped under __________ group.

a) Amides b) Anilines and Nitro-phenols
c) Substituted urea d) None **Ans : b**

229. Chick weed is used as __________ pollution indicator in the air.

a) NO_2 b) SO_4
c) CO_2 d) CO **Ans : a**

230. The longevity of Convolvulus arvensis seed is about ________ years.

a) 10 b) 12
c) 22 d) 18 **Ans : c**

231. The author of the book Principles in Weed Management is ____________.

a) Crafts b) O.P. Gupta
c) Aldrich and Kramer d) V.S.Rao **Ans : c**

232. The author of the book Scientific Weed Management is ____________.

a) Crafts b) O.P. Gupta
c) Aldrich and Kramer d) V.S.Rao **Ans : b**

233. The author of the book Modern Weed Control is ____________.

a) Crafts b) O.P. Gupta
c) Aldrich and Kramer d) V.S.Rao **Ans : a**

234. The author of the book Principles of Weed Control is ____________.

a) Crafts b) O.P. Gupta
c) Aldrich and Kramer d) V.S.Rao **Ans : d**

235. Parthenium hysterophorus is native of _______________.

a) India b) China
c) Africa d) Mexico **Ans : d**

236. The most important weed in Berseem is ____________.

a) Chicory b) Melilotus
c) Amaranthus d) Cyperus

Ans : a

237. Pigweed is capable of absorbing large amount of ____________ which helps for competing with the crop plants.

a) Nitrate b) Phosphate
c) Potassium d) All

Ans : a

238. The Herbicidal property of 2, 4-D was known during the year ____________ .

a) 1941 b) 1942
c) 1943 d) 1944

Ans : d

239. ____________ cropping is a conventional practice of controlling parasitic weeds.

a) Agropyron b) Brush weeds
c) Cyperus d) Parthenium

Ans : d

240. __________ weed is used as leafy vegetable.

a) Amaranthus b) Brush weeds
c) Cyperus d) Parthenium

Ans : a

241. Amaranthus species produces ____________ seeds / plant.

a) 20000 b) 15000
c) 10000 d) 5000

Ans : a

242. In Cynodon dactylon_the vegetative propagation is through ____________.

a) Seeds b) Suckers
c) Rhizomes d) Underground stems

Ans : d

243. __________ and ________________ (Scientists) were the first who found the use of 2,4-D as herbicide.

a) Henmer and Tukey b) Zimmerman and Hitchcock
c) Marth and Mitchell d) None

Ans : a

244. Simazine is grouped under __________ group of herbicide.

a) Substituted ureas b) Triazines
c) Amides d) Nitriles

Ans : b

245. Critical stage of weed competition in cotton crop is ____________.

a) Seedling b) Vegetative
c) Reproductive d) All

Ans : a

246. Chemical name of Dalapon is ____________.
 a) 2, 2, Dichloropropionic acid b) 2, 4, Dichloropropionic acid
 c) 2, 6, Dichloropropionic acid d) None
 Ans : a

247. Cotton crop is highly sensitive to the herbicide ____________.
 a) 2, 4, D b) Grammaxone
 c) MCPA d) MCPB
 Ans : a

248. Longer the interval between emergence of weed and crop ____________ is the competition.
 a) Lesser b) Higher
 c) Equal d) None
 Ans : a

249. Weed competition causes a reduction in the grain yield by 30% in __________ rice and 90% in ________ rice.
 a) Lowland b) Upland
 c) Upland & Lowland d) Lowland & Upland
 Ans : d

250. The milk will get an undesirable flavour when the cattle grazes ____________ weed on the grassland
 a) Sorghum helpens b) Argemone maxicana
 c) Sweet clover d) All
 Ans : a

251. Striga is a partial root parasite on
 a) Lucern b) Tomato
 c) Potato d) Sorghum
 Ans : d

252. Orobanche is a total root parasite on
 a) Lucern b) Tobacco
 c) Potato d) Sorghum
 Ans : b

253. Cassytha filiformis is a partiall stem parasite on
 a) Lucern b) Orange
 c) Mango d) Sorghum
 Ans : b

254. Loranthus longifolius is a partiall stem parasite on
 a) Lucern b) Orange
 c) Mango d) Sorghum
 Ans : c

255. The weeds that belong to the family .... are known as sedges
 a) Gramineae b) Convolvulaceae
 c) Euphorbiaceae d) Cyperaceae
 Ans : d

256. Critical period of weed competition for transplanted rice is .... days after transplanting

a) 30-50 b) 15-45
c) 20-60 d) 25-60 **Ans : a**

257. Butachlor and propanil belong to .... group of herbicide

a) Ureas b) Carbamates
c) Aliphatic d) Amides **Ans : d**

258. Alternate host of gram caterpillar is one of the following:

a) Amaranthus b) Crotalaria sp.
c) Agropyron repens d) Cenchrus sp. **Ans : a**

259. Skin allergy is caused by one of the following:

a) Agropyron b) Brush weeds
c) Cyperus d) Parthenium **Ans : d**

260. One of the following weeds belong to the Leguminaecae family

a) Elesusine indica b) Melilotus alba
c) Panicum repens d) Cyperus deformis **Ans : b**

261. Which one of the following weedicide is recommended for wheat + mustard

a) 2, 4-D b) Pendimethaline
c) Isoproturon d) MCPB **Ans : : c**

262. Which of the following herbicides is used for chemical tillage

a) 2, 4, D b) Grammaxone
c) MCPA d) MCPB **Ans : b**

263. Which one of the following weedicides is available in wettable powder form

a) 2, 4-D b) Pendimethaline
c) Isoproturon d) MCPB **Ans : c**

264. Which of the following statement is false

a) 2, 4-D is specific to kill monocots
b) MCPA is specific for killing dicots
c) Aquatic weed "chara chinensis" can be controlled by 2, 4-D
d) Glyphosate kills woody plants **Ans : a**

265. . ...is the pre-emergence herbicide used for controlling weeds in jute

a) Benthiocarp b) MSMA
c) DSMA d) Nitrofen **Ans : d**

266. An example of worst perennial weed
a) Paspalum conjugatum
b) Eleusine indica
c) Eichhornia crassipes
d) Chenopodium album
Ans : a

267. Weeds which will grow well on alkali soil
a) Barnyard grass
b) Nut grass
c) Quack grass
d) Chenopodium album
Ans : d

268. Herbicide which can be applied before planting/sowing
a) Glyphosate
b) Alachlor
c) Atrazine
d) Fluchloralin
Ans : d

269. One of the main difference between rice and Echinochloa crusgalli (weed) is
a) Leaves are rough and hairy (rice) and leaves are smooth (weed)
b) Stems are erect and produce tillers (rice) and do not produce tillers (weed)
c) Have auricles and ligules (rice) and no auricles and ligules (weed)
d) No difference till flowering
Ans : c

270. ....under the trade name Collego is normally applied to rice and soybean
a) Bioherbicide
b) Chemical herbicide
c) Mycoherbicide
d) Natural plant origin herbicide
Ans : c

271. Avena fatua is a problematic weed in
a) Pulses
b) Maize
c) Cotton
d) Wheat
Ans : d

272. The best time of application of herbicide in transplanted flooded paddy fields would be
a) To apply herbicide 1-2 weeks after transplanting
b) To apply herbicide 1-2 days after transplanting
c) To apply herbicide just before puddling
d) Tto apply herbicide soon after puddling but before the start of trans planting operation
Ans : b

273. Most commonly used herbicide in sugarcane is
a) 2- 4-D ester and lasso
b) Gromoxone & Butachlor
c) Sencore isoproturon
d) Atrataf and Tafazine
Ans : d

274. Harmonal herbicides are
a) The one which when applied at recommended rates also promote crop growth besides killing the weeds
b) The ones which when applied at very low concentrations induce various responses in plants
c) Total herb-killers and kill all type of growing plants
d) None of these
Ans : b

275. Amongst the 3 commercial preparation of 2, 4-D available in the market the most volatile and likely to cause extensive damage by drift to sensitive crop like cotton in nearby fields are

a) Sodium salts of 2, 4-D b) Amine salts of 2, 4-D
c) Esters of 2, 4-D d) All are equal **Ans : a**

276. Amongst the various herbicides available in the market the most commonly used herbicide in transplanted rice is

a) Isoproturon b) Butachlor
c) Fluchlorin d) Atrataf **Ans : b**

277. Find out the quantity of simazine WP to be sprayed in one-hectare area (if the rate of application per hectare = 3.00 kg al/ha & active ingredients 80%)

a) 3.75 b) 3.5
c) 3.25 d) 3.00 **Ans : a**

278. If the recommended dose of oxydiazon (0.25% ai) to rice is 0.5 kg ai/ha, how much amount of commercial product you will apply in the crop

a) 1 kg b) 2 kg
c) 1.5 kg d) 2.5 kg **Ans : b**

279. Indicator of Uncultivated/neglected, alkaline, low fertility soil is

a) Wild carrot b) Chickweed
c) Clovers d) Dandelion **Ans : a**

280. Indicator of cultivated/tilled, high fertility soil is

a) Wild carrot b) Chickweed
c) Clovers d) Dandelion **Ans : b**

281. Indicator of Low nitrogen in soil

a) Wild carrot b) Chickweed
c) Clovers d) Dandelion **Ans : c**

282. Indicator of cultivated/tilled, clay, acidic soil is

a) Wild carrot b) Chickweed
c) Clovers d) Dandelion **Ans: d**

283. Indicator of wet, acidic soil is

a) Rumex sp. b) Chickweed
c) Clovers d) Dandelion **Ans : a**

284. Indicator of High potassium

a) Fumaria officinalis b) Chickweed
c) Clovers d) Dandelion **Ans : a**

285. Indicator of High fertility, cultivated/ tilled soil is
a) Sececio vulgaris b) Chickweed
c) Clovers d) Dandelion **Ans : a**

286. Which of the following weed brought reduction in land value
a) Cynodon dectylon b) Cyperus rotundus
c) Setaria glauca d) Melilotus indica **Ans :b**

287. Indicator of Low potassium soil is
a) Achillea millefolium b) Chickweed
c) Clovers d) Dandelion **Ans : a**

288. Who coined the term Allelopathy
a) Jethro tull b) Mohr
c) Misterlisch d) Molisch **Ans :d**

289. Maximum number of seeds is produced by which of the following weeds
a) Amaranthus b) Parthenium
c) Utrica sp d) Argemone mexicana **Ans : a**

290. What is the trade name of Alachlor?
a) Basalin b) Stomp
c) Lasso d) Machete **Ans : c**

291. Which paint company introduced 2, 4-D?
a) Dupont b) Cynamid
c) Shaw wallace d) None **Ans : d**

292. Cotton is susceptible to drift of ....
a) 2, 4-D b) Stomp
c) Lasso d) Machete **Ans : a**

293. What is the trade name of paraquat?
a) Machete b) Stomp
c) Lasso d) Gramoxone **Ans : d**

294. What is the scientific name of Bermuda grass?
a) Cynodon dactylon b) Cyperus rotundus
c) Brecharia mutica d) None **Ans : a**

295. Which group of herbicide is having long herbicidal residue problem in soil?
a) Trianilines b) Ureas
c) Amide d) None **Ans : a**

296. Origin of *Lantana camara* is ______

a) Central America b) Euresia
c) Africa d) Europe
Ans : a

297. Coupling of herbicide as molecule with cell constituents in living plant is termd as

a) Cell elongation b) Association
c) Conjugation d) Metabolism
Ans : c

298. ______________is a common safner used in agriculture

a) NA b) GA
c) NAA d) IBA
Ans : a

299. Weeds completing their life cycle in a season/year are known as

a) Annual weeds b) Biennial weeds
c) Perennial weeds d) None
Ans : a

300. ....is an annual weed

a) Bidens pilosa b) Chenopodium album
c) Phalaris minor d) All
Ans : d

301. Which one of the following is non selective herbicide

a) Alachlor b) Butachlor
c) Paraquat d) Atrazine
Ans : c

302. Where is NRCWS located at

a) Jaipur b) Jodhpur
c) Jabalpur d) Jamshedpur
Ans : c

303. What is the percentage loss caused by weeds in Indian agriculture?

a) 45 b) 55
c) 75 d) 25
Ans : a

304. The world's worst weed is

a) Cynodon dactylon b) Cyperus rotundus
c) Brecharia mutica d) None
Ans : b

305. Common name of herbicide is assigned by which organization?

a) ISO b) ISRO
c) BSI d) None
Ans : a

306. Which paint company introduced 2, 4-D?
a) Sherwin Williams paint company b) Berger
c) Nerolac d) None **Ans : a**

307. Which one of the following pairs is not correctly matched

Herbicide group		*Common name*
a) Phenoxy acids	-	2,4,5-T
b) Ureas	-	Isoproturon
c) Acetamide	-	Butachlor
d) Carbamates	-	Dinoseb

Ans : d

308. In India % use of herbicide out of total agro-chemicals is equivalent to............%
a) 38 b) 48
c) 28 d) 58 **Ans : a**

309. What is the common name of Cynodon dactylon?
a) Motha b) Para grass
c) Bermuda grass d) None **Ans : c**

310. Irritation in colocasia is caused due to presence of which chemical?
a) Oxalic acid b) Crystals of calcium oxolate
c) Boric acid d) Mallic acid **Ans : b**

311. What is the trade name of oxadiazon
a) Monstar b) Bluestar
c) Gramaxone d) Ronstar **Ans : d**

312. What is the shape of stem of Borreria hispida
a) Square b) Round
c) Triangular d) None **Ans : a**

313. What is the origin of Echinochloa crusgalli?
a) Tropical America b) Central America
c) Europe d) Asia **Ans : a**

314. In which year Dinoseb was discovered?
a) 1938 b) 1948
c) 1928 d) 1933 **Ans : d**

315. What is the trade name of Diquat?
a) Rogor b) Rhonstar
c) Reglone d) Roundup **Ans : c**

316. Orobanche can be survive on which crop
a) Tobacco b) Sorghum
c) Maize d) All **Ans : a**

317. Origin of Cyperus rotundus is ______

a) Tropical America b) Central America
c) Euresia d) Asia **Ans : c**

318. Amaranthus spp is a ________ plant

a) C3 b) C4
c) Both d) None **Ans : b**

319. __________is used in slash and burn cultivation to compete with Lantana camara

a) Chromolaena odorata b) Cascuta
c) Orobanche d) Striga **Ans : a**

320. Allelopathy is derived from words

a) Greek b) Latin
c) Persian d) None **Ans : a**

321. The break down of herbicides inside the plant into non-toxic compounds is termed as

a) Conjugation b) Reverse metabolism
c) Metabolism d) Detoxification

322. ______________ is the most problematic weed present in 92 countries.

a) Cynodon dactylon b) Cyperus rotundus
c) Brecharia mutica d) None **Ans : b**

323. _______________ was the first person to use the word weed in the present day of meaning.

a) G.J.Mandel b) J.Tull
c) Shoemacher d) G.Tull **Ans : b**

324. Trade name of a combination of Atrazine + Simazine is ____________.

a) Printex b) Weedmar
c) Dhoom d) Gramaxone **Ans : a**

325. Which herbicide is first used in India for the selective control of weed in wheat?

a) Isoproturon b) 2,4-D
c) Pendimethalin d) All **Ans : b**

326. 2, 4-D is used for ____________ weeds

a) Grassy b) Broad leaved
c) Both d) None **Ans : b**

327. ______________ is a Pseudo cereal

a) Buck wheat
b) Dicoccum wheat
c) Duram wheat
d) All

Ans : a

328. Trade name of propanil is __________

a) Stam F-34
b) Stomp
c) Avadex
d) Treflan

Ans : a

329. Trade name of trifluralin is __________.

a) Stam F-34
b) Stomp
c) Avadex
d) Treflan

Ans : d

330. Metsulfuron methyl is a herbicide belonging to ___________ group

a) Urea
b) Sulfonylurea
c) Uracil
d) Carbamates

Ans : b

331. Name the weed which was first biologically controlled

a) Parthenium
b) Lantana
c) Amaranthus
d) Water hycinth

Ans : b

332. The toxic constituent produced by Lantana camara

a) Lontadens
b) Lentacil
c) Bromocil
d) None

Ans : a

333. Boreria hispida belongs to which family

a) Gramineae
b) Leguminoceae
c) Verbanaceae
d) Liliaceae

Ans : c

334. ....is an adsorbent

a) Activated charcoal
b) Nepthalic anhydride
c) Anhydrous ammonia
d) None

Ans : a

335. ....is an adsorbent and safener

a) Activated charcoal
b) Nepthalic anhydride
c) Anhydrous ammonia
d) None

Ans : b

336. Which is the oldest group among the chemical classification of herbicide

a) Triazines
b) Amide
c) Nitrophenols
d) Sulfonylurea

Ans : c

337. Dodder is a total stem parasite on

a) Lucern
b) Tomato
c) Potato
d) Sorghum

Ans : a

338. Which one of the following is non selective herbicide

a) Alachlor b) Butachlor
c) Paraquat d) Atrazine Ans : c

339. Which one of the following is not narrow leaved weeds

a) Cynodon dectylon b) Cyperus rotundus
c) Setaria glauca d) Melilotus indica Ans : d

340. Which one of the following is not kharif season weed

a) Chenopodium album b) Amarauthus viridisi
c) Echinocloa crusgalli d) Commelina benghalensis Ans : a

341. Which one of the following was introduced in the country to eradicate prickly pear (*Opuntia dilleni*) weed

a) Chrysomella spp. b) Dactylopius tomentosus
c) Zygogramma bicolarata d) Neochetina spp. Ans : a

342. Which one of the following microbial agents is being commercially exploited as biocontrol agent

a) Penicillium notatum b) Bacillus subtilis
c) Trichoderma viridiae d) Sclerotium rolfsii Ans : c

343. Plant which can grow well under acidic soil conditions

a) Acedophiles b) Halophytes
c) Heliophytes d) Sciophytes Ans : a

344. Which one of the following causes more wastage of herbicide by drift

a) High volume sprayer b) Ultra-low volume sprayer
c) Hand sprayer d) Low volume sprayer Ans : d

345. In functional allelopathy

a) Toxic substances are released as such from the plant
b) A precursor is released which is converted into active substances by some microorganism
c) There is no question of release of any toxic substance
d) Release of nitrogen from nodule of legume take place. Ans : b

346. If the rate of application per hectare is 3.00 kg ai., the quantity of simazine WP (80% ai.) required to be sprayed in 0.33 hectare area would be

a) 0.50 kg b) 0.75 kg
c) 1.25 kg d) 1.87 kg Ans : c

347. The herbicides containing carbon and hydrogen in their molecule are called

a) Arsenic b) Acid
c) Organic herbicide d) Salt Ans : c

348. The theoretical yield of parent acid from an active ingredient is termed as

a) Active ingredient
b) Technical ingredient
c) Acid equivalent
d) Acid value

Ans : c

349. The weeds which grow well characteristically in acidic soils having pH between 4.5 and 6.5

a) Acidophiles
b) Basophiles
c) Sciophiles
d) None

Ans : a

350. The plants which are undesirable regardless of time and place are known as

a) Absolute weed
b) Alien weeds
c) Relative weeds
d) None

Ans : a

351. The active component of a formulation/ product is called as

a) Active ingredient
b) Technical ingredient
c) Acid equivalent
d) Acid value

Ans : a

352. When a weed is allowed to move from place of its origin to a new area and it establishes itself there, it becomes an introduced weed in its new environment. Such weeds are known as

a) Alien weeds
b) Invader weeds
c) Obligate weeds
d) Facultative weeds

Ans : a

353. Anthrophytes are

a) Alien weeds
b) Invader weeds
c) Obligate weeds
d) Facultative weeds

Ans : a

354. The inhibition of the growth of plants species other than producer species through production of chemicals is called

a) Allelo-inhibition
b) Allelopathy
c) Phyto inhibition
d) None

Ans : a

355. Any direct or indirect harmful effect that one plant has on another or mutually on each other through the production of chemical compounds that escape into the environment is called as

a) Allelo-inhibition
b) Allelopathy
c) Phyto inhibition
d) None

Ans : b

True or False

356. Simazine and Atrazine are the most commonly used herbicides for soybean. **Ans : false**

357. Cypenus difformis belongs to family compositeae **Ans : false**

358. Orobanche is a parasitic weed on tobacco **Ans : true**

359. The scientific name of congress grass is Parthenium hysterophorus **Ans : true**

360. Herbicide mixture is always more effective than a single herbicide application. **Ans : false**

361. Cypenus difformis is a kharif annual weeds. **Ans : true**

362. All monocot weeds belong to family poaceae. **Ans : false**

363. All unwanted plants are weeds but all weeds are not unwanted plants. **Ans : true**

364. Simazine and Atrazine are the most commonly used herbicides for maize and sugarcane **Ans : true**

365. 2, 4-D mostly used in grasses, is a systemic non selective herbicide **Ans : false**

❑❑❑

10 Chapter

Cropping System

Multiple Choice Questions (MCQ's)

1. Multiple cropping refers to
   a) Growing two or more crops in the same field in a year
   b) Growing two or more crops in sequence on the same field in a year
   c) Growing two crops in the same field in a year
   d) All the above **Ans : a**

2. Sequential cropping refers to
   a) Growing two or more crops in the same field in a year
   b) Growing two or more crops in sequence on the same field in a year
   c) Growing two crops in the same field in a year
   d) All the above **Ans : b**

3. Double cropping refers to
   a) Growing two or more crops in the same field in a year
   b) Growing two or more crops in sequence on the same field in a year
   c) Growing two crops in a year in sequence
   d) All the above **Ans : c**

4. Growing two or more crops simultaneously in the same field is
   a) Multiple cropping b) Double cropping
   c) Intercropping d) None of the above **Ans : c**

5. Cultivation of crop regrowth after harvest is
   a) Multiple cropping b) Double cropping
   c) Intercropping d) Ratoon cropping **Ans : d**

6. Growing two or more crops simultaneously with no distinct row agreement is
   a) Mixed intercropping b) Row intercropping
   c) Strip Intercropping d) Relay intercropping **Ans : a**

7. Growing two or more crops simultaneously with one or more crops in distinct row agreement is
   a) Mixed intercropping b) Row intercropping
   c) Strip Intercropping d) Relay intercropping **Ans : b**

8. Growing two or more crops simultaneously during part of life cycle of each is
   a) Mixed intercropping b) Row intercropping
   c) Strip Intercropping d) Relay intercropping **Ans : d**

9. Strip intercropping
   a) Growing two or more crops simultaneously with no distinct row agreement
   b) Growing two or more crops simultaneously during part of life cycle of each
   c) Growing two or more crops simultaneously with one or more crops in distinct row agreement
   d) None **Ans : d**

10. Cropping system is
    a) Cultivation of crops in a farm
    b) Cropping patterns used on a farm
    c) Cropping systems used on a farm and its interaction with farm resources, other farm enterprises and available technology which determine their makeup.
    d) None **Ans : c**

11. Land equivalent ratio is
    a) Cultivation of crops in a farm
    b) Cropping patterns used on a farm
    c) The sum of the fractions of the yield of the intercrops, relative to their sole crop yields.
    d) None **Ans : c**

12. What is crop rotation?
    a) Growing more than one crop at a time
    b) Growing of, crop's one after other to maintain soil fertility
    c) Growing of an associate crop in between the rows of main crop
    d) Growing of crops together in strips **Ans : b**

13. The basic principal of taking crop rotation is
    a) To get higher crop production
    b) To get higher returns per unit area of the soil
    c) To maintain the fertility status of the soils
    d) To keep the weeds under control **Ans : c**

14. Intercropping is a system of cropping to make the maximum use of-
    a) Solar energy b) Water resources
    c) Soil resource d) All above **Ans : d**

15. Relay cropping is defined as
    a) A system of cropping where one crop hands over the land to the next crop in a quick succession
    b) A system of cropping where two or three crops are grown annually
    c) A system of cropping where two or three crops are grown together by mixing the seeds.
    d) A system of cropping where an associate crop is grown in between the rows of principal crop. **Ans : a**

16. Practicing agriculture without the use of chemicals and fertilizers so as to protect ecological balance is known as
    a) Inorganic farming b) Natural farming
    c) Extensive farming d) Intensive farming **Ans : b**

17. The most effective cropping system for returning mineral elements to the soil is
    a) Monocropping b) Crop rotation
    c) Double cropping d) Relaycropping **Ans : b**

18. Cropping intensity is
    a) The extent of the use of land for purposes other than cropping
    b) The extent of the use of land for cropping purposes during a given year
    c) Net cropped area
    d) Gross cropped area **Ans : b**

19. The crops which are grown during summer are known as
    a) Rabi b) Kharif
    c) Zaid d) None of the above **Ans : c**

20. In a sub-tropical area if the average rainfall is 400-500 mm, which of the following crop is most suitable to cultivate ------
    a) Paddy b) Maize
    c) Sorghum d) Cotton **Ans : c**

21. Mixed farming referred only to:
    a) Growing of crop together by mixing the seeds
    b) Growing of cash crops and food grains together
    c) Crop production combined with dairy farming at farm.
    d) Growing of fruits and vegetable mixed **Ans : c**

22. Cropping intensity means :
    a) Percentage ratio of gross cropped area to net cropped area
    b) Percentage ratio of net cropped area to gross cropped area
    c) Percentage ratio of the number of crops in rotation to period of one rotation
    d) Percentage ratio of gross cropped area to the net cropped area **Ans : a**

23. Legume crop should be included in rotation; because
    a) It increase the total pulse production
    b) Emphasis has been given on pulses and oilseeds in the 20-point programme
    c) It helps in maintenance of soil fertility
    d) It increases cropping intensity **Ans : c**

24. Which of the following refers *to* cover, crop
    a) Rapid growing crops grown, to protect the soil from rain and wind action.
    b) Tall growing crops which protects the shade loving crops
    c) Growing of a crop in strips all around the main crop
    d) Crops grown at right angle to the direction of prevailing winds **Ans : a**

25. Any direct or indirect harmful effect that one plant has on another through the release of chemical substances in the root environment is called
    a) Allelopathy b) Annidation
    c) Toxic effect d) None **Ans : a**

26. Which of the following is the pre-requisite for multiple cropping.
    a) Irrigation facility b) Intensive use of fertilizers
    c) Availability of HYV d) All of the above **Ans : a**

27. The ratio of number of hectare days required in mono culture to the number of hectare days used in intercropping to produce identical quantities of each of the component crop is called
    a) ATER b) NAR
    c) CGR d) None **Ans : a**

28. A farming system in which arable crops are grown with trees or shrubs established mainly to hasten soil fertility restoration and enhance soil productivity is called
    a) Alley cropping b) Agroforestry
    c) Lay farming d) None **Ans : a**

29. Agronomic efficiency is expressed as
    a) Kg/kg b) Kg/day
    c) Kg/ha d) None **Ans : a**

30. A system in which land is managed for the concurrent production of agricultural and forest crops and for the rearing of domesticated animal.
    a) Agro-silvi-pastoral system b) Agri-horti-silvi pastoral system
    c) Horti-silvi pastoral system d) None **Ans : a**

31. Agro-silvi-pastoral system is a combination of agrisilviculture and.
    a) Silvi-pastoral system b) Agri-horti system
    c) Horti pastoral system d) None **Ans : a**

32. The unit of crop produced per unit of nutrient added is called
    a) Agronomic efficiency b) Nutrient use efficiency
    c) Nutrient uptake d) None **Ans : a**

33. The term 'Cropping Scheme' refers to
    a) Allotment of crops to the pieces of land on a farm
    b) Allotment of crop rotations to the pieces of land in a farm
    c) Planning and budgeting of the crop production in a farm
    d) Planning of schedules for different farm operations **Ans : b**

34. Trap Crops" refer *to*
    a) Crops grown around the main crop which are more susceptible to insect pest and diseases than the main crop
    b) Crops whi.ch are resistant to pest and diseases
    c) Crops efficient in trapping solar radiation.
    d) Crops efficient hi trapping rainfall. **Ans : a**

35. A short duration crop in between two main crops is termed as-
    a) Cash Crop b) Catch crop
    c) Companion crop d) Ephemeral **Ans : b**

36. The practice of growing four crops in a year from same piece of land in such a way that one crop is sown immediately after or just before the harvest of the previous crop is called as
    a) Double cropping b) Relay cropping.
    c) Intensive cropping d) Multiple cropping **Ans : b**

37. In a cropping sequence of maize-wheat-moong (summer) the cropping intensity is

a) 100% b) 200%
c) 300% d) 400% Ans : c

38. Mixed cropping or growing of more than one crops together in the same field helps farmers in

a) Providing better and continuous cover on land
b) Protecting soil against beating rain drops
c) Assured income to the farmers
d) All these ways Ans: d

39. Crop seasons in India are named as

a) Kharif b) Zaid Kharif
c) Rabi, Zaid Rabi d) All these Ans : d

40. From being of May to middle of October is

a) Kharif season b) Zaid Kharif season
c) Zaid Rabi season d) Rabi season Ans : a

41. From August to January is the period of

a) Kharif season b) Zaid Kharif season
c) Zaid Rabi season d) Rabi season Ans : b

42. From middle of October to middle of April is the period of

a) Kharif season b) Zaid kharif season
c) Rabi season d) Zaid rabi season Ans : c

43. The period of Zaid rabi season in India is from

a) Beginning of May to middle of October
b) August to January
c) Middle of October to middle of April
d) Beginning of February to beginning of May Ans : d

44. Which of the following two years crop rotations are practiced by Indian farmers in different parts of the country

a) Rice-pulses
b) Jowar or maize-wheat or gram
c) Cotton or jowar-groundnut or jowar
d) All these Ans : d

45. Which of the following crop rotations practiced by Indian farmers cover three years period

a) Wheat-maize-sugarcane b) Sugarcane-wheat-cotton
c) Wheat-wheat-turia d) All these Ans : d

46. Which of the following crop mixtures are not important for farmers of Punjab
    a) Jowar and sugar, bajra and moong
    b) Maize and urad, wheat and lentil or gram
    c) Jowar and moth, bajra and cowpea
    d) Jowar and cowpea bajra and guar **Ans : b**

47. Important crop mixtures grown by farmers of Rajasthan are
    a) Maize with ragi or urad or cowpea or velvet bean or guar or senji, jowar with moong, bajra with moth
    b) Wheat with gram or lentil or barley or rapeseed, barley with gram
    c) Cotton with til or senji or moth or melons, gram with toria
    d) All these **Ans : d**

48. Growing four crops sequentially in a year in the same field is
    a) Multiple cropping b) Double cropping
    c) Quadruple cropping d) None of the above **Ans : c**

49. Important crop mixtures grown by farmers of Uttar Pradesh are
    a) Wheat and mustard, wheat and barley, wheat and linseed, wheat and gram, wheat and lentil, barley and mustard, barley, mustard and gram, barley and gram, barley and peas
    b) Jowar and arhar, jowar arhar and urd, jowar and lobia, jowar and moong, jowar, arhar and moong, jowar and sessamum, jowar and guar, jowar, arhar and til
    c) Bajra, arhar and urd, bajra and tur, bajra, arhar and til, bajra, arhar and groundnut, bajra and groundnut sessamum and tur, cotton and tur, maize and urd, groundnut and tur peas and mustard, groundnut and sessamum, kodon and arhar, gram and linseed)
    d) All these **Ans : d**

50. Growing two or more crops one after the other on the same field in a year is
    a) Multiple cropping b) Double cropping
    c) Sequential cropping d) None of the above **Ans : c**

51. Which of the following crop mixtures are not grown by farmers of West Bengal
    a) Jute and cowpea, jute and khesari
    b) Paddy and khesari
    c) Wheat and linseed, wheat and gram
    d) Paddy and jute **Ans: a**

52. Intensification of cropping both in time and space takes place in
   a) Multiple cropping
   b) Double cropping
   c) Sequential cropping
   d) None of the above

Ans : a

53. Which of the following crop mixtures are important for farmers of Delhi area
   a) Jowar and moong, wheat and gram, maize and pulses, wheat and mustard
   b) Jowar and guar, wheat and barley, gram and barley, maize and sessamum
   c) Bajra and tur, Bajra and tobacco, Wheat and potato
   d) Both a & b.

Ans : d

54. Intensification of cropping in time dimensions takes place in
   a) Multiple cropping
   b) Double cropping
   c) Sequential cropping
   d) None of the above

Ans : c

55. Which of the following crop mixtures are not important for farmers of Himachal Pradesh
   a) Wheat and barley, barley and gram
   b) Wheat and gram, maize and sessamum
   c) Wheat and mustard, maize and pulses
   d) None of these

Ans: d

56. Intensification of cropping both in time and space takes place in
   a) Multiple cropping
   b) Intercropping
   c) Double cropping
   d) Both a & b

Ans : d

57. Most important crop mixture grown by farmers on Tripura is
   a) Cotton, sessamum and autumn paddy,
   b) Paddy and chillies
   c) Wheat and gram
   d) Wheat and peas

Ans : a

58. Which of the following crop mixtures are not important for farmers of Manipur
   a) Paddy and chillies
   b) Cotton, sessamum and autumn paddy
   c) Wheat and gram
   d) Wheat and peas

Ans : b

59. In relay cropping
    a) Second crop is sown after the harvest of previous crop
    b) Second crop is sown before the harvest of previous crop
    c) Second crop is sown next year
    d) None of the above **Ans : b**

60. Wheat and barley are not grown as mixed crops in
    a) Jammu & Kashmir b) Rajasthan
    c) Haryana
    d) Uttar Pradesh and Himachal Pradesh **Ans : c**

61. Cotton is grown as a mixed crop with paddy in
    a) Gujarat b) Maharashtra
    c) Assam d) All these states **Ans : d**

62. Important crop mixtures grown in Gujarat are
    a) Jowar and tur or urad, bajra and moong or tur
    b) Wheat and gram, paddy and tur or cotton, maize and tur
    c) Both a & b
    d) None of these **Ans: c**

63. Important crop mixtures grown in Maharashtra are
    a) Jowar with tur or moth or safflawer or moong or urad or cotton or groundnut or cotton or and tur or groundnut and tur or safflower and wheat or gram, rabi jowar with gram or moong
    b) Bajra with moth or cstor or moth and hulga or tur or urad or niger or moth and moong
    c) Paddy with gram or castor or cowpea
    d) Both a & b. **Ans : d**

64. Important crop mixtures grown in Karnataka are
    a) Jowar with ragi or linseed or bajra and tur or safflower or gram or tur and grounfnut or gram and linseed or urad or horsegram or tur and horsegram or cotton and horsegram
    b) Bajra with horsegram or ragi or tur or maize or Dolichos, ragi with Kodra or above or rapeseed or tur or jowar and avare or Italian millet or Dolichos or sessamum or horsegram or cotton
    c) Cotton with navane or groundnut or Kodon or chillies or tur, wheat with safflower or safflower and jowar, gram with safflower
    d) All these **Ans : d**

65. An appropriate combination of different farm enterprises and the means available to the farmer to raise them successfully is called
    a) Farming system b) Cropping system
    c) Cropping pattern d) None **Ans : a**

66. The yearly sequence and spatial arrangement of crops and fallow in an area is called

a) Cropping pattern b) Cropping system
c) Crop rotation d) None Ans : b

67. The recurrent succession of crops on the same piece of land either in a year or over a period of time is called

a) Farming system b) Cropping system
c) Crop rotation d) None Ans : c

68. In mixed farming the main objective is

a) Subsistence b) Higher profitability
c) Ecological balance d) None Ans : a

69. In farming system the main objective is

a) Subsistence b) Higher profitability
c) Ecological balance d) Both b& c Ans : d

70. The proportion of area under different crops at a point of time in a unit area is called

a) Cropping pattern b) Cropping system
c) Crop rotation d) None Ans : a

71. In low rainfall area (<750mm/annum)..................is practiced

a) Monocropping b) Double cropping
c) Multiple cropping d) Intercropping Ans : d

72. When rainfall is >750mm/annum..................is practiced

a) Monocropping b) Double cropping
c) Multiple cropping d) Intercropping Ans : b

73. In intercropping the interaction between different crops may be

a) Competitive b) Non competitive
c) Complementary d) Any one of the above Ans : d

74. Allelopathy is

a) Harmful effect b) Beneficial effect
c) Both d) None Ans : a

75. Annidation in time takes place in intercropping of

a) Sorghum with redgram b) Groundnut with redgram
c) Maize with greengram d) All the above Ans : d

76. The complementary interaction between component crops both in space and time is called
a) Allelopathy b) Annidation
c) Toxic effect d) None
Ans : b

77. In multistoried cropping the principal of .... is used
a) Allelopathy b) Annidation in space
c) Annidation in time d) None
Ans : b

78. When two crops of widely varying duration are planted and their peak demends of light and nutrients are likely to occur at different periods, thus reducing competition is called
a) Allelopathy b) Annidation in space
c) Annidation in time d) None
Ans : c

79. Planting of shade trees in coffee plantation is an example of
a) Allelopathy b) Annidation in space
c) Annidation in time d) None
Ans : b

80. RYT is mostly used to measure yield advantage in
a) Pasture land b) Fallow land
c) Crop land d) None
Ans : a

81. Conversion of yield of different intercrops into equivalent yield of one crop on monetary basis is called
a) CEY b) RYT
c) LER d) None
Ans : a

82. Intercropping of setaria with redgram is an example of
a) Allelopathy b) Annidation in space
c) Annidation in time d) None
Ans : b

83. The yield advantage measured on the basis of unit area and unit population is called
a) CEY b) RYT
c) LER d) None
Ans : b

84. RYT of < 1 indicates
a) Yield reduction b) Yield enhancement
c) No change d) None
Ans : a

85. The sum of area under different crops and harvested in a single year divided by total cultivated area is called
a) CEY b) RYT
c) LER d) MCI
Ans : d

86. Multiple cropping index (MCI) is expressed as
   a) Percentage b) Hectares
   c) Quintals d) None
   Ans : a

87. Cultivation of long duration crops will result in ....in cropping intensity index (CII)
   a) increase b) decrease
   c) no change d) None
   Ans : b

88. The gross cropped area divided by net cultivated area is called
   a) MCI b) RYT
   c) LER d) CII
   Ans : d

89. The zone with maximum yield and spread index is called
   a) Agro ecological zone b) Efficient cropping zone
   c) Deficient cropping zone d) None
   Ans : b

90. The area with similar soil, vegetation and population density characteristics, resulting in a similar type of cropping system is called
   a) Agro ecological zone b) Efficient cropping zone
   c) Deficient cropping zone d) None
   Ans : a

91. The yield index of a crop is calculated by dividing
   a) Yield of state / yield of country x100
   b) Average yield of state / average yield of country x100
   c) Area of state/Area of country x100
   d) Average area of state / average area of country x100
   Ans : b

92. The spread index of a crop is calculated by dividing
   a) Yield of state / yield of country x100
   b) Average yield of state / average yield of country x100
   c) Area of state/Area of country x100
   d) Average area of state / average area of country x100
   Ans : d

93. Intercropping of sorghum with redgram is an example of
   a) Allelopathy b) Annidation in space
   c) Annidation in time d) None
   Ans : c

94. Which of the following crop mixtures are not grown by farmers in Orissa
   a) Gram with safflower or jowar, ragi with bajra or Kodra or rapeseed
   b) Mustard with linseed or horsegram
   c) Gram with linseed
   d) Ragi with maize
   Ans : a

95. A farming system in which arable crops are grown with trees or shrubs established mainly to hasten soil fertility restoration and enhance soil productivity is called

a) Alley cropping b) Agroforestry
c) Lay farming d) None **Ans : a**

96. A system in which land is managed for the concurrent production of agricultural and forest crops and for the rearing of domesticated animal.

a) Agro-silvi-pastoral system b) Agri-horti-silvi pastoral system
c) Horti-silvi pastoral system d) None **Ans : a**

97. Agro-silvi-pastoral system is a combination of agrisilviculture and

a) Silvi-pastoral system b) Agri-horti system
c) Horti pastoral system d) None **Ans : a**

98. Jowar and paddy are grown as mixed crops in

a) Uttar Pradesh (Rohilkhand) b) Madhya Pradesh (Raipur)
c) Orissa (balangir) d) Tamil Nadu (Coastal region) **Ans : a**

99. Individual crop species which are part of the multiple cropping system are called

a) Component crops b) Multiple crops
c) Copanion crops d) All **Ans : a**

100. Example of simultaneous poly culture is

a) Mixed cropping b) Intercropping
c) Relay planting d) All **Ans : d**

❑❑❑

11 Chapter

Cereal Crop Production

Multiple Choice Questions (MCQ's)

Rice

1. The centre of origin of rice is

   a) India and Burma　　b) America
   c) S. Africa　　d) None　　**Ans : a**

2. The second most cultivated crop of the world is

   a) Rice　　b) Wheat
   c) Maize　　d) Jowar　　**Ans : a**

3. The most cultivated crop in India is

   a) Rice　　b) Wheat
   c) Maize　　d) Jowar

4. In India the productivity of rice is highest in the state

   a) Punjab　　b) Haryana
   c) W. Bengal　　d) None　　**Ans : b**

5. The inflorescence in rice is

   a) Siliqua　　b) Panicle
   c) Spikelets　　d) Ear　　**Ans : b**

6. Area under rice cultivation in India is approximately

   a) 39 m ha　　b) 45m ha
   c) 50m ha　　d) 29m ha　　**Ans : b**

7. Number of species included under genus oryza

a) 23	b) 21	
c) 22	d) 24	Ans : d

8. Rice varieties cultivated in India belongs to genus

a) Asiatica	b) Japonica	
c) Javanica	d) Indica	Ans : d

9. Optimum temperatures towards maturity of rice ranges between

a) 21-37°C	b) 25-30°C	
c) 20-25°C	d) None	Ans : c

10. ....is a long duration variety of rice

a) Govind	b) Bala	
c) Garima	d) None	Ans : c

11. ....is an early maturing semi dwarf variety of rice

a) Govind	b) Bala	
c) Garima	d) None	Ans : b

12. Oryza sativa is having .... chromosomes

a) 24	b) 1	
c) 22	d) 3	Ans : a

13. Oryza sativa is

a) Diploid	b) Tetraploid	
c) Hexaploid	d) None	Ans : d

14. Number of sub species included under oryza sativa

a) 24	b) 1	
c) 22	d) 3	Ans : d

15. Rice varieties cultivated in India belongs to genus

a) Asiatica	b) Japonica	
c) Javanica	d) Indica	Ans : d

16. Rice varieties cultivated in Japan belongs to genus

a) Asiatica	b) Japonica	
c) Javanica	d) Indica	Ans : b

17. Optimum temperatures towards maturity of rice ranges between

a) 21-37°C	b) 25-30°C	
c) 20-25°C	d) None	Ans : c

18. Optimum temperatures throughout the growth of rice ranges between
    a) 21-37°C b) 25-30°C
    c) 20-25°C d) None Ans : a

19. Optimum temperatures for blooming of rice ranges between
    a) 21.5-37°C b) 25-30.5°C
    c) 26.5-29.°C d) None Ans : c

20. Minimum temperature for sprouting of rice is
    a) 15°C b) 22-23°C
    c) 10°C d) None Ans : c

21. Minimum temperature for flowering of rice is
    a) 15°C b) 25-30°C
    c) 10°C d) None Ans : b

22. The shoot of rice is called
    a) Stem b) Trunk
    c) Culm d) None Ans : c

23. Shoot of rice is
    a) Hollow b) Solid
    c) Hollow except at nodes d) Solid except at nodes Ans : c

24. Leaf of rice is
    a) Sessile b) Petiolate
    c) Both d) None Ans : a

25. In rice.................. is very prominent
    a) Auricle b) Ligule
    c) Lemma d) Palea Ans : b

26. Panicle of rice is borne on
    a) Uppermost node b) Lowermost node
    c) Middle nodes d) None Ans : a

27. Hull of rice comprised of
    a) Auricle b) Lemma
    c) Palea d) Both b & c Ans : d

28. The Dapog method of raising rice nursery was introduced in India from
    a) Burma b) Japan
    c) USA d) Philippines Ans : d

29. The area need to raise rice seedlings by dapog method, sufficient for transplanting one hectare area is-
a) 25-30 sqm b) 50-60 sqm
c) 250-300 sq m d) 400-500 sqm. Ans : b

30. Dapog seedlings will be ready for transplanting within days.
a) 21-24 b) 18-20
c) 11-14 d) 8-10 Ans : c

31. In Orissa Summer or spring season rice is known as
a) Aus b) Arnall
c) Dalua d) Boro Ans : c

32. Dalua crop is sown in
a) Nov-Dec b) June-July
c) May-June d) March-April Ans : a

33. Boro crop is sown in
a) Nov-Dec b) June-July
c) May-June d) March-April Ans : a

34. Aman crop is sown in
a) Nov-Dec b) June-July
c) May-June d) March-April Ans : b

35. Aus crop is sown in
a) Nov-Dec b) June-July
c) May-June d) March-April Ans : c

36. Dalua crop is harvested in
a) Nov-Dec b) June-July
c) May-June d) March-April Ans : d

37. Aman crop is harvested in
a) Nov-Dec b) June-July
c) May-June d) March-April Ans : a

38. Aus crop is harvested in
a) Nov-Dec b) June-July
c) May-June d) Sept-Oct Ans : d

39. Which of the following management practices is more problematic in direct seeded rice cultivation
a) Fertilizer management b) Weed management
c) Water management d) Bushening Ans : b

40. Which of the following ratios between nursery area and transplantation is correct when rice seedling are raised by "Dapog method"

a) 1: 50 b) 1: 25
c) 1:100 d) 1:250 **Ans : d**

41. In early sixtees chinese scientists discovered a sponteneous mutant in rice in the indica variety. It was

a) Dee-gee-woo-gen b) Tsai -yuen-chung
c) Norin-8 d) Peta **Ans : a**

42. Long duration Indica rice type is

a) Photo insensitive b) Photosensitive
c) Weakly photo sensitive d) None of the above **Ans : b**

43. Long duration rice variety is

a) Tella Hamsa b) IR-8
c) Mashuri d) Jaya **Ans : c**

44. The most commonly practised method of growing rice seedling nursery in the northern Indian plains is

a) Growing seedlings on raised beds
b) Growing dry and un sprouted seeds in puddled flat seed beds
c) Growing rice nursery by DAPOG method)
d) Sowing pre-sprouted seeds in well puddled seed beds in standing water **Ans : a**

45. The varieties which belong to species *Oryza glaberrima* are found in

a) Europe b) Asia
c) America d) Africa **Ans : d**

46. The rice varieties grown in India belongs to

a) Indica b) Japonica
c) Javanica d) Asiatica **Ans : a**

47. Rice grown in Indonesia belongs to

a) *O. glaberrima* b) *O. sativa indica*
c) *O. sativa japonica* d) *O. sativa javanica* **Ans : d**

48. Non traditional areas of rice cultivation include

a) Punjab and Haryana b) Uttar Pradesh and Bihar
c) West Bengal and Orissa d) Tamil Nadu and Andhra Pradesh **Ans : a**

49. The rice inflorescence is known as

a) Panicle b) Spikelets
c) Ear d) Siliqua **Ans : a**

50. The dehulled rice grain is known as
a) White rice b) Brown rice
c) Red rice d) Grey rice **Ans : b**

51. Rice crop needs
a) Warm and humid climate b) Hot and humid climate
c) Dry and hot conditions d) Cold and dry climate **Ans : b**

52. Photoperiodically, rice is a type of plant
a) Long day b) Short day
c) Day neutral d) Intermediate **Ans : b**

53. The optimum pH of soil for rice cultivation is varies in range
a) 4 – 6 b) 5 - 6.5
c) 6 – 7 d) 7 – 9 **Ans : b**

54. IR-8 variety of rice is introduced in India from
a) Taiwan b) Philippines
c) Indonesia d) USA **Ans : b**

55. Which of the following variety is a dwarf mutant
a) IR-8 b) Jaya
c) Pusa-2-21 d) Jagannath **Ans : d**

56. The seed rate of rice cultivation in broadcasting and drilling, is sufficient
a) 100, 60 kg/ha b) 60, 100 kg/ha
c) 45, 100 kg/ha d) 100,45 kg/ha **Ans : a**

57. For transplanting one hectare area of rice how much area is sufficient for nursery raising
a) 100 m^2 b) 500 m^2
c) 1000 m^2 d) 1500 m^2 **Ans : c**

58. Dapong method of raising nursery of rice has been introduced in India from
a) Indonesia b) Taiwan
c) Israel d) Philippines **Ans : d**

59. In wet bed method of nursery raising of rice, seedling would be ready for transplanting at an age of
a) 14 days b) 15 – 20 days
c) 20 - 25 days d) 25 - 30 days **Ans : c**

60. In rice, Dapog seedlings would be ready for transplanting
a) 11 - 14 days b) 14 – 17 days
c) 17 – 20 days d) 20 –23 days **Ans : a**

61. Dapog method requires .... sq.m area for raising nursery for 1 ha area
   a) 10-15 b) 15-20
   c) 20-25 d) 25 -30 **Ans : d**

62. Propanil (Stam F-34) herbicide should be applied in rice after transplanting
   a) 2 - 4 Dat b) 6 - 8 DAT
   c) 10 - 14 DAT d) 2 weeks **Ans : b**

63. Propanil (Stam F-34) herbicide should be applied in rice at......... kg ai/ha
   a) 2 b) 3
   c) 4 d) 5 **Ans : b**

64. The Kresek occurs in early stage of plant growth of rice in
   a) BLB b) Bacterial leaf streak
   c) Tungro virus d) False Smut **Ans : a**

65. Tungro virus of rice is transmitted by
   a) Stem borer b) Leaf roller
   c) Green leaf hopper d) Gundhi bug **Ans : c**

66. Dead heart and white head damage to rice is caused by
   a) Gall midge b) Leaf roller
   c) Army worm d) Stem borer **Ans : d**

67. Silvery shoot or onion leaf symptom of rice tillers
   a) Rice hispa b) Gall midge
   c) Leaf roller d) Gundhi bug **Ans : b**

68. Grassy stunt virus disease of rice is transmitted by
   a) Green leaf hopper b) Brown plant hopper
   c) White fly d) Gundhi bug **Ans : b**

69. Gundhi bug insect attack on rice during
   a) Germination b) Flowering
   c) Milking stage d) Harvesting stage **Ans : c**

70. Test weight of basmati rice is
   a) 15 g b) 21 g
   c) 24 g d) 30 g **Ans : b**

71. *Oryza sativa* is a diploid species having chromosome
   a) 18 b) 24
   c) 28 d) 42 **Ans : b**

72. Gene responsible for dwarfing characters in rice is
 a) Tift 23A b) Dee-gee-woo-gen
 c) Norin-10 d) Opaque-2 **Ans : b**

73. Golden rice is a rich source of
 a) Vitamin A b) Vitamin B
 c) Ascorbic acid d) Vitamin K **Ans : a**

74. Idea of super rice was given by
 a) GH Shull b) GS Khush
 c) VL Chopra d) Yoshida **Ans : b**

75. Dwarf rice is introduced in India in the year
 a) 1964 b) 1965
 c) 1963 d) 1969 **Ans : c**

76. Maximum area under rice is in
 a) India b) USA
 c) Russia d) China **Ans : a**

77. The first rice variety having short plant height, lodging resistance and higher grain yield was
 a) Dee-gee-woo-gen b) Norin-10
 c) Lerma Rojo 64A d) Sonara-64 **Ans : a**

78. Govind an early maturing semi dwarf variety of rice developed by selection from crossig
 a) IR-20 x IR-24 b) Peta x Dee-gee-woo-gen
 c) IR 262 x TKM-6 d) Taichung 65 x Maying Ebos 80/2 **Ans : a**

79. IR-8 an photo insensitive variety of rice developed by selection from crossig
 a) IR-20 x IR-24 b) Peta x Dee-gee-woo-gen
 c) IR 262 x TKM-6 d) Taichung 65 x Maying Ebos 80/2 **Ans : b**

80. Rice crop yielding 50 quintals of grain per hectare removes N, P_2O_5 and K_2O kg/ha, respectively
 a) 100 – 150, 70 – 80 and 125 – 150
 b) 90 – 100, 60- 70 and 100 – 110
 c) 120 – 40 – 80
 d) 80 – 30/40 – 20 **Ans : a**

81. The recommended dose of N, P_2O_5 and K_2O kg/ha for deshi rice are
 a) 60 – 40 – 40 b) 120 – 40 – 40
 c) 30 – 20 – 30 d) 40 – 30 – 40 **Ans : a**

82. The recommended dose of N, P_2O_5 and K_2O kg/ha for dwarf rice are
a) 100 - 60 -40 b) 120 - 40 - 40
c) 80 - 60 - 40 d) 40 - 30 -40 **Ans : a**

83. The N fertilizer use efficiency in rice can be increased by using
a) S-coated urea b) Urea super granules
c) BGA d) Both a & b **Ans : d**

84. IR-20 variety of rice developed by selection from crossig
a) IR-20 x IR-24 b) Peta x Dee-gee-woo-gen
c) IR 262 x TKM-6 d) Taichung 65 x Maying Ebos 80/2 **Ans : c**

85. Mahsuri an late maturing variety of rice developed by crossig
a) IR-20 x IR-24 b) Peta x Dee-gee-woo-gen
c) IR 262 x TKM-6 d) Taichung 65 x Maying Ebos 80/2 **Ans : d**

86. Rice variety Mahsuri has been introduced in India from
a) Phillipines b) Taiwan
c) Malaysia d) Indenosia **Ans : c**

87. Rice variety IR-20 has been evolved at
a) Phillipines b) Taiwan
c) Malaysia d) Indenosia **Ans : a**

88. 'Akiochi' disease in rice is due to the toxicity caused by
a) Zinc b) Iron
c) Phosphorus d) Hydrogen sulphide **Ans : b**

89. 'Khaira' disease in rice is due to deficiency of
a) Zinc b) Iron
c) Phosphorus d) Hydrogen sulphide **Ans : a**

90. Which one of the following pairs is not correctly matched
a) Bajra- Tift 23 D_2A b) Maize- T Cytoplasm
c) Rice-Rht1 Rht2 d) Rice-Norin 10 **Ans : d**

91. Which one of the following pairs is not correctly matched
a) Paddy-gall fly: Silver leaf b) Rice bug: Repugnant smell
c) Rice stem borer: Black ears d) Rice zig-zag leaf hopper: Orange leaf **Ans : c**

92. Which vegetable oil is good for heart patient
a) Groundnut oil b) Mustard oil
c) Soybean oil d) Sunflower oil **Ans : d**

93. Nitrogen is applied into transplanted rice in the proportion of
   a) 50% at basal+ nil at tillering+50% at panicle emergence stage,
   b) Nil at basal+50% at tillering+50% at panicle emergence stage
   c) 25% at basal+50% at tillering+25% at panicle emergence stage
   d) 50% at basal+25% at tillering+25% at panicle emergence stage. **Ans : d**

94. Dee-gee-woo-gen a dwarf indica variety from
   a) Phillipines b) Taiwan
   c) Malaysia d) Indenosia **Ans : b**

95. Peta a tall variety from
   a) Phillipines b) Taiwan
   c) Malaysia d) Indenosia **Ans : d**

96. The optimum plant population per hectare of sorghum is
   a) 50,000 plants b) 1,00,000 plants
   c) 1,50,000 plants d) 2,00,000 plants **Ans : b**

97. Which one of the following is a green house gas
   a) Oxygen b) Ammonia
   c) Methane d) Chlorine **Ans : c**

98. Khaira disease in rice is caused due to
   a) Fungal infection
   b) Zinc deficiency
   c) Excessive application of potassium
   d) Bacterial infection **Ans : b**

99. An annual weed commonly found in rice field is
   a) Allagalliis arvellsis b) Ec1lillochloa crusgalli
   c) Amaralltht/s spillosl/s d) Phalaris minor **Ans : b**

100. Seed rate of transplanted basmati rice is
   a) 25-30 kg/ha b) 35-40 kg/ha
   c) 45-60 kg/ha d) 55-70 kg/ha **Ans : a**

101. Mat type nursery is related to
   a) Tobacco crop b) Paddy crop
   c) Onion crop d) Brinjal **Ans : b**

102. Zero-till seed drill is used for sowing of
   a) Rice in maize-rice cropping system
   b) Rice in mung rice cropping system
   c) Rice in rice-rice cropping system
   d) None of these **Ans : c**

103. The panicle initiation stage in rice plants comes
   a) Immediate after transplanting b) After maximum tillering stage
   c) After boot leaf stage d) At any time after transplanting **Ans : c**

104. Which gas is released from paddy fields
   a) CH_4 b) H_2S
   c) CO_2 d) NH_3 **Ans : a**

105. Varieties of rice cultivated in India belong to
   a) Oryza sativa b) Oryza glaberrima
   c) Oryza perennis d) All these species **Ans : a**

106. Seed of rice is called
   a) Siliqua b) Caryopsis
   c) Berry d) Droop **Ans : b**

107. Rice contains 7-8% protein which is located in
   a) Inside the endosperm
   b) Husk
   c) Aleurone layer (outside the endosperm)
   d) Embryo **Ans : c**

108. The shortest duration rice variety is known to nature matures in
   a) 85-90 days b) 60-75 days
   c) 60 days d) 70 days **Ans : d**

109. Which of the following soils are suitable for rice cultivation
   a) Riverine alluvium, red yellow and red loamy
   b) Hill and submontane soils, tarai soils and laterite soils
   c) Coastal alluvium soils, red sandy soils, mixed red and black soils and medium and shallow black soils
   d) All these soils **Ans : d**

110. When rice is sown directly in dry field and crop is raised without availability of standing water, this system of cultivation is known as
   a) Dry system b) Semi-dry system
   c) Boro system d) Koder system **Ans: a**

111. When rice is sown directly in dry field but rain water is compounded when crop is about one and a half to two months old and thereafter enough water is available to the crop, this system of rice cultivation is called
   a) Dry system b) Semi-dry system
   c) Wet system d) Boro system **Ans : b**

112. When rice grown under wet conditions by sowing sprouted seeds or transplanting seedlings in puddle soil with 3-5 cm deep water, this is called

a) Semi-wet system
b) Puddle system
c) Wet system
d) Stagnant or standing water system

Ans : c

113. The interculture operation of semi-dry rice cultivation which helps to control weeds and adjust crop plant population is called

a) Beushening
b) Dry sowing
c) Dribbling
d) None

Ans : a

114. Optimum depth of puddling for rice growing in clay and clay-loam soils is

a) 5 cm
b) 10 cm
c) 3-4 cm
d) 5-6 cm

Ans : b

115. The objective of puddling operation in rice cultivation is to

a) Providing soft seedbed for seedlings to establish faster
b) Increase nutrient availability by minimizing leaching losses by creating reduced condition
c) Incorporated weeds and stubbles into the soil and minimize weed problem.
d) Achieve all these

Ans : d

116. Advantage of raising transplanted rice crop is

a) reduced cost of maintenance of seedlings as nurseries occupy only 10% of the main area
b) Better and more effective use of plant protection measures in nursery
c) Seedling treatment for nutrient deficiency and for protection against pests and diseases can be done before transplanting
d) All the above advantages

Ans : d

117. For raising direct seeded crop of rice the amount of seed required to broadcast in one hectare is

a) 80-100 kg
b) 60-70 kg
c) 40-50 kg
d) 100-125 kg

Ans : a

118. For raising direct seeded crop of rice by dibbling amount of seed required for one hectare is

a) 80-100 kg
b) 60-70 kg
c) 40-50 kg
d) 100-125 kg

Ans : b

119. For raising nursery for one hectare of field the amount of seed required is

a) 80-100 kg b) 60-70 kg
c) 40-50 kg d) 100-125 kg **Ans : c**

120. When rice seedlings are raised in puddle beds, it is called

a) Wet nursery b) Puddle nursery
c) Dapog nursery d) Nursery **Ans : a**

121. When rice nursery is grown in raised beds of well prepared dry soils (without stagnant water), it is called

a) Raised nursery b) Dry nursery
c) Dapog nursery d) Field nursery **Ans : b**

122. When rice seedlings are grown on concrete floor or on a raised bed of soil covered with polythene sheets, this method of raising nursery is called

a) Raised nursery b) Dry nursery
c) Dapog nursery d) Concrete nursery **Ans : c**

123. In nursery raised by 'Dapog' method, seedlings become ready for transplanted at the age of

a) 14 days b) 21 days
c) 25 days d) 30 days **Ans : a**

124. Kharif rice seedlings are transplanted at the age of

a) 20-25 days b) 30-40 days
c) 10-14 days d) 40-45 days **Ans : a**

125. Rabi rice seedlings are transplanted at the age of

a) 40-45 days b) 30-40 days
c) 20-25 days d) 10-14 days **Ans : b**

126. Which of the following rice varieties are not drought tolerant

a) N-22, Bala, Brown Gora, White Gora, Black Gora, rasi
b) Lal Nakanda, CR 143-2-2
c) Kanchan, Kiran, Bhavani
d) CO 14, Vani, K 333 **Ans : d**

127. Blast resistant varieties of rice are

a) CO 14, K 332 b) K 333, Jaya
c) Rasi, Vani d) All these **Ans : d**

128. Rice varieties resistant to bacterial blight are

a) IET-4141 b) BJ-1
c) TKM 6 d) All these **Ans : d**

129. Rice varieties resistant to tungru virus are
  a) Kataribhog, PTB 2 b) O. nivarl
  c) IR 362 d) TKM 6 **Ans : a**

130. Rice varieties resistant to grassy-stunt virus are
  a) O. nivari, IR 362 b) Kataribhog
  c) PTB 2 d) TKM 6 **Ans : a**

131. Rice varieties resistant to green leaf hopper
  a) Vani, PTB 2, PTB 7 b) RP 5-12, Pankhari 203
  c) PTB 18 d) All these **Ans : d**

132. Rice varieties resistant to gall midge are
  a) Eswarakora, Surek b) CR 94, MRI 523
  c) PTB 18, PTB 21, P d) All these **Ans : d**

133. Rice varieties resistant to brown plant hopper are
  a) PTB 21, PTB 33, IET 5118
  b) IET 5119, T-1425, N-1432
  c) Ennapatta, T-1421, Sonamukhi, Pandi, ARC 6650
  d) All these **Ans : d**

134. Rice varieties resistant to white backed plant hopper are
  a) T-1471, Lal Basmat b) Eswaramangalam
  c) Kodiyan d) All these **Ans : d**

135. Stem borer tolerant varieties of rice are
  a) TKM 6, IET 2815 b) W 1263, IET 2845
  c) IET 2812 d) All these **Ans : d**

Bajra

136. The centre of origin of Bajra is
  a) India and Burma b) America
  c) W. Africa d) N. Africa **Ans : c**

137. In terms of productivity Bajra ranks ....in India
  a) Third b) Fifth
  c) Fourth d) Sixth **Ans : b**

138. The inflorescence in Bajra is
  a) Siliqua b) Panicle
  c) Spikelets d) Ear **Ans : b**

139. In terms of area under cultivation in India Bajra ranks
a) Third b) Fifth
c) Fourth d) Sixth Ans : c

140. Bajra varieties cultivated in India belongs to genus
a) Typhoides b) Nigritarum
c) Globosum d) Indica

141. Optimum temperatures for normal vegetative growth of Bajra ranges between
a) 21-37°C b) 28-35°C
c) 20-28°C d) None Ans : c

142. The recommended seed rate for bajra is _______ kg/ha
a) 8-10 b) 4 - 5
c) 15-18 d) 20-25 Ans : b

143. The best observed pH for bajra cultivation is ________
a) 5.0.5.5 b) 6.0-7.5
c) 7.5 - 8.5 d) 8.5-9.5 Ans : b

144. The optimum plant population for bajra is around
a) One lacs b) Two lacs
c) Three lacs d) Four lacs Ans : a

145. Wherever one irrigation is available, the bajra crop should irrigated at
a) Late booting to early flowering b) Tillering stage to early booting
c) Flowering stage d) Milking stage Ans : a

146. The identified critical stage of irrigation in bajra is
a) Tillering b) Flowering
c) Both of these d) Maturity Ans : c

147. Pearl millet BK 560 - 230 is a cross from
a) 5141 A x J 104 b) 5141 A x D 230
c) 5141 A x 23 DB d) 5141 A x 23 D2 B Ans : a

148. The male sterile parent of bajra hybrid HHB 45 is
a) MS 5141 A b) MS 5054 A
c) Tift 23 A d) MS 5071 A Ans : a

149. The first inter-specific hybridization in bajra was carried out during the year
a) 1941 b) 1954
c) 1965 d) 1975 Ans : a

150. The male-sterile line in bajra was discovered by
a) G.W.Burton b) N.E.Bourloge
c) F.W.Gardner d) None **Ans : a**

151. The male-sterile line was discovered during the year
a) 1950 b) 1975
c) 1985 d) 1995 **Ans : a**

152. The first hybrid HB1 was released during the year
a) 1965 b) 1975
c) 1985 d) 1995 **Ans : a**

153. The first male-sterile line in hybrid bajra is
a) Tift 23A b) J-104
c) 5141 A d) Bil-38 **Ans : a**

154. HB-1 bajra was developed by crossing
a) Tift 23A x J-104 b) 5141 A x J-104
c) 5054 A x J-104 d) Tift 23A x Bil-38 **Ans : a**

155. The inflorescence of pearl millet is usually called
a) Panicle b) Ear
c) Rachis d) Rosette **Ans : a**

156. The Bajra grain is known as
a) Caryopsis b) Pod
c) Kernel d) All **Ans : a**

157. Bajra cop needs
a) Warm climate b) Hot and humid climate
c) Dry and hot conditions d) Cold and dry climate **Ans : a**

158. Photoperiodically, Bajra is a type of plant
a) Long day b) Short day
c) Day neutral d) Intermediate

159. The sowing time of Bajra is
a) April b) July
c) June d) May **Ans : b**

160. The optimum pH of soil for Bajra cultivation is varies in range
a) 4 – 6 b) 5 – 6.5
c) 6 – 7 d) 7 – 9 **Ans : c**

161. Test weight of Bajra is

a) 15-20 g b) 8-12 g
c) 20-25 g d) 12-15 g Ans : b

162. The scientific name of bajra is

a) *P. typhoides* b) *P.purpureum*
c) *P. spherococcum* d) *P. spelta* Ans : a

163. The recommended dose of N, P_2O_5 and K_2O kg/ha for hybrid bajra is

a) 60 - 40 - 40 b) 100 - 60 - 40
c) 30 - 20 - 30 d) 40 - 30 - 40 Ans : b

164. The recommended dose of N, P_2O_5 and K_2O kg/ha for rainfed bajra is

a) 60 - 40 - 40 b) 100 - 60 - 40
c) 30 - 20 - 30 d) 40 - 30 - 40 Ans : a

165. Bajra belongs to the family

a) Cyperacene b) Gramineae
c) Solanaceae d) Malvaceae Ans : b

166. Which of the following pairs of crop and critical stage of irrigation is not correctly matched

a) Bajra: Ear head b) Cotton: Pre-flowering
c) Wheat: Crown root initiation d) Groundnut: pod development Ans : a

167. Which one of the following pairs is not correctly matched

a) Bajra- Tift 23 D_2A b) Maize- T Cytoplasm
c) Wheat-Rht1 Rht2 d) Bajra-Norin 10 Ans : d

168. The optimum plant population per hectare of bajra is

a) 50,000 plants b) 1,00,000 plants
c) 1,50,000 plants d) 2,00,000 plants Ans : b

169. Maximum yield of bajra is obtained at a plant geometry of

a) 45 x 15 cm b) 60 x 15 cm
c) 45 x 25 cm d) 60 x 20 cm Ans : a

170. Which one of the following is not correctly matched Crop Test weight in gram

a) Pearlmillet 5-7
b) Wheat 20-22
c) Rice 22-27
d) Sunflower 60-62 Ans : a

171. Which of the following can be transplanted
a) Groundnut b) Bajra
c) Mustard d) Soybean
Ans : b

172. Which cereal is tillering in nature
a) Maize b) Bajra
c) Sorghum d) All
Ans : b

173. In kera method of bajra sowing the seed is drilled
a) By hand behind the plough in the furrows
b) By funnel behind the plough in the furrows
c) By seed drill
d) All
Ans : a

174. In pora method of bajra sowing the seed is drilled
a) By hand behind the plough in the furrows
b) By funnel behind the plough in the furrows
c) By seed drill
d) All
Ans : b

175. The protein content in bajra is around
a) 12% b) 8%
c) 14% d) 15%
Ans : a

Barley

176. The centre of origin of barley is
a) Asia and Ethiopia b) America
c) S. Africa d) None
Ans : a

177. In India the highest producer of barley is
a) Punjab b) U.P.
c) W. Bengal d) None
Ans : b

178. The inflorescence in barley is
a) Siliqua b) Panicle
c) Spike d) Ear
Ans : c

179. Area under barley cultivation in India is approximately
a) 3.9 m ha b) 1.84 m ha
c) 5.0m ha d) 2.9m ha
Ans : b

180. Optimum temperatures towards maturity of barley ranges between
a) 21-37°C b) 28-30°C
c) 20-25°C d) None
Ans : b

181. Which of the following management practices is more problematic in barley cultivation

a) Fertilizer management b) Weed management
c) Water management d) Bushening **Ans : b**

182. The optimum temperature range for sowing of barley is

a) 10-12°C b) 15-20°C
c) 12-15°C d) 25-30°C **Ans : b**

183. Most suitable temperature for growth of barley is

a) 32°C b) 25°C
c) 37°C d) 12-15°C **Ans : d**

184. Barley crop planted at a spacing of 20 x 10 cm will have .... plants/ha

a) 5,00, 000 b) 1,00,000
c) 1,25, 000 d) 2,50,000 **Ans : a**

185. The recommended seed rate for barley is _______ kg/ha

a) 8-10 b) 12-15
c) 15-18 d) 75-90 **Ans : d**

186. The identified critical stage of irrigation in barley is

a) Tillering b) Flag leaf stage
c) Both of these d) Maturity **Ans : c**

187. Which one is the most critical stage of irrigation in Barley

a) TIliering b) CRI
c) Jointing d) Flowering **Ans : a**

188. In two rowed barley ....spikelet is fertile

a) Central b) Left
c) Right d) All **Ans : a**

189. In six rowed barley ....spikelet is fertile

a) Central b) Left
c) Right d) All **Ans : d**

190. Barley crop needs

a) Warm and humid climate b) Hot and humid climate
c) Dry and hot conditions d) Cold and dry climate **Ans : d**

191. Photoperiodically, barley is a type of plant

a) Long day b) Short day
c) Day neutral d) Intermediate **Ans : a**

192. The seed rate of barley cultivation in broadcasting and drilling, is sufficient

a) 100, 60 kg/ha b) 60, 100 kg/ha
c) 45, 100 kg/ha d) 100,45 kg/ha **Ans : a**

193. The normal seed rate of barley is

a) 50 kg/ha b) 75 kg/ha
c) 100 kg/ha d) 125 kg/ha

194. Barley crop yielding 50 quintals of grain per hectare removes N, P_2O_5 and K_2O kg/ha, respectively

a) 80-120, 60-70 and 100-125 b) 90-100, 60- 70 and 100-110
c) 120-40-80 d) 80-30/40-20 **Ans : a**

195. First irrigation to the barley crop should be given at

a) CRI stage b) Tillering stage
c) Jointing stage d) Dough stage **Ans : b**

196. Which of the following is the ideal temperature range for tillering stage in barley

a) 10-15^{0}C b) 16-20^{0}C
c) 20-23^{0}C d) 23-25^{0}C **Ans : a**

197. The depth of sowing of barley is

a) 1-2 cm b) 3-5 cm
c) 6-8 cm d) 5-6 cm **Ans : b**

198. Botanical name of barley is

a) Hordeum vulgare b) Sorghum bicolar
c) Panicum miliacium d) None of these **Ans : a**

199. Nitrogen is applied to barley in the proportion of

a) 50% at basal+ 50% at first irrigation,
b) Nil at basal+50% at tillering+50% at panicle emergence stage
c) 25% at basal+50% at tillering+25% at panicle emergence stage
d) 50% at basal+25% at tillering+25% at panicle emergence stage **Ans : a**

200. The shoot of barley is called

a) Stem b) Trunk
c) Culm d) None **Ans : c**

201. ....is a variety of barley suitable for chapati making

a) Amber b) RD 137
c) Neelam d) Rajkiran **Ans : c**

202. ....is a nematode resistant variety of barley
a) Amber b) RD 137
c) Neelam d) Rajkiran **Ans : d**

203. ....is a six row variety of barley
a) Amber b) RD 137
c) Neelam d) Rajkiran **Ans : a**

204. Shoot of barley is
a) Hollow b) Solid
c) Hollow except at nodes d) Solid except at nodes **Ans : c**

205. Leaf of barley is
a) Sessile b) Petiolate
c) Both d) None **Ans : a**

206. Spike of barley is borne on
a) Uppermost node b) Lowermost node
c) Middle nodes d) None **Ans : a**

207. Six rowed barley is
a) *Hordeum vulgare* b) *Hordeum distinction*
c) *Hordeum irregulare* d) All **Ans : a**

208. *Hordeum distinction* is
a) Two rowed b) Six rowed
c) Four rowed d) b & c **Ans : a**

209. *Hordeum irregulare* is
a) Two rowed b) Six rowed
c) Four rowed d) b & c **Ans : a**

210. Two rowed barley is
a) *Hordeum vulgare* b) *Hordeum distinction*
c) *Hordeum irregulare* d) b & c **Ans : d**

211. Ratna variety of barley tolerant to saline alkaline conditions released from
a) Phillipines b) Taiwan
c) Malaysia d) IARI **Ans : d**

212. ....is a salt tolerant variety of barley
a) Amber b) RD 137
c) Neelam d) Rajkiran **Ans : a**

213. .... variety of barley is suitable for malting
a) Amber b) RD 137
c) Neelam d) Rajkiran **Ans : b**

214. For control of wild oat in barley it is recommended to apply
a) Avadex b) 2,4-D
c) Whip super d) b & c **Ans : a**

215. For control of broad leaved weeds in barley it is recommended to apply
a) Avadex b) 2,4-D
c) Whip super d) b & c **Ans : b**

Maize

216. ....is a double cross hybrid variety of maize
a) Deccan hybrid b) Ganga safed-2
c) Hi starch d) All these **Ans : a**

217. ....is a three way cross hybrid variety of maize
a) Deccan hybrid b) Ganga 5
c) Hi starch d) All these **Ans : c**

218. ....is a single cross hybrid variety of maize
a) Paras b) Ganga 5
c) Ganga 4 d) All these **Ans : a**

219. ....is a high lysine composite variety of maize
a) Shakti b) Rattan
c) Protina d) All these **Ans : d**

220. Very early maturing varieties of maize which mature in 65-75 days are
a) Sathi b) Kathri
c) Teen Pakhia d) All these **Ans : d**

221. Which of the following maize varieties are classified as 'medium early-maturing' type
a) Basi b) KT 41
c) Jaunpur, Rudrapur Local d) All these **Ans : d**

222. Which of the following maize varieties is late maturing type
a) Malan b) Basi
c) KT 41 d) Rudrapur Local **Ans : a**

223. 'Ganga-1, 'Ganga-101' Ranjit, Deccan, Ganga 5, Ganga safed 2, Hi-strach, Ganga 4, Himalayan 123, Ganga 3, VL 54 are high yielding hybrid varieties of

a) Wheat b) Maize
c) Rice d) Sorghum Ans : b

224. Vijay, Amber, Sona, Kisan, Jawahar and Vikram are composite varieties of

a) Wheat b) Rice
c) Maize d) Sorghum Ans : c

225. Farmers using composite varieties of maize can save their own seed for the following year provided

a) They take adequate precaution to keep their seed fields at least 300 metres away from other maize fields
b) They select the best plants and not ears
c) They bulk together the seed from at least 500-1000 plants to represent the population
d) They observe all these precautions Ans : d

226. Which of the following composite varieties of maize are very rich in essential amino acids, particularly lysine and tryptophan

a) Shakti, Rattan and Protina b) Vijay and Amber
c) Sona and Kisan d) Jawahar and Vikram Ans : a

227. After thinning the number of plants per hectare at 50 cm x 20 cm is :

a) 50000 b) 75000
c) 100000 d) 125000 Ans : c

228. The centre of origin of maize is

a) India and Burma b) Mexico
c) S. Africa d) None Ans : b

229. The inflorescence in maize is

a) Siliqua b) Panicle
c) Spikelets d) Ear Ans : c

230. Number of species included under genus Zea

a) 1 b) 2
c) 3 d) 4 Ans : a

231. Optimum temperatures towards maturity of maize ranges between

a) 21-37°C b) 25-30°C
c) 20-25°C d) None Ans : c

232. Which of the following is the leading state in rabi maize production

a) Bihar b) Uttar Pradesh
c) Punjab d) Gujarat Ans : a

233. Which of the following is the leading state in maize production

a) Bihar b) Uttar Pradesh
c) Punjab d) Gujarat Ans : c

234. Of the following varieties of maize which one is the high lysine maize composite.

a) Himalaya 123 b) Hi-starch
c) Vijay d) Shakti Ans : b

235. In composite maize, which of the following statement is true

a) Seed can be sown year after year with almost same level of performance.
b) Fresh seed is to be obtained for every 'sowing, to get high yields
c) Seed is a mixture of a' few hybrid lines which are mixed before sowing.
d) Seed is a mixture of a few local and hybrid strains. Ans : a

236. Male sterile lines are used in India in production of new hybrids of

a) Wheat b) Maize
c) Bajra d) Sorghum Ans : b

237. Brace roots are found in

a) Rice b) Wheat
c) Maize d) Barley Ans : c

238. Yield of any maize hybrid decreases in ____________ generation by about 20 to 40 per cent in comparison to its ____________ yields.

a) F1 and F2 b) F2 and F1
c) F2 and F3 d) None Ans : b

239. Temperature requirement for proper germination of maize is

a) 18 - 25°C b) 25 - 33°C
c) 33 - 35°C d) 16 - 18°C Ans : a

240. Temperature requirement for proper grain filling of maize is

a) 28 - 30°C b) 20 - 33°C
c) 33 - 35°C d) 16 - 18°C Ans : a

241. The female flower of maize is known as
a) Spadix b) Tassel
c) Panicle d) Umbel
Ans : a

242. The male flower of maize is known as
a) Cob b) Tassel
c) Panicle d) Umbel
Ans : b

243. The normal seed rate of maize is
a) 25 kg/ha b) 75 kg/ha
c) 100 kg/ha d) 125 kg/ha
Ans : a

244. Hybrid maize crop requires.......... kg N, P_2O_5 and K_2O / ha, respectively
a) 100 - 120, 50 - 60 and 40 -50 b) 90 - 100, 60- 70 and 100 - 110
c) 120 - 40 - 80 d) 80 - 40 - 20
Ans : a

245. The recommended dose of N, P_2O_5 and K_2O kg/ha for composite maize are
a) 80 - 40 - 40 b) 120 - 40 - 40
c) 30 - 20 - 30 d) 40 - 30 - 40
Ans : a

246. Nitrogen is applied to maize in the proportion of
a) 1/3 as basal+ 1/3 at Knee height+1/3% at tasseling,
b) 1/2 as basal+ nil at Knee height+1/2% at tasseling,
c) Nil as basal+ 1/2 at Knee height+1/2% at tasseling,
d) 1/2 as basal+ 1/2 at Knee height
Ans : d

247. The optimum plant population per hectare of Maize is
a) 50,000 plants b) 60000-80000 plants
c) 1,50,000 plants d) 2,00,000 plants
Ans : b

248. Maximum yield of maize is obtained at a plant geometry of
a) 45 x 20 cm b) 60 x 25 cm
c) 90 x 30 cm d) 105 x 30 cm
Ans : b

249. In which of the following, composite and synthetic cultivars are used
a) Rice b) Wheat
c) Maize d) Cotton
Ans : c

250. Opaque-2 maize composites are richer as compared to normal maize in
a) Tryptophan b) Lysine
c) Tryptophan & Lysine d) Protein
Ans : c

251. Flint corn is commonly grown in India is also known as

a) *Zea mays* indentata b) *Zea mays* indurate
c) *Zea mays* everta d) *Zea mays* saccharata Ans: b

252. The most suitable temperature for maize growth is

a) 21°C b) 26°C
c) 29°C d) 32^0C **Ans : c**

253. First hybrid maize Ganga-1 was developed in India in

a) 1957 b) 1961
c) 1964 d) 1965 **Ans : b**

254. Most common herbicide used in maize is

a) Atrazine b) Alachlor
c) Simazine d) Metribuzin **Ans : c**

255. Which cereal is non tillering in nature

a) Maize b) Wheat
c) Barley d) Paddy **Ans : a**

256. Which one of the following is a hybrid variety of maize

a) Vikram b) Amber
c) Kissan d) Sangam **Ans : d**

257. Who introduced the technique of production of double cross hybrid maize

a) EM East b) DF Jones
c) GH Shull d) Mendel **Ans : b**

258. Which one of the following is a three way cross hybrid variety of maize

a) Hi-Starch b) Ganga-5
c) Deccan hybrid d) Sangam **Ans : a**

259. Required plant population of maize for fodder may be obtained by using a seed rate of (kg/ha)

a) 40-45 b) 75-100
c) 90-110 d) 100-125 **Ans : a**

260. Required plant population of maize for cob may be obtained by using a seed rate of (kg/ha)

a) 20-25 b) 75-100
c) 90-110 d) 100-125 **Ans : a**

261. Dent corn is

a) *Zea mays indentata* b) *Zea mays indurata*
c) *Zea mays saccharata* d) *Zea mays amylacea* **Ans : a**

262. Flint corn is
a) *Zea mays indentata* b) *Zea mays indurata*
c) *Zea mays saccharata* d) *Zea mays amylacea* **Ans : b**

263. Sweet corn is
a) *Zea mays indentata* b) *Zea mays indurata*
c) *Zea mays saccharata* d) *Zea mays amylacea* **Ans : c**

264. Flour corn is
a) *Zea mays indentata* b) *Zea mays indurata*
c) *Zea mays saccharata* d) *Zea mays amylacea* **Ans : d**

265. Pop corn is
a) *Zea mays everta* b) *Zea mays indurata*
c) *Zea mays saccharata* d) *Zea mays amylacea* **Ans : a**

266. Waxy corn is
a) *Zea mays indentata* b) *Zea mays ceratina*
c) *Zea mays saccharata* d) *Zea mays amylacea* **Ans : b**

267. Pop corn is a product of
a) *Maize* b) *Wheat*
c) *Rice* d) *Barley* **Ans : a**

268. Seminal roots in maize originates from
a) Radicle b) Plumule
c) Endosperm d) Nodes **Ans : a**

269. Crown roots in maize originates from
a) Radicle b) Plumule
c) Lower nodes above soil d) Lower nodes below soil **Ans : d**

270. Brace roots in maize originates from
a) Radicle b) Plumule
c) Lower nodes above soil d) Lower nodes below soil **Ans : c**

271. Seminal roots in maize are ....in numbers
a) 30-50 b) 3-5
c) 10-15 d) 20-30 **Ans : b**

272. Roots help to establish the plant firmly to ground in maize is
a) Brace roots b) Seminal roots
c) Crown roots d) All **Ans : a**

273. For weed control in maize simazine can be applied as

a) 1-1.25kg a.i./ha as pre emergence
b) 2-2.5kg a.i./ha as pre emergence
c) 1-1.25kg a.i./ha as post emergence
d) All

Ans : a

274. Area under maize cultivation in India is approximately

a) 39 m ha b) 6m ha
c) 50m ha d) 29m ha

Ans : b

275. In India the productivity of maize is highest in the state

a) Punjab b) Haryana
c) W. Bengal d) None

Ans : a

Millets

276. The centre of origin of finger millet is

a) India b) America
c) S. Africa d) Europe

Ans : a

277. Scientific name of finger millet is

a) Elusine coracana b) Echinocloa crusgalli
c) Panicum miliaceum d) Paspalum scrobiculatum

Ans : a

278. Scientific name of Cheena is

a) Elusine coracana b) Echinocloa crusgalli
c) Panicum miliaceum d) Paspalum scrobiculatum

Ans : c

279. Scientific name of kodo is

a) Elusine coracana b) Echinocloa crusgalli
c) Panicum miliaceum d) Paspalum scrobiculatum

Ans : d

280. Scientific name of sawan is

a) Elusine coracana b) Echinocloa *frumentacea*
c) Panicum miliaceum d) Paspalum scrobiculatum

Ans : b

281. Scientific name of foxtail millet is

a) Setaria italica b) Echinocloa crusgalli
c) Panicum miliaceum d) Paspalum scrobiculatum

Ans : a

282. Scientific name of teff millet is

a) Setaria italica b) Eragrostis tef
c) Panicum miliaceum d) Paspalum scrobiculatum

Ans : b

283. Scientific name of fonio millet is

a) Digitaria exilis b) Echinocloa crusgalli
c) Panicum miliaceum d) Paspalum scrobiculatum

Ans : a

284. Scientific name of foxtail millet is

a) Coix lachryma b) Echinocloa crusgalli
c) Panicum miliaceum d) Paspalum scrobiculatum **Ans : a**

285. The recommended seed rate for finger millet is _______ kg/ ha

a) 8-10 b) 12-15
c) 15-18 d) 20-25 **Ans : a**

286. The recommended dose of N, P_2O_5 and K_2O kg/ha for finger millet is

a) 60 – 40 – 30 b) 120 – 40 – 40
c) 30 – 20 – 30 d) 40 – 30 – 40 **Ans : a**

287. The seed of Sawan is called as

a) Caryopsis b) Panicle
c) Spikelets d) Siliqua **Ans : a**

288. The centre of origin of sawan is

a) Central Asia b) America
c) S. Africa d) Europe **Ans : a**

289. The inflorescence in sawan is

a) Caryopsis b) Panicle
c) Spikelets d) Ear **Ans : b**

290. Optimum temperatures towards maturity of sawan ranges between

a) 21-37°C b) 25-30°C
c) 20-25°C d) None **Ans : c**

291. The recommended seed rate for sawan is _______ kg/ha

a) 8-10 b) 12-15
c) 15-18 d) 20-25

292. Transplanting of seedlings is commonly practiced in

a) Sawan b) Finger millet
c) Kodo d) Pearl millet **Ans : b**

293. Most commonly followed method of planting sawan in India is

a) Direct seeding in rows
b) Transplanting on raised seed beds
c) Direct seeding in puddled beds
d) Transplanting in standing water in puddled fields **Ans : a**

294. Sawan crop needs
a) Warm and humid climate
b) Hot and humid climate
c) Dry and hot conditions
d) Cold and dry climate
Ans : a

295. The recommended dose of N, P_2O_5 and K_2O kg/ha for sawan are
a) 60 – 40 – 40
b) 120 – 40 – 40
c) 30 – 20 – 30
d) 40 – 30 – 20
Ans : d

296. The depth of sowing of sawan is
a) 1-2 cm
b) 3-4 cm
c) 4-5 cm
d) 5-6 cm
Ans : b

297. Botanical name of Ragi is
a) Eleusine coracana
b) Echinochloa frumentacea
c) Echinocloa crusgalli
d) None of these
Ans : a

298. The optimum plant population per hectare of sawan is
a) 1 lacs
b) 2 lacs
c) 3 lacs
d) 4 lacs
Ans : d

299. Maximum yield of sawan is obtained at a plant geometry of
a) 45 x 20 cm
b) 60 x 30 cm
c) 20 x 10 cm
d) 105 x 30 cm
Ans : c

300. The centre of origin of proso millet is
a) India
b) America
c) S. Africa
d) Europe
Ans : a

301. The inflorescence in proso millet is
a) Caryopsis
b) Panicle
c) Spikelets
d) Ear
Ans : b

302. The recommended seed rate for proso millet is _______ kg/ha
a) 8-10
b) 12-15
c) 15-18
d) 20-25
Ans : a

303. Most commonly followed method of planting proso millet in India is
a) Direct seeding in rows
b) Transplanting on raised seed beds
c) Direct seeding in puddled beds
d) Transplanting in standing water in puddled fields
Ans : a

304. Proso millet crop needs

a) Warm and humid climate b) Hot and humid climate
c) Dry and hot conditions d) Cold and dry climate Ans : a

305. The recommended dose of N, P_2O_5 and K_2O kg/ha for sawan are

a) 60 – 40 – 40 b) 120 – 40 – 40
c) 30 – 20 – 30 d) 40-60 – 30 – 20 Ans : d

306. The test weight of teff millet is ______

a) 8-10g b) 12-15g
c) 0.15-0.2g d) 20-25 Ans : c

Oat

307. The centre of origin of oat is

a) Asia minor b) America
c) S. Africa d) None

308. The inflorescence in oat is

a) Siliqua b) Panicle
c) Spike d) Ear Ans : b

309. The optimum temperature range for sowing of oat is

a) 10-12°C b) 15-20°C
c) 12-15°C d) 25-30°C Ans : b

310. Most suitable temperature for growth of oat is

a) 32°C b) 20-25°C
c) 37°C d) 12-15°C

311. Oat crop planted at a spacing of 20 x 10 cm will have .... plants/ha

a) 5,00, 000 b) 1,00,000
c) 1,25, 000 d) 2,50,000

312. The recommended seed rate for oat is ______ kg/ha

a) 50-60 b) 12-15
c) 15-18 d) 75-90 Ans : a

313. The identified critical stage of irrigation in oat is

a) Tillering b) Flag leaf stage
c) Flowering d) Maturity Ans : c

314. Which one is the most critical stage of irrigation in Oat
a) TIliering b) CRI
c) Jointing d) Flowering **Ans : d**

315. Oat crop needs
a) Warm and humid climate b) Hot and humid climate
c) Dry and hot conditions d) Cold and dry climate **Ans : d**

316. The seed rate of oat cultivation for seed and fodder, is sufficient
a) 100, 60 kg/ha b) 60, 100 kg/ha
c) 45, 100 kg/ha d) 100,45 kg/ha **Ans : b**

317. The normal seed rate of Oat is
a) 50 kg/ha b) 60 kg/ha
c) 100 kg/ha d) 125 kg/ha **Ans : b**

318. Botanical name of oat is
a) Hordeum vulgare b) Sorghum bicolar
c) Panicum miliacium d) Avena sativa **Ans : d**

319. Grey speck disease in oat it is due to
a) mn b) zn
c) cu d) mg **Ans : a**

320. Wild oat is
a) *Avena fatua* b) *Sorghum bicolar*
c) *Panicum miliacium* d) *Avena sativa*

321. Nitrogen is applied to oat in the proportion of
a) 50% at basal+ 50% at first irrigation,
b) Nil at basal+50% at tillering+50% at panicle emergence stage
c) 25% at basal+50% at tillering+25% at panicle emergence stage
d) 50% at basal+25% at tillering+25% at panicle emergence stage. **Ans : a**

322. The shoot of oat is called
a) Stem b) Trunk
c) Culm d) None **Ans : c**

323. Shoot of oat is
a) Hollow b) Solid
c) Hollow except at nodes d) Solid except at nodes **Ans : c**

324. Leaf of oat is
a) Sessile b) Petiolate
c) Both d) None **Ans : a**

325. Spike of oat is borne on

a) Uppermost node b) Lowermost node
c) Middle nodes d) None **Ans : a**

326. Grey speck disease in oat it is due to

a) Virus b) Bacteria
c) Fungus d) Nutritional **Ans : d**

327. Avena sativa is

a) Diploid b) Tetraploid
c) Hexaploid d) Hploid **Ans : c**

328. For control of broad leaved weeds in oat it is recommended to apply

a) Avadex b) 2,4-D
c) Whip super d) b & c **Ans : b**

Sorghum

329. The centre of origin of sorghum is

a) India and Africa b) America
c) S. Africa d) None **Ans : a**

330. The inflorescence in sorghum is

a) Siliqua b) Head
c) Spikelets d) Ear

331. Area under sorghum cultivation in India is approximately

a) 39 m ha b) 16m ha
c) 50m ha d) 29m ha **Ans : b**

332. Optimum temperatures for growth of sorghum ranges between

a) 21-37°C b) 26-30°C
c) 20-25°C d) None **Ans : b**

333. The recommended seed rate for sorghum is _______ kg/ha

a) 8-10 b) 10-12
c) 15-18 d) 20-25 **Ans : b**

334. Higher yield of sorghum could be achieved by:

a) Use of higher dose of phosphate
b) Adequate amount of N
c) No nitrogen application
d) Closer planting **Ans : a**

335. Sorghum yield is largely reduced mainly due to:
a) Cold wind at the flowering stage
b) Weeds
c) Lack of adequate fertilizer application
d) Failure to apply irrigation in flowering stage **Ans : b**

336. High yielding hybrid varieties of jowar grown under optimal conditions can yield upto
a) 2000 kg/hectare b) 3000 kg/hectare
c) 5000-6000 kg/hectare d) 1500 kg/hectare **Ans : c**

337. Which of the following are not traditional varieties of jowar
a) Saoner, Ramkel, Aispuri, maldandis 35-1,
b) Dogadi, bilichigan, flugar white,
c) Hagari, Yenigar, budhperio,
d) CSH-1, CSH-2, CSH-4, CSV-4, CSV-9, CSV-10 **Ans : d**

338. Which of the following are improved and high yielding varieties of sorghum
a) CSH-1, CSH-2, CSV-1, CSV-2, CSV-4
b) Soaner, Ramkel, Maldandis-2
c) Dagadi, Bilichigan,
d) Hagari, Yenigar, Chasatio **Ans : a**

339. In India sorghum is predominantly cultivated on
a) Medium and deep black soils b) Alluvial soils
c) Sandy soils d) Clay soils

340. Which of the following management practices is more problematic in sorghum cultivation
a) Fertilizer management b) Weed management
c) Water management d) Bushening **Ans : b**

341. Sorghum crop planted at a spacing of 40 x 15 cm will have – plants/ha
a) 50, 000 b) 1,00,000
c) 1,25, 000 d) 1,66,666 **Ans : d**

342. Sorghum is considered as camel crop because of
a) Deep root system b) Resistance to drought
c) Shallow root system d) Nutrient exhaustiveness **Ans : b**

343. For Kharif season sorghum, the recommended optimum seed rate is kg/ha
a) 8-10 b) 10-12
c) 15-20 d) 20-25 **Ans : b**

344. The thinning of sorghum for seed is to maintain a plant to plant distance of ______________ cm.

a) 5-10 b) 10-15
c) 15-20 d) 20-25 Ans : c

345. The Directorate of Sorghum Research e is located at_______

a) Cuttak b) Coimbtore
c) Hyderabad d) Bangalore Ans : c

346. Most commonly followed method of planting sorghum in India is

a) Direct seeding in rows
b) Transplanting on raised seed beds
c) Direct seeding in puddled beds
d) Transplanting in standing water in puddled fields Ans : a

347. Highly tolerant variety of sorghum for striga

a) SPV-IO4 b) SPV-222
c) SAR-2 d) CSH-5 Ans : d

348. The hybrid seed of CSH-l can be produced by crossing

a) 2077 A x Swarna b) 2077 A x IS 3691
c) 2077 A x CS 3541 d) MSCK 60A x IS84 Ans : c

349. Hybrid seed of CSH-2 can be produced by crossing

a) MSCK 60-A x CS 351 b) MSCK 60-A x Swarna
c) MSCK *60-A* x IS 3691 d) MSCK *60-A)* x IS 3891 Ans : c

350. CSH is a cross between

a) 1036 Z x Swarna b) 2077 A x CS 3541
c) 1076 A x Swarna d) 1076 B x Swarna Ans : b

351. The inflorescence of sorghum is called

a) Head b) Ear
c) Rachis d) Rosette Ans : a

352. Test weight of sorghum is

a) 20-25 gram b) 25-30 gram
c) 30-35 gram d) 70-72 gram Ans : b

353. Soil sickness is caused by

a) Maize b) Jowar
c) Linseed' d) Both b & c Ans : b

354. Required plant population of jowar may be obtained by using a seed rate of (kg/ha)

a) 9-10 b) 12-15
c) 15-18 d) 18-20 Ans : b

355. CSH-l was first sorghum hybrid released in
a) 1961 b) 1962
c) 1963 d) 1964 Ans : d

Wheat

356. ....is a suitable variety of wheat under rainfed conditions
a) C 306 b) K 65
c) Narbada 4 d) All these Ans : d

357. ....is a durum variety of wheat
a) PBW 34 b) Malvika
c) Jairaj d) All these Ans : d

358. ....is a single gene dwarf variety of wheat
a) HD 2177 b) HD 2189
c) Narbada 4 d) All these Ans : b

359. ....is a double gene dwarf variety of wheat
a) HD 2177 b) HD 2189
c) Narbada 4 d) All these Ans : a

360. The centre of origin of wheat is
a) S-W Asia b) America
c) S. Africa d) Mexico Ans : a

361. In India the productivity of wheat is highest in the state
a) Punjab b) Haryana
c) W. Bengal d) None Ans : b

362. The inflorescence in wheat is
a) Siliqua b) Panicle
c) Spike d) Ear Ans : c

363. Area under wheat cultivation in India is approximately
a) 39 m ha b) 45m ha
c) 50m ha d) 29m ha Ans : d

364. Optimum temperatures towards maturity of wheat ranges between
a) 21-37°C b) 25-30°C
c) 30-35°C d) None Ans : c

365. The recommended seed rate for wheat is _______ kg/ha
a) 80-100 b) 12-15
c) 40-60 d) 20-25 Ans : a

366. Optimum temperatures for growth and development of wheat ranges between

a) 21-37°C b) 25-30°C
c) 10-15°C d) None **Ans : c**

367. Optimum temperature for sprouting of wheat is

a) 15°C b) 20-25°C
c) 4-5°C d) None Ans: b

368. Temperature having detrimental effect on growth of wheat is

a) 15°C b) 25-30°C
c) 30-35°C d) None **Ans : c**

369. The shoot of wheat is called

a) Stem b) Trunk
c) Culm d) None **Ans : c**

370. Shoot of wheat is

a) Hollow b) Solid
c) Hollow except at nodes d) Solid except at nodes **Ans : c**

371. Leaf of wheat is

a) Sessile b) Petiolate
c) Both d) None **Ans : a**

372. In wheat.................. is very prominent

a) Auricle b) Ligule
c) Lemma d) Palea **Ans : b**

373. Panicle of wheat is borne on

a) Uppermost node b) Lowermost node
c) Middle nodes d) None **Ans : a**

374. Hull of wheat comprised of

a) Auricle b) Lemma
c) Palea d) Both b & c **Ans : d**

375. Higher yield of wheat could be achieved by

a) Use of higher dose of phosphate
b) Adequate amount of N
c) No nitrogen application
d) Closer planting **Ans : b**

376. The second most cultivated crop of the world is

a) Wheat b) Rice
c) Maize d) Jowar **Ans : a**

377. Wheat varieties cultivated in India belongs to species

a) Aestivum b) Durum
c) Dicocum d) Indica Ans : a

378. High yielding dwarf varieties of wheat were developed by-

a) Dr. Norman E. Borlaug b) Dr. B.P. Pal
c) Dr. M.S. Swaminathan d) None of these Ans : a

379. The triticum aestivum is __________ wheat

a) Hexaploid b) Diploid
c) Tetraploid d) Monoploid Ans : a

380. The optimum temperature range for sowing of wheat is

a) 10-15.C b) 15-20.C
c) 20-25'C d) 25-30'C Ans : c

381. The Directorate of Wheat Research Institute is located at_______

a) Karnal b) Coimbtore
c) Hyderabad d) Bangalore Ans : a

382. The optimum plant population for wheat is around

a) 1-2 lacs b) 2-3 lacs
c) 3-4 d) 4-5 lacs Ans : d

383. Wherever one irrigation is available, the wheat crop should irrigated at

a) CRI stage b) Tillering stage
c) Flowering stage d) Milking stage Ans : a

384. Which one is the most critical stage of irrigation in wheat

a) TIliering b) CRI
c) Jointing d) Flowering Ans : b

385. Which of the following statements refers to the concept of critical stage of irrigation:

a) When irrigation is not given reduction of yield is maximum
b) When water absorption rate is maximum
c) When water loss by the plant is maximum
d) All of the above Ans : a

386. Wheat variety__________ has given consistently superior performance in Barani (rainfed) tracts of the country.

a) N.P. 875 b) Sonalika
c) C-306 d) P.V. 18 Ans : c

387. The normal seed rate of wheat is

a) 50 kg/ha b) 75 kg/ha
c) 100 kg/ha d) 125 kg/ha Ans : c

388. Wheat crop yielding 50 quintals of grain per hectare removes N, P_2O_5 and K_2O kg/ha, respectively

a) 100-150, 70-80 and 125-150 b) 90-100, 60-70 and 100-110
c) 120-40-80 d) 80-30/40-20 Ans : a

389. The recommended dose of N, P_2O_5 and K_2O kg/ha for deshi wheat are

a) 60-40-40 b) 120-40-40
c) 30-20-30 d) 40-30-40 Ans : a

390. The recommended dose of N, P_2O_5 and K_2O kg/ha for dwarf wheat are

a) 100 - 60 -40 b) 120 - 40 - 40
c) 80 - 60 - 40 d) 40 - 30 -40 Ans : a

391. The N fertilizer use efficiency in wheat can be increased by using

a) S-coated urea b) Urea super granules
c) BGA d) Both a & b Ans : d

392. Triticale is a cross between

a) Wheat x rye b) Oat x barley
c) Wheat x barley d) None of these Ans : a

393. Phalaris minor belongs to the family

a) Cyperacene b) Gramineae
c) Solanaceae d) Malvaceae Ans : b

394. The nitrogen losses in wheat can be reduced by

a) Basal application b) Top dressing
c) Split application d) Both b & c Ans : d

395. First irrigation to the wheat crop should be given at

a) 21day stage b) 45 day stage
c) 60 day stage d) 80 day stage Ans : a

396. Phalaris minor in wheat can be controlled by using

a) 2,4-D b) Isoproturon
c) Atrazine d) Gramaxone Ans : b

397. Test weight of Phalaris minor is

a) 2 g b) 4 g
c) 22 g d) 40 g Ans : a

398. Directorate of Wheat Research (DWR) is located at
a) Hyderabad b) New Delhi
c) Karnal d) Bikaner Ans : c

399. Which one of the following is the characteristic of stripe rust of wheat
a) Brown, elongated ruptured pustules on stem and leaf-sheath
b) Yellow, small, spherical to oval ruptured uredo-pustules in rows of leaf
c) Brown, spherical to oval, ruptured uredo-pustules on leaf
d) Black telio pustules, scattered on under surface of the leaf Ans : c

400. Which of the following is the ideal temperature range for tillering stage in wheat
a) 10-15^{0}C b) 16-20^{0}C
c) 20-23^{0}C d) 23-25^{0}C Ans : b

401. Species of wheat not cultivated at present in India is
a) Triticum aestivum b) Triticum dicoccum
c) Triticum durum d) Triticum sphaerococum Ans : d

402. Wheat species known as Triticum sphaerococcum has been put out of cultivation because of
a) Low productivity and high susceptibility of disease
b) Bad taste and bad cooking quality
c) Quick spoilage in the shortage
d) All these reasons Ans : a

403. Triticum aestivum, the common bread wheat which is the most important wheat species in India, occupies the following proportion of India's total wheat area
a) 14% b) 85%
c) 95% d) 79% Ans : b

404. Which of the following species of wheat are cultivated all over India from the sea level upto high snow cover red regions of Himalaya is
a) Triticum dicoccum b) Triticum durum
c) Triticumaestivum d) All these Ans : c

405. Triticum durum, the macaroni wheat, occupies the following proportion of India's cultivated area under wheat
a) 14% b) 1%
c) 85% d) 20% Ans : a

406. Wheat species known as Triticum dicoccum is cultivated on a very restricted scale in
a) Gujarat b) Maharashtra
c) Andhra and Tamilnadu d) All these states Ans : d

407. Cultivation of macaroni wheat is restricted to
a) Western India b) Central & Southern India
c) North-eastern India d) North-Western India Ans : b

408. Which one of the following is correctly matched

	Crop		*Seed rate*
a)	Wheat	-	100
b)	Urd	-	40
c)	Maize	-	10
d)	Mustard	-	30

Ans : a

409. North-Western Plains of India wheat is harvested in
a) Mid-April to beginning of May b) March to April
c) February end to March d) February Ans : a

410. North Eastern Indian Plains wheat is sown in
a) Mid-October to mid November b) Mid-November to mid-December
c) October d) April-May Ans : b

411. In North-Eastern Indian Plains wheat is harvested
a) Mid-April to beginning of May b) March–April
c) September-October d) February end to March end Ans : b

412. In Central Zone of India wheat is harvested
a) Mid-April to beginning of May b) March-April
c) September-October d) February end to March end Ans : d

413. In peninsular Region of India wheat is harvested
a) February (second half) b) Mid-April to beginning of May
c) March-April d) February end to March end Ans : a

414. In which of the following regions of India is the rainfed wheat sown in October and the irrigated wheat sown in November
a) North-Western Plains Region b) Peninsular Region
c) Central region d) All these regions Ans : d

415. In Northern Hill Region of India wheat is sown in
a) October b) Mid-November to mid-December
c) December-January d) Mid-October to November end Ans : a

416. In Northern Hill region of India wheat is harvested in
a) March-April b) May-June
c) February end to March end d) Mid-April to beginning of May Ans : c

417. In high Himalayan ranges and valleys wheat is sown

a) April-May b) May-June
c) September-October d) June-July Ans : a

418. In high mountain ranges and valleys in Himalayas wheat is harvested in

a) April-May b) May-June
c) September-October d) March-April Ans : d

419. Although wheat can grow well on a variety of soils, the best suited soil for wheat is well drained

a) Loam b) Clayey loam
c) Both these d) Alluvial Ans : c

420. Wheat is grown as a mixed crop with

a) Barley, mustard b) Safflower, sugarcane
c) Gram, lentil d) Any of these Ans : d

421. Modern dwarf varieties of wheat are sown at the depth

a) 5-6 cm b) 8-9 cm
c) 7-8 cm d) 2-3 cm Ans : a

422. Depth of sowing of tall varieties of wheat is

a) 5-6 cm b) 8-9 cm
c) 2-3 cm d) 7-8 cm Ans : d

423. Seed rate of wheat depending upon variety varies

a) 75 kg to 125 kg/ha b) 100 kg to 200 kg/ha
c) 50 kg to 60 kg/ha d) 40 kg to 70 kg/ha Ans : a

424. Loose snut of wheat can be controlled by seed treatment with

a) Solar energy b) Hot water
c) Vitavax d) Any of these treatments Ans : d

425. Two to 3 tonnes of farm yard manure, 40 kg of nitrogen and 20 kg of phosphorus are recommended for wheat crop

a) Irrigated areas b) Mixed cropping with mustard
c) Mixed cropping with barley d) Rainfed areas Ans: d

426. Under rainfed conditions, wheat crop is supplied with 40 kg of nitrogen and 20 kg of P_2O_5 before or at sowing time at the depth

a) 3-4 cm below soil surface b) 3-4 cm below the seed
c) 10 cm below seed d) 5 cm below seed Ans: b

427. Under irrigated conditions, wheat is supplied with full doses of phosphorus and potash and half dose of nitrogen at the time of sowing at the depth of

a) 5 cm below the seed b) 3-4 cm below the seed
c) 10 cm below soil surface d) 5 cm below soil surface **Ans : a**

428. Girija, Arjur, Pratap, Kalyansona, Sonalica, Chotilerma, C 306 are varieties of

a) Barley b) Wheat
c) Oat d) Gram **Ans : b**

429. The chromosome numbers in T. aestivum is

a) 21 b) 7
c) 14 d) 42 **Ans : a**

430. The chromosome numbers in T. monococcum is

a) 21 b) 7
c) 14 d) 42 **Ans : b**

431. The chromosome numbers in T. durum is

a) 21 b) 7
c) 14 d) 42 **Ans : c**

432. *Triticum aestivum* is

a) Diploid b) Trtraploid
c) Hexaploid d) Haploid **Ans : c**

433. Common bread wheat is

a) Diploid b) Trtraploid
c) Hexaploid d) Haploid **Ans : c**

434. Common bread wheat is

a) T. aestivum b) T. durum
c) T. vulgare d) T. dicoccum **Ans : c**

435. Emmer wheat is

a) T. aestivum b) T. durum
c) T. vulgare d) T. dicoccum **Ans : d**

436. Macaroni wheat is

a) T. aestivum b) T. durum
c) T. vulgare d) T. dicoccum **Ans : b**

437. Which of the following is most common weed in wheat

a) Phalaris minor b) Avena fatua
c) Cyperus rotundus d) All **Ans : a**

438. Broad leaved weeds in wheat can be controlled by applying

a) Isoproturon b) 2,4-D

c) Metsulfurom methyl d) All **Ans : d**

439. The most cultivated crop of the world is

a) Rice b) Wheat

c) Maize d) Jowar **Ans : b**

440. The second most cultivated crop in India is

a) Rice b) Wheat

c) Maize d) Jowar **Ans : b**

441. The crown roots in wheat appear

a) Above soil surface

b) Below soil surface but above seed

c) Below soil surface and below seed

d) Below seed **Ans : b**

❑❑❑

12 Chapter

Oilseed Crop Production

Multiple Choice Questions (MCQ's)

Groundnut

1. ....is abunch type variety of groundnut

a) Jawa b) HD 2189
c) Narbada 4 d) All these **Ans : b**

2. Which of the following countries occupies first position in the world with regard to area and production of groundnut

a) India b) U.S.A
c) China d) Burma **Ans : a**

3. Total area under groundnut in India is

a) 18.9 million hectares b) 7.5 million hectares
c) 17.8 million hectares d) 6.0 million hectares **Ans : b**

4. Total annual production of groundnut in India is about

a) 17.8 million tones b) 18.9 million tones
c) 9.0 million tones d) 7.5 million tones **Ans : c**

5. Seventy per cent of area and 75% of the production of groundnut in India is concentrated in

a) Gujarat and Andhra Pradesh b) Tamilnadu
c) Karnataka d) All these states **Ans : d**

6. Which of the following soil types are best suited for cultivation of groundnut
   a) Sandy loam soils
   b) Loam soils
   c) Black soils with good drainage
   d) All these soils

   Ans : d

7. Stiff clay soils are not suitable for cultivation of groundnut crop because
   a) Root elongation is hampered due to unaerobic conditions in the rhizosphere
   b) Flowering delayed
   c) Pod development is hampered
   d) Growth arrested

   Ans : c

8. A manure-cum-fertilizer dose of 6.25 tonnes of farmyard manure, 10-20 kg of nitrogen, 20-40 kg of phosphorus and 20-40 kg of potassium per hectare is recommended for
   a) Rainfed groundnut crop b) Irrigated groundnut crop
   c) Mustard crop d) Rapeseed crop

   Ans : a

9. A manure-cum-fertilizer dose of 12.5 tonnes of farmyard manure, 20-40 kg of nitrogen, 40-90 kg of phosphorus and 20-40 kg of potassium (K_2O) per hectare is recommended for
   a) Rainfed groundnut crop b) Irrigated groundnut crop
   c) Irrigated mustard crop d) Irrigated rapeseed crop

   Ans : b

10. Application of 500 kg of gypsum hectare is recommended for growing crop at
    a) Sowing time b) 15 days before sowing
    c) Pegging stage d) The time of irrigation

    Ans : c

11. Weeds in groundnut crop can be controlled by pre-emergence application at the rate of 3 litres per hectare of
    a) Tok- E-25 b) Lasso
    c) Either of these d) Neither of these

    Ans : c

12. GAUGI-1, AH 32, Junagarh 113, Punjab-I, AH 334, GAUG-10 are varieties of groundnut recommended for
    a) Andhra Pradesh b) Gujarat
    c) Tamilnadu d) Punjab

    Ans : b

13. The centre of origin of groundnut is
    a) Brazil b) America
    c) S. Africa d) None

    Ans : a

14. The inflorescence in groundnut is
a) Raceme b) Panicle
c) Spikelets d) Ear
Ans : a

16. Optimum temperatures towards maturity of groundnut ranges between
a) 31-37°C b) 25-30°C
c) 20-22°C d) None
Ans : c

17. The recommended seed rate for spreading type groundnut is ______ kg/ha
a) 80-90 b) 50-60
c) 70-80 d) 60-70
Ans : d

18. The recommended seed rate for bunch type groundnut is ______ kg/ha
a) 80-100 b) 50-60
c) 70-80 d) 60-70
Ans : a

19. Higher yield of groundnut could be achieved by:
a) Use of higher dose of phosphate
b) Adequate amount of N
c) No nitrogen application
d) Closer planting
Ans : a

20. Bunch type varieties of groundnut belongs to sub species
a) Fastigata b) Procumbens
c) Javanica d) Indica
Ans : a

21. Spreading type varieties of groundnut belongs to sub species
a) Fastigata b) Procumbens
c) Javanica d) Indica
Ans : b

22. The largest producer of groundnut in the world is ________
a) India b) China
c) China d) Pakistan
Ans : a

23. Groundnut besides good quantity of oil also contains about ....per cent of protein
a) 40 b) 30
c) 20 d) 10
Ans : c

24. The oil content in groundnut seeds is about __________ per cent
    a) 45 b) 30
    c) 20 d) 10 **Ans : a**

25. Among the oilseed crops groundnut ranks__________ in India
    a) First b) Second
    c) Third d) Fourth **Ans : a**

25. Groundnut seed normally contains protein and oil per cent __________ respectively.
    a) 20 and 45 b) 25 and 50
    c) 45 and 26 d) 43 and 22 **Ans : a**

26. The gynophore of groundnut is commonly referred to as the
    a) Monadelph b) Pod
    c) Ovate d) Peg **Ans : d**

27. The most suitable soil for groundnut is________
    a) Sandy and sandy loam b) Silty loam and silty clay loam
    c) Clayey loam d) Loamy clay **Ans : a**

28. Deep ploughing of soil in a groundnut field should be normally avoided because it
    a) Delays harvesting b) Delays germination
    c) Makes harvesting difficult d) Delays peg and pod development **Ans : c**

29. In India amongst all the oilseed crops groundnut accounts for more than-
    a) 40% acreage and 60% production
    b) 60% acreage and 40% production
    c) 20% acreage and 40% production
    d) 60% acreage and 80% production **Ans : a**

30. The recommended planting geometry for groundnut is ________ cm
    a) 25 x 15 b) 30 x 10
    c) 45 x 15 d) 50 x 20 **Ans : c**

31. The optimum plant population for groundnut is around
    a) One lacs b) Two lacs
    c) Three lacs d) Four lacs **Ans : b**

32. The Cultivation of which crop requires earthing
    a) Groundnut b) Paddy
    c) Cotton d) Wheat **Ans : a**

33. The groundnut seed is known as
a) Kernel b) Pod
c) Caryopsis d) None
Ans : a

34. The groundnut fruit is known as
a) Kernel b) Pod
c) Caryopsis d) None
Ans : b

35. Groundnut crop needs
a) Warm and humid climate b) Hot and humid climate
c) Dry and hot conditions d) Cold and dry climate

36. Temperature requirement for proper growth and development of groundnut is
a) 21- 26.5^{0}C b) 16 - 20^{0}C
c) 25 - 30^{0}C d) 30 - 35^{0}C
Ans : a

37. ....is essential for root development and proper growth of groundnut
a) Calcium b) Sulphur
c) Boron d) None
Ans : a

38. ....is essential for synthesis of oil and protein
a) Calcium b) Sulphur
c) Boron d) None
Ans : b

39. Pop pod of groundnut is due to the deficiency of
a) Calcium b) Sulphur
c) Boron d) None
Ans : a

40. Groundnut crop requires N, P_2O_5 and K_2O kg/ha, respectively
a) 100-150, 70-80 and 125-150 b) 90-100, 60-70 and 100-110
c) 30-60-40 d) 80–30/40–20
Ans : c

Soybean

41. Which of the following countries occupies first position in the world with regard to area and production of soybean
a) India b) U.S.A
c) China d) Burma
Ans : c

42. Maximum area and production of soybean in India is concentrated in
a) Gujarat and Andhra Pradesh b) Tamilnadu
c) Karnataka d) Madhya Pradesh
Ans : d

43. Which of the following soil types are best suited for cultivation of soybean
    a) Sandy loam soils b) Loam soils
    c) Black soils with good drainage d) All these soils **Ans : d**

44. Weeds in soybean crop can be controlled by pre-emergence application at the rate of 3 litres per hectare of
    a) Tok- E-25 b) Lasso
    c) Either of these d) Neither of these **Ans : c**

45. The inflorescence in soybean is
    a) Raceme b) Panicle
    c) Spikelets d) Ear **Ans : a**

46. Optimum temperatures towards growth of soybean ranges between
    a) 31-37°C b) 26.5-30°C
    c) 20-22°C d) None **Ans : b**

47. The recommended seed rate for soybean for grain purpose is _______ kg/ha
    a) 20-30 b) 50-60
    c) 70-80 d) 60-70 **Ans : a**

48. The recommended seed rate for fodder soybean is _______ kg/ha
    a) 80-100 b) 50-60
    c) 70-80 d) 60-70 **Ans : a**

49. Soybean besides good quantity of oil also contains about ....per cent of protein
    a) 40 b) 30
    c) 20 d) 10 **Ans : a**

50. The oil content in soybean seeds in about __________ per cent
    a) 45 b) 30
    c) 20 d) 10 **Ans : c**

51. Among the oilseed crops soybean ranks_________ in India
    a) First b) Second
    c) Third d) Fourth **Ans : d**

52. Soybean seed normally contains protein and oil per cent __________ respectively.
    a) 20 and 45 b) 25 and 50
    c) 40 and 20 d) 43 and 22 **Ans : c**

53. The fruit of soybean is commonly referred to as the
a) Monadelph b) Pod
c) Ovate d) Peg **Ans : b**

54. The most suitable soil for soybean is________
a) Loam b) Silty loam and silty clay loam
c) Clayey loam d) Loamy clay **Ans : a**

55. The recommended planting geometry for kharif soybean is _______ cm
a) 25 x 15 b) 30 x 10
c) 45 x 5 d) 50 x 20 **Ans : c**

56. The recommended planting geometry for summer soybean is _______ cm
a) 25 x 15 b) 30 x 10
c) 30 x 5 d) 50 x 20 Ans : c

57. The recommended depth of sowing for soybean is _______ cm
a) 2 -3 b) 5 x 6
c) 3 - 4 d) 6 x 8 **Ans : c**

58. The optimum plant population for soybean is around
a) One lacs b) Two lacs
c) Three lacs d) Four lacs Ans : c

59. Soybean crop needs
a) Warm and moist climate b) Hot and humid climate
c) Dry and hot conditions d) Cold and dry climate **Ans : a**

60. Temperature requirement for proper growth and development of soybean is
a) 26.5 -30^{0}C b) 16 - 20^{0}C
c) 25 - 30^{0}C d) 30 - 35^{0}C **Ans : a**

61. The suitable rhizobium strain for soybean is
a) Japonicum b) Phaseoli
c) Soja d) Glycine **Ans : a**

62. The post emergence herbicide for control of broad leaved weeds in soybean is
a) Isoproturon b) Glyphosate
c) Chlorimuron ethyl d) None **Ans : c**

63. Soybean crop requires N, P_2O_5 and K_2O kg/ha, respectively
a) 100-150, 70-80 and 125-150 b) 90-100, 60-70 and 100-110
c) 30-60-40 d) 80-30/40-20 Ans : c

Sesamum

64. Which of the following countries occupies first position in the world with regard to area and production of sesamum
a) India b) U.S.A
c) China d) Burma Ans : a

65. Total area under sesamum in India is about
a) 5 million hectares b) 3 million hectares
c) 8 million hectares d) 6.0 million hectares Ans : b

66. Total annual production of sesamum in India is about
a) 17.8 lakh tones b) 18.9 lakh tones
c) 5.0 lakh tones d) 7.5 lakh tones Ans : c

67. Maximum productivity of sesamum in India is in
a) Orissa b) Tamilnadu
c) Karnatak d) W. Bengal Ans : a

68. Which of the following soil types are best suited for cultivation of sesamum
a) Sandy loam soils b) Loam soils
c) Black soils with good drainage d) All these soils Ans : d

69. Weeds in sesamum crop can be controlled by pre-emergence application of
a) Basalin a) Lasso
b) Either of these c) Neither of these Ans : c

70. Maximum acerage of sesamum in India is in
a) Orissa b) Tamilnadu
c) Karnatak d) W.Bengal Ans : d

71. The centre of origin of sesamum is
a) India b) America
c) S. Africa d) None Ans : a

72. In India the productivity of sesamum is highest in the state
a) Punjab b) Orissa
c) W. Bengal d) None Ans : b

73. The inflorescence in sesamum is

a) Cymose b) Panicle
c) Spikelets d) Ear Ans : a

74. Optimum temperatures for growth and development of sesamum ranges between

a) 31-37°C b) 25-27°C
c) 20-25°C d) None Ans :b

75. The recommended seed rate for sesamum is _______ kg/ha

a) 8-10 b) 6-8
c) 15-18 d) 20-25 **Ans : b**

76. Sesamum yield is largely reduced mainly due to:

a) Cold wind at the flowering stage
b) Attack of aphids
c) Lack of adequate fertilizer application
d) Failure to apply irrigation in flowering stage **Ans : a**

77. Chromosome numbering *Sesamum indicum* is

a) 2n=26 b) 2n=32
c) 2n=64 d) 2n=42 **Ans : a**

78. Chromosome numbering *Sesamum prostatum* is

a) 2n=26 b) 2n=32
c) 2n=64 d) 2n=42 **Ans : b**

79. Chromosome numbering *Sesamum Occidentale* is

a) 2n=26 b) 2n=32
c) 2n=64 d) 2n=42 **Ans : a**

80. The largest producer of sesamum in the world is ______________

a) India b) China
c) China d) Pakistan **Ans : a**

81. Sesamum besides good quantity of oil also contains about ......per cent of protein

a) 40 b) 30
c) 20 d) 10 **Ans : c**

82. The oil content in sesamum seeds in about _________ per cent

a) 40-50 b) 30 -40
c) 20-30 d) 10-20 **Ans : a**

83. Among the oilseed crops sesamum ranks________ in India

a) First b) Second
c) Third d) Fourth **Ans : c**

84. Sesamum seed normally contains protein and oil per cent ________ respectively.

a) 25 and 40 b) 20 and 48
c) 45 and 26 d) 43 and 22 **Ans : b**

85. The fruit of sesamum is commonly referred to as the

a) Monadelph b) Pod
c) Ovate d) Capsule **Ans : d**

86. The most suitable soil for sesamum is________

a) Sandy loam b) Silty loam and silty clay loam
c) Clayey loam d) Loamy clay **Ans : a**

87. The Identified critical stages of irrigation for sesamum is______

a) 4-5 leaf stage b) Flowering
c) Pod development d) All **Ans : d**

88. In case of only one irrigation for sesamum it should be applied at______

a) 4-5 leaf stage b) Reproductive
c) Seed development d) All **Ans : b**

89. The recommended planting geometry for sesame is _______ cm

a) 25 x 15 b) 30 x 10
c) 45 x 15 d) 50 x 20 **Ans : b**

90. The optimum plant population for Sesamum is around

a) One lacs b) Two lacs
c) Three lacs d) Four lacs **Ans : c**

91. The Cultivation of which crop requires thinning

a) Sesamum b) Paddy
c) Cotton d) Wheat **Ans : a**

92. The sowing time of kharif sesamum is

a) Nov. - Dec) b) June-July
c) May-June d) March – April **Ans : b**

Mustard

93. Botanical name of rai is
    a) *Brassica juncea*
    b) *Eruca sativa*
    c) *Brassica campestris* var. *dichotoma*
    d) *Brassica nigra* **Ans : a**

94. Botanical name of taramira is
    a) *Brassica juncea*
    b) *Eruca sativa*
    c) *Brassica campestris* var. *dichotoma*
    d) *Brassica nigra* **Ans : b**

95. Botanical name of brown sarson is
    a) *Brassica juncea*
    b) *Eruca sativa*
    c) *Brassica campestris* var. *dichotoma*
    d) *Brassica nigra* **Ans : c**

96. Botanical name of Banarsi rai is
    a) *Brassica juncea*
    b) *Eruca sativa*
    c) *Brassica campestris* var. *dichotoma*
    d) *Brassica nigra* **Ans : d**

97. Oil content in mustard varies between
    a) 10-20% b) 30-46%
    c) 20-30% d) 46-60% **Ans : d**

98. The centre of origin of rai is
    a) China b) America
    c) S. Africa d) None **Ans : a**

99. The centre of origin of brown sarson is
    a) China b) Afganistan
    c) S. Africa d) None **Ans : b**

100. In India the productivity of mustard is highest in the state
    a) Punjab b) Haryana
    c) W. Bengal d) None **Ans : b**

101. The inflorescence in mustard is
    a) Siliqua b) Receme
    c) Spikelets d) Ear **Ans : b**

102. The fruit of mustard is
a) Siliqua
b) Receme
c) Caryopsis
d) Ear
Ans : a

103. Area under mustard cultivation in India is approximately
a) 39 m ha
b) 6m ha
c) 50m ha
d) 29m ha
Ans : b

104. Optimum temperatures for growth and development of mustard ranges between
a) 30-37°C
b) 25-30°C
c) 21-25°C
d) None
Ans : c

105. The recommended seed rate for mustard is _______ kg/ha
a) 8-10
b) 6-8
c) 15-18
d) 20-25
Ans : b

106. Higher yield of mustard could be achieved by
a) Use of higher dose of phosphate
b) Adequate amount of N
c) Use of sulphur
d) All
Ans : d

107. Mustard yield is largely reduced mainly due to
a) Cold wind at the flowering stage
b) Attack of aphids
c) Lack of adequate fertilizer application
d) None
Ans : b

108. The largest producer of rapeseed and mustard in the world is _________
a) India
b) China
c) China
d) Pakistan
Ans : c

109. The crops of group rapeseed and mustard (except. taramlra) belongs to the family_________
a) Cruciferae
b) Leguminosae
c) Composite
d) Euphorbiaceae
Ans : a

110. The fruit of rapeseed and mustard is known as ___________
a) Pod
b) Caryopsis
c) Siliquae
d) Dutum
Ans : c

111. The..........crop is more prone to frost and cold injury
a) Toria
b) Paharirai
c) Sarson
d) Banarsi Rai
Ans : a

112. The thinning of mustard is to maintain a plant to plant distance of _____________ cm.

a) 5-10 b) 10-15
c) 15-20 d) 20-25 **Ans : b**

113. The mustard crop is considered ready for harvesting when its pod turns in colour.

a) Greenish -yellow b) Dark green
c) Yellowish green d) Yellowish brown **Ans : d**

114. Mustard was planted at a plant to plant spacing of 12.5 cm with a population of 1,60,000 plants/ha, its row spacing will be

a) 30 cm b) 50 cm
c) 40 cm d) 60 cm **Ans : b**

115. Mustard has...............% protein content

a) 10-15 b) 15-25
c) 25-30 d) 30-40 **Ans : c**

116. Among the oilseed crops mustard ranks_________ in India

a) First b) Second
c) Third d) Fourth **Ans : b**

117. Mustard seed normally contains protein and oil per cent __________ respectively.

a) 25 and 40 b) 22 and 43
c) 45 and 26 d) 43 and 22 **Ans : a**

118. Fluchloralin herbicide should be applied in mustard

a) 1 - 2 DAS b) 6 - 8 DAS
c) 10 - 14 DAS d) 2 weeks **Ans: a**

119. Isoproturon should be applied in mustard

a) 15-20 DAS b) 6 - 8 DAS
c) 10 - 14 DAS d) 3 weeks after saving **Ans : d**

120. Isoproturon herbicide should be applied in mustard at......... kg ai/ha

a) 0.5 b) 0.75
c) 1.5 d) 2 **Ans : b**

121. The recommended dose of N, P_2O_5 and K_2O kg/ha for rainfed mustard are

a) 60 - 20 - 20 b) 120 - 40 - 40
c) 80 - 20 - 30 d) 40 - 30 - 40 **Ans : a**

122. The recommended dose of N, P_2O_5 and K_2O kg/ha for irrigated mustard are
   a) 100 - 50 -40   b) 120 - 40 - 40
   c) 80 - 60 - 40   d) 40 - 30 -40   **Ans : a**

123. Which one is the recommended optimum plant population for mustard
   a) 2,00,000 plants   b) 66,000 plants
   c) 80,000 plants   d) 90,000 plants   **Ans : a**

124. Which of the following cruciferous oil seeds are self fertile
   a) Brassica juncea (Brown mustard or rai, or raya or laha) and B campestris var. sarson (yellow sarson)
   b) Brassica campestris var. dichotoma (brown sarson) and B cumpestris var. toria (toria, lahi or maghi labhi)
   c) Eruca satica (taramira or tara) and Brassica nigra (banarasi rai)
   d) Brassica alba (white mustard)   **Ans : a**

125. Which of the following cruciferous oil seeds is known as mustard, in trade
   a) Brassica campestris var. sarson and B campestris var. dichotoma
   b) Brassica juncea
   c) Brassica campestris var. toria
   d) Eruca sativa   **Ans : b**

126. Which of the following oil seeds are classified as rapeseed in trade
   a) Brassica alba
   b) Brassica nigra
   c) Brassica campestris var. sarson, B campestris var. dichotoma, B campestris var. toria and Eurca sativa
   d) Brassica juncea   **Ans : c**

127. Which of the following cruciferous oil seeds is not cultivated in India
   a) Brassica alba (white mustard)   b) Brassica juncea (rai)
   c) Brassica nigra (Banarasi rai)   d) Eruca sativa (tamarira)   **Ans : a**

128. BTHI, M 18 and M 27 are the varieties of
   a) Yellow sarson   b) Brown sarson
   c) Raya   d) Toria   **Ans: b**

129. Varuna, Type-1, Type 10, Type 42, type 151, YS Pb 24, and M3 are the varieties of
   a) Yellow sarson   b) Brown sarson
   c) Raya   d) Toria   **Ans : a**

130. Which of the following cruciferous oil seeds is used as a spice
a) *Brassica juncea* (rai)
b) *Eruca sativa* (taramira)
c) *Brassica campestris* var. *dichotoma* (brown sarson)
d) *Brassica nigra* (Banarsi rai)
Ans : d

131. Sangam, Br. 36, ITSA and B-54 are varieties of
a) Yellow sarson b) Brown sarson
c) Raya d) Toria
Ans : d

132. Prakash, Faridkot selection, K-1 and B-85 are the varieties of
a) Yellow sarson b) Brown sarson
c) Raya d) Toria
Ans : c

133. Test weight of mustard is
a) 3-5 gram b) 25-30 gram
c) 30-35 gram d) 70-72 gram
Ans : a

Castor

134. The centre of origin of castor is
a) Burma b) America
c) Africa d) None
Ans : c

135. In India the productivity of castor is highest in the state
a) Punjab b) Haryana
c) Gujarat d) None
Ans : c

136. In India the area of castor is highest in the state
a) Punjab b) Haryana
c) Gujarat d) Andhra Pradesh
Ans : d

137. In India maximum production of castor is in the state
a) Punjab b) Haryana
c) Gujarat d) None
Ans : c

138. The inflorescence in castor is
a) Raceme b) Panicle
c) Spikelets d) Ear
Ans : a

139. Area under castor cultivation in India is approximately
a) 7 lakh ha b) 10 lakh ha
c) 12 lakh ha d) 29 lakh ha
Ans : a

140. Optimum temperatures for growth and development of castor ranges between
a) 31-37°C b) 25-30°C
c) 20-26°C d) None
Ans : c

141. The recommended seed rate for castor is ______ kg/ha
a) 10-15 b) 6-10
c) 15-18 d) 20-25 Ans : a

142. Castor belongs to__________ family
a) Euphorbiaceae b) Leguminosae
c) Cruciferеae d) Compositae Ans : a

143. The usual plant to plant spacing recommended for castor is
____
a) 30-50 cm b) 15-30 cm
c) 50-75 cm d) 75-90 cm Ans : a

144. Castor besides good quantity of protein also contains about per cent of oil
a) 40-50 b) 30-40
c) 20-30 d) 10-20 Ans : a

145. Which of the followings contains the highest oil content
a) Pigeonpea b) Soybean
c) Greengram d) Castor Ans : d

146. Among the oilseed crops castor ranks________ in India
a) Eighth b) Sixth
c) Third d) Fourth Ans : a

147. The optimum plant population for castor is around
a) 30000-40000 b) 100000
c) 40000-70000 d) 70000-90000 Ans : c

148. Castor crop needs
a) Warm and humid climate b) Hot and humid climate
c) Dry and hot conditions d) Cold and dry climate Ans : c

149. Photoperiodically, castor is a type of plant
a) Long day b) Short day
c) Day neutral d) Intermediate Ans : a

150. The recommended dose of N, P_2O_5 and K_2O kg/ha for deshi castor are
a) 60 – 40 – 40 b) 120 – 40 – 40
c) 30 – 20 – 30 d) 40 – 30 – 40 Ans : a

151. The recommended dose of N, P_2O_5 and K_2O kg/ha for hybrid castor are
a) 100 – 60 –40 b) 120 – 40 – 40
c) 80 – 60 – 60 d) 40 – 30 –40 Ans : c

152. Ricinus communis is originated in

a) India b) Eastern Africa
c) Tropical America d) Tropical Asia **Ans : b**

153. In castor nipping is done to

a) Encourage branching b) Remove axillary buds
c) Both d) None **Ans : b**

154. The fruit of castor is

a) Caryopsis b) Capsule
c) Siliqua d) None **Ans : b**

155. The turkey red oil is prepared by treating castor oil with

a) H_2SO_4 b) HCL
c) $BaCl_2$ d) None **Ans : a**

Sunflower

156. Which of the following countries occupies first position in the world with regard to area and production of sunflower

a) India b) U.S.S.R. (*Former*)
c) China d) Burma **Ans : b**

157. Total area under sunflower in India is

a) 18.9 million hectares b) 7.5 million hectares
c) 17.8 million hectares d) 4.0 lakh hectares **Ans : d**

158. Maximum area and production of sunflower in India is concentrated in

a) Gujarat and Andhra Pradesh b) Tamilnadu
c) Maharastra and Karnataka d) All these states **Ans : c**

159. Which of the following soil types are best suited for cultivation of sunflower

a) Sandy loam soils b) Loam soils
c) Black soils with good drainage d) All these soils **Ans : d**

160. Weeds in sunflower crop can be controlled by pre-emergence application at the rate of 3 litres per hectare of

a) Tok- E-25 b) Lasso
c) Either of these d) Neither of these **Ans : c**

161. The centre of origin of sunflower is

a) Mexico b) America
c) S. Africa d) None **Ans : a**

162. The inflorescence in sunflower is

a) Capitulum b) Panicle
c) Ear d) Raceme
Ans : a

163. Optimum temperatures for sunflower ranges between

a) 31-37°C b) 28-32°C
c) 20-22°C d) None
Ans : b

164. The recommended seed rate for sunflower is ____________ kg/ha

a) 10-15 b) 50-60
c) 20-30 d) 60-70
Ans : a

165. Sunflower besides good quantity of oil also contains about ....per cent of protein

a) 40 b) 30
c) 20 d) 10
Ans : c

166. The oil content in sunflower seeds is about _________ per cent

a) 45 b) 30
c) 20 d) 10
Ans : a

167. Sunflower seed normally contains protein and oil per cent _________ respectively.

a) 20 and 45 b) 25 and 50
c) 45 and 26 d) 43 and 22
Ans : a

168. The recommended planting geometry for sunflower is _______ cm

a) 25 x 15 b) 30 x 10
c) 45 x 30 d) 50 x 20
Ans : c

169. The optimum plant population for mustard is around

a) 1.3 lacs b) 2 lacs
c) 3 lacs d) 4 lacs
Ans : a

170. Sunflower crop needs

a) Warm and humid climate b) Hot and humid climate
c) Dry and hot conditions d) Cold and dry climate
Ans : a

171. Sunflower is a............... pollinated crop

a) Cross b) Self
c) Bisexual d) None
Ans : a

172. Sunflower crop producing 14 q seed exhausts N, P_2O_5 and K_2O kg/ha, respectively

a) 175, 70 and 125
b) 90 - 100, 60- 70 and 100 - 110
c) 30 - 60 - 40
d) 80 - 30/40 - 20

Ans : a

173. Sunflower is a............... crop

a) Short day
b) Long day
c) Day neutral
d) None

Ans : c

174. Sunflower crop requires N, P_2O_5 and K_2O kg/ha, respectively

a) 100-150, 70-80 and 125-150
b) 90-100, 60-70 and 100-110
c) 80-60-40
d) 80-30/40-20

Ans : c

Safflower

175. Which of the following countries occupies first position in the world with regard to area and production of safflower

a) India
b) U.S.A.
c) China
d) Burma

Ans : a

176. 72 per cent of area and 73% of the production of safflower in India is concentrated in

a) Gujarat and Andhra Pradesh
b) Tamilnadu
c) Karnataka
d) Maharastra

Ans : d

177. Which of the following soil types are best suited for cultivation of safflower

a) Sandy loam soils
b) loam soils
c) Heavy soils
d) all these soils

Ans : c

178. Stiff clay soils are suitable for cultivation of safflower crop because

a) Favours root elongation
b) Early Flowering
c) Better moisture retention
d) Better growth

Ans : c

179. A manure-cum-fertilizer dose of 5-10 tonnes of farmyard manure,20-50kg of nitrogen, 20-50 kg of phosphorus and 20-30 kg of potassium per hectare is recommended for

a) Rainfed safflower crop
b) Irrigated safflower crop
c) Mustard crop
d) Rapeseed crop

Ans : a

180. A manure-cum-fertilizer dose of 10 tonnes of farmyard manure, 75 kg of nitrogen, 50-75 kg of phosphoric acid and 30-40 kg of potassium (K_2O) per hectare is recommended for

a) Rainfed safflower crop
b) Irrigated safflower crop
c) Transplanted safflower crop
d) Irrigated rapeseed crop

Ans : b

181. Weeds in safflower crop can be controlled by pre-emergence application at the rate of 1.5 kg ai per hectare of

a) Tok- E-25 b) Fluchloralin
c) Either of these d) Neither of these **Ans : b**

182. The centre of origin of safflower is

a) Arabia b) Americ
c) S. Africa d) None **Ans : a**

183. The inflorescence in safflower is

a) Head b) Panicle
c) Spikelets d) Ear **Ans : a**

184. Optimum temperatures towards maturity of safflower ranges between

a) 31-37°C b) 24-32°C
c) 20-22°C d) None **Ans : b**

185. Optimum temperatures for proper germination of safflower is

a) 31-37°C b) 24-32°C
c) 20-22°C d) 15°C **Ans : d**

186. Optimum temperatures for proper growth and development of safflower is

a) 31-37°C b) 24-32°C
c) 20-21°C d) 15°C **Ans : c**

187. The recommended seed rate for safflower is _______ kg/ha

a) 80-90 b) 50-60
c) 70-80 d) 10-15 **Ans : d**

188. Higher yield of safflower could be achieved by

a) Use of higher dose of phosphate
b) Adequate amount of N
c) Sulphur application
d) All **Ans : d**

189. Cultivated safflower varieties belongs to species

a) Tinctorius b) Palaestinus
c) Oxyacantha d) Flavescens **Ans : a**

190. Cultivated safflower has .... chromosomes

a) 2n=20 b) 2n=24
c) 2n=44 d) 2n=64 **Ans : b**

191. Wild safflower found in India is

a) Tinctorius
b) Palaestinus
c) Oxyacantha
d) Flavescens

Ans : c

192. Wild safflower found in Negev Desert is

a) Tinctorius
b) Palaestinus
c) Oxyacantha
d) Flavescens

Ans : b

193. The largest producer of safflower in the world is ________

a) India
b) China
c) China
d) Pakistan

Ans : a

194. Safflower besides good quantity of oil also contains about ....per cent of protein

a) 40
b) 30
c) 20
d) 10

Ans : a

195. The oil content in safflower seeds is about __________ per cent

a) 25-35
b) 35-45
c) 10-15
d) 10-15

Ans : a

196. Safflower seed normally contains protein and oil per cent __________ respectively.

a) 20 and 45
b) 25 and 50
c) 45 and 20
d) 40 and 30

Ans : d

197. The most suitable soil for safflower is________

a) Sandy and sandy loam
b) Silty loam and silty clay loam
c) Clayey loam
d) Loamy clay

Ans : c

198. India amongst safflower growing countries accounts for more than-

a) 60% acreage and 48% production
b) 60% acreage and 30% production
c) 20% acreage and 40% production
d) 60% acreage and 80% production

Ans : a

199. The recommended planting geometry for safflower is _______ cm

a) 25 x 15
b) 30 x 10
c) 45 x 15
d) 50 x 20

Ans : c

200. The optimum plant population for safflower is around

a) One lacs
b) Two lacs
c) Three lacs
d) Four lacs

Ans : b

201. Which oil contains maximum linoleic acid
a) Safflower b) Groundnut
c) Cotton d) Mustard **Ans : a**

202. The cultivation of which crop requires cool and dry weather
a) Safflower b) Paddy
c) Cotton d) Maize **Ans : a**

203. Safflower oil contains ....% linoleic acid
a) 55-80 b) 25-35
c) 35-45 d) 45-55 **Ans : a**

204. Which oil is recommended for patients suffering from heart diseases
a) Safflower b) Groundnut
c) Cotton d) Mustard **Ans : a**

205. Carthamin is extracted from
a) Safflower b) Groundnut
c) Cotton d) Mustard **Ans : a**

206. The safflower crop besides oil is grown for centuries for extraction of
a) Kernel b) Dye
c) Resins d) None **Ans : b**

207. Safflower crop needs
a) Warm and humid climate b) Hot and humid climate
c) Dry and hot conditions d) Cool and dry climate **Ans : d**

208. ....is essential for high oil content and quality of safflower
a) Calcium b) Sulphur
c) Boron d) None **Ans : b**

209. ....is essential for synthesis of oil and protein
a) Calcium b) Sulphur
c) Boron d) None **Ans : b**

Niger

210. Maximum area and production of Niger in India is concentrated in
a) Gujarat and Andhra b) Tamilnadu
c) Karnataka d) M.P., Orissa and Maharastra **Ans : d**

211. Which of the following soil types are best suited for cultivation of niger

a) Sandy loam soils
b) Loam soils
c) Clayey loam
d) All the above

Ans : d

212. A manure-cum-fertilizer dose of 5 tonnes of farmyard manure, 20 kg of nitrogen, 20 kg of phosphorus and 1 kg of potassium per hectare is recommended for

a) Rainfed Niger crop
b) Irrigated Niger crop
c) Mustard crop
d) Rapeseed crop

Ans : a

213. .... is a parasitic weed in niger crop

a) Cascuta
b) Striga
c) Orobanche
d) None

Ans : a

214. Dodder in niger crop can be controlled by pre-emergence application at the rate of 2.0 kg ai per hectare of

a) Tok- E-25
b) Fluchloralin
c) Pronamide
d) Neither of these

Ans : c

215. The centre of origin of niger is

a) Ethiopia
b) America
c) S. Africa
d) None

Ans : a

216. The inflorescence in niger is

a) Raceme
b) Panicle
c) Spikelets
d) Capitulum

Ans : d

217. Optimum temperatures for proper growth and development of niger is

a) 31-37°C
b) 18-23°C
c) 20-21°C
d) 15°C

Ans : b

218. The recommended seed rate for niger is _______ kg/ha

a) 80-90
b) 5-8
c) 70-80
d) 10-15

Ans : b

219. Cultivated niger varieties belongs to species

a) Abyssinica
b) Palaestinus
c) Oxyacantha
d) Flavescens

Ans : a

220. The oil content in niger seeds in about _________ per cent

a) 25-35
b) 35-40
c) 10-15
d) 10-15

Ans : b

221. The most suitable soil for niger is_______
a) Sandy loam and Clayey loam b) Silty loam and silty clay loam
c) Loamy clay d) None
Ans : c

222. The recommended planting geometry for niger is ______ cm
a) 25 x 15 b) 30 x 15
c) 45 x 15 d) 50 x 20
Ans : b

223. The optimum plant population for niger is around
a) One lacs b) Two lacs
c) Three lacs d) Four lacs
Ans : c

224. Which oil contains maximum linolinic acid
a) Niger b) Groundnut
c) Cotton d) Mustard
Ans : a

225. The cultivation of which crop requires cool and dry weather
a) Niger b) Paddy
c) Cotton d) Maize
Ans : a

226. Niger oil contains ....% linolinic acid
a) 55-80 b) 25-35
c) 35-45 d) 45-55
Ans : d

227. Fiber from niger is extracted from
a) Root b) Seed
c) Stem d) None
Ans : c

228. The niger crop besides oil is grown for centuries for extraction of
a) Fiber b) Dye
c) Resins d) None
Ans : a

229. Niger crop needs
a) Warm and humid climate b) Hot and humid climate
c) Dry and hot conditions d) Cool and dry climate
Ans : d

Linseed

230. Which of the following countries occupies first position in the world with regard to area and production of linseed
a) India b) U.S.A.
c) China d) Burma
Ans : a

231. 70% of the production of linseed in India is concentrated in
a) Madhaya Pradesh & Uttar Pradesh
b) Tamilnadu
c) Karnataka
d) Maharastra
Ans : a

232. Which of the following soil types are best suited for cultivation of linseed
a) Clay loam soils b) Loam soils
c) Heavy soils d) All the above
Ans : a

233. A fertilizer dose of 50 kg of nitrogen, 25 kg of phosphorus and 25 kg of potassium per hectare is recommended for
a) Rainfed Linseed crop b) Irrigated Linseed crop
c) Mustard crop d) Rapeseed crop
Ans : a

234. A fertilizer dose of 90 kg of nitrogen, 40 kg of phosphorus and 40 kg of potassium per hectare is recommended for
a) Rainfed linseed crop b) Irrigated linseed crop
c) Transplanted linseed crop d) Irrigated rapesed crop
Ans: b

235. Weeds in linseed crop can be controlled by post-emergence application at the rate of 0.5 kg ai per hectare of
a) MCPB b) Glyphosate
c) 2,4-D d) Neither of these
Ans : a

236. The centre of origin of linseed is
a) Arabia b) America
c) S. Africa d) Mediterranean
Ans : d

237. The inflorescence in linseed is
a) Raceme b) Cymose
c) Spikelets d) Ear
Ans : b

238. Optimum temperatures towards seed formation of linseed ranges between
a) 31-37°C b) 15-20°C
c) 20-22°C d) None
Ans : b

239. Optimum temperatures for proper germination of linseed is
a) 31-37°C b) 25-30°C
c) 20-22°C d) 15°C
Ans :b

240. The recommended seed rate for line sowing of linseed is _______ kg/ha
a) 20 b) 30
c) 40 d) 10
Ans : d

241. The recommended seed rate for sowing of linseed by broadcasting is _______ kg/ha

a) 20-30 b) 30-40
c) 40-50 d) 10-15 Ans : a

242. The recommended seed rate for sowing of utera crop of linseed is _______ kg/ha

a) 20-30 b) 30-40
c) 40-50 d) 10-15 Ans : b

243. Linseed is commonly known as......... in western countries

a) Coir b) Flax
c) Fiber d) All Ans : b

244. Scientific name of linseed is

a) *Linum usitatissimum* b) *Linum angustifolium*
c) *Linum grandiflorum* d) None Ans : a

245. Linseed has .... chromosomes

a) 2n=20 b) 2n=30
c) 2n=44 d) 2n=64 Ans : b

246. The largest producer of linseed in the world is ________

a) India b) China
c) China d) Pakistan Ans : a

247. Linseed besides good quantity of oil also contains about ....per cent of protein

a) 40 b) 30
c) 20 d) 10 Ans : c

248. The oil content in linseed seeds in about ____________ per cent

a) 25-35 b) 35-45
c) 10-15 d) 10-15 Ans : b

249. Linseed seed normally contains protein and oil per cent __________ respectively.

a) 20 and 40 b) 25 and 50
c) 45 and 20 d) 40 and 30 Ans : a

250. The most suitable soil for linseed is________

a) Sandy loam b) Silty loam
c) Clayey loam d) Loamy clay Ans : c

251. India amongst linseed growing countries accounts for more than-

a) 27% acreage and 20% production
b) 60% acreage and 30% production
c) 20% acreage and 40% production
d) 60% acreage and 80% production

Ans : a

252. The recommended planting geometry for linseed is _______ cm

a) 25 x 5
b) 30 x 10
c) 45 x 15
d) 50 x 20

Ans : a

253. The optimum plant population for linseed is around

a) 8 lacs
b) 10 lacs
c) 5 lacs
d) 4 lacs

Ans : a

254. Linseed oil contains ....% linolinic acid

a) 20-30
b) 30-40
c) 40-50
d) 50-60

Ans : d

255. The crop cultivated for food, oil, fiber and feed is

a) Linseed
b) Groundnut
c) Cotton
d) Mustard

Ans : a

256. The linseed fruit is known as

a) Capsule
b) Head
c) Caryopsis
d) None

Ans : a

257. The linseed crop besides oil is grown for centuries for extraction of

a) Fiber
b) Dye
c) Resins
d) None

Ans : a

❑❑❑

13 Chapter

Pulse Crop Production

Multiple Choice Questions (MCQ's)

Blackgram

1. The centre of origin of blackgram is
   a) India b) America
   c) S. Africa d) None **Ans : a**

2. In India the productivity of blackgram is highest in the state
   a) Punjab b) Madya Pradesh
   c) Orissa d) Andhra Pradesh **Ans : d**

3. In India the area of blackgram is highest in the state
   a) Punjab b) Madhya Pradesh
   c) Orissa d) Andhra Pradesh **Ans : c**

4. In India ....state is having maximum production of blackgram
   a) Punjab b) Madhya Pradesh
   c) Orissa d) Andhra Pradesh **Ans : d**

5. Blackgram is also known as
   a) Mung b) Urd
   c) Khesari d) None **Ans : b**

6. Blackgram contains about............% protein
   a) 60 b) 24
   c) 1.3 d) None **Ans : b**

7. Blackgram contains about............% carbohydrate
   a) 60 b) 44
   c) 1.3 d) None **Ans : a**

8. Blackgram contains about............% fat
   a) 6.0 b) 2.4
   c) 1.3 d) 1 **Ans : c**

9. Blackgram is the richest source of ....among pulses
   a) Lysine b) Phosphoric acid
   c) Peptine d) None **Ans : b**

10. Scientific name of blackgram is
    a) Vigna mungo b) Vigna radiata
    c) Phaseolus radiata d) None **Ans : a**

11. Blackgram belongs to family
    a) Gramineae b) Leguminoseae
    c) Compositeae d) None **Ans : b**

12. Crop grown for food, fodder, manuring and soil conservation is
    a) Urd b) Groundnut
    c) Soybean d) None **Ans : a**

13. Blackgram originated from ....a wild plant
    a) Phaseolus sublobatus b) Phaseolus radiata
    c) Vigna radiata d) None **Ans : a**

14. Early maturing, black coloured bold seeded varieties of blackgram belongs to
    a) Vigna mungo var. niger b) Vigna mungo var. viridis
    c) Vigna mungo var. mutans d) Vigna mungo var. helpense **Ans : a**

15. Late maturing, green coloured small seeded varieties of blackgram belongs to
    a) Vigna mungo var. niger b) Vigna mungo var. viridis
    c) Vigna mungo var. mutans d) Vigna mungo var. helpense **Ans : b**

16. The inflorescence in blackgram is
    a) Axillary receme b) Panicle
    c) Spikelets d) Ear **Ans : a**

17. Area under blackgram cultivation in India is approximately
    a) 39 m ha b) 2.4m ha
    c) 50m ha d) 29m ha
    Ans : b

18. The recommended seed rate for blackgram is _______ kg/ha
    a) 8-10 b) 12-15
    c) 15-18 d) 20-25
    Ans : b

19. Blackgram yield is largely reduced mainly due to:
    a) Heavy rains at the flowering stage
    b) Attack of aphids
    c) Lack of adequate fertilizer application
    d) Failure to apply irrigation in flowering stage
    Ans : a

20. ....is a kharif/summer crop raised under residual soil moisture in dry lands
    a) Cotton b) Maize
    c) Chickpea d) Black gram
    Ans : d

21. Blackgram crop planted at a spacing of 25 x 10 cm will have ....lakh plants/ha
    a) 1 b) 2
    c) 3 d) 4
    Ans : d

22. Which of the following is the leading state in blackgram production
    a) Bihar b) Uttar Pradesh
    c) A.P. d) Gujarat
    Ans : c

23. The place of origin of blackgram is
    a) Burma b) India
    c) Bangladesh d) Ceylon
    Ans : b

23. For Kharif season blackgram, the recommended optimum seed rate is kg/ha
    a) 8-10 b) 12-15
    c) 15-20 d) 20-30
    Ans : b

24. For summer season blackgram, the recommended optimum seed rate is kg/ha
    a) 8-10 b) 12-15
    c) 15-20 d) 20-30
    Ans : d

25. Blackgram yield is largely reduced mainly due to:
    a) Heavy rains at the flowering stage
    b) Attack of aphids
    c) Frost
    d) All
    Ans : d

26. Khariff blackgram should be seeded at a row to row spacing of ....cm

a) 30-45 b) 15-20
c) 20-25 d) 25-30 **Ans : a**

27. Summer blackgram should be seeded at a row to row spacing of ....cm

a) 30-45 b) 15-20
c) 30-35 d) 20-30 **Ans : d**

28. The fruit of blackgram is known as ___________

a) Pod b) Caryopsis
c) Siliquae d) Dutum **Ans : a**

29. Blackgram was planted at a plant to plant spacing of 10 cm with a population of 3,33,333 plants/ha, its row spacing will be

a) 30 cm b) 50 cm
c) 40 cm d) 60 cm **Ans : a**

30. The protein content in blackgram seeds in about _________ per cent

a) 40 b) 30
c) 24 d) 10 **Ans : c**

31. Most commonly followed method of planting blackgram in India is

a) Direct seeding in rows
b) Transplanting on raised seed beds
c) Direct seeding in puddled beds
d) Transplanting in standing water in puddled fields **Ans : a**

32. The type of germination in urdbean is known as

a) Epigeal b) Hypogeal
c) Hypoepigeal d) Epihypogeal **Ans : d**

33. Which one of the following is not correctly matched

Crop		*Seed rate*
a) Wheat	-	100
b) Urd	-	50
c) Maize	-	25
d) Mustard	-	10

Ans : b

34. Blackgram is originated in

a) India b) Tropical America
c) China d) Indonesia **Ans : a**

35. Which one *of* the following pulse crops is used as a pulse, a fodder and a green manure crop
a) Groundnut b) Urd
c) Cowpea d) Pea
Ans : b

36. Blackgram needs a..............climate
a) Hot and humid b) Cool and humid
c) Hot and dry d) None
Ans : a

37. It is recommended to apply NPK @................kg/ha to urd
a) 100:60:40 b) 20:40:40
c) 60:40:40 d) None
Ans : b

38. It is recommended to apply................ @ 1kg a.i. /ha to urd
a) Stomp b) Basalin
c) Isoproturon d) None
Ans : b

39. The cohesiveness and tenderness of cooked blackgram depends largely on
a) The percentage of amylose
b) The percentage of amylopectin
c) The proportion of amylose and protein
d) The proportion of amylose and amylopectin
Ans : c

Gram

40. The centre of origin of gram is
a) S-W Asia b) America
c) S. Africa d) None
Ans : a

41. The inflorescence in gram is
a) Axilary receme b) Panicle
c) Spikelets d) Ear
Ans : a

42. Area under gram cultivation in India is approximately
a) 3.9 m ha b) 6.5m ha
c) 5.0m ha d) 65m ha
Ans : b

43. Optimum temperatures towards maturity of gram ranges between
a) 21-37°C b) 25-30°C
c) 20-25°C d) None
Ans : b

44. The recommended seed rate for gram is _______ kg/ha
a) 8-10 b) 75-100
c) 15-18 d) 25-25
Ans : b

45. Higher yield of gram could be achieved by:
   a) Use of higher dose of phosphate
   b) Adequate amount of N
   c) No nitrogen application
   d) Closer planting
   Ans : a

46. The leading producer of gram is
   a) Burma b) India
   c) Bangladesh d) Ceylon
   Ans : b

47. The country having maximum acreage of gram is
   a) Burma b) India
   c) Bangladesh d) Ceylon
   Ans : b

48. Gram is commonly known as
   a) Redgram b) Greengram
   c) Blackgram d) Bengalgram
   Ans : d

49. The most Important pulse crop in India is
   a) Chickpea b Cowpea
   c) Pigeonpea d) Greengram
   Ans : a

50. The recommended seed rate for kabuli gram is _______ kg/ ha
   a) 8-10 b) 100-125
   c) 15-18 d) 20-25
   Ans : b

51. The largest producer of chickpea in the world is ________
   a) India b) China
   c) Nepal d) Pakistan
   Ans : a

52. Nipping in gram means a process of __________
   a) Removal of all buds present in the axil of leaves
   b) Removal of leaves
   c) Removal of terminal bud
   d) Buming of leaves
   Ans : c

53. The basic aim of nipping in gram is
   a) To reduce the plant height
   b) To enlarge branching
   c) To divert energy and nutrients from flower head to leaves
   d) To protect the plants against lodging
   Ans : b

54. What is the ideal temperature for sowing of Gram
   a) 10°C to 25°C b) 15°C to 20.C
   c) 10°C to 15°C d) 25.C to 30.C
   Ans : a

55. The optimum plant population for gram is around
a) One lacs b) Two lacs
c) Three lacs d) Four lacs Ans : c

56. Wherever one irrigation is available, the gram crop should irrigated at
a) CRI stage b) Tillering stage
c) Preflowering stage d) Milking stage Ans : c

57. The dehulled gram grain is known as
a) White gram b) Brown gram
c) Red gram d) Grey gram Ans : b

58. Gram crop needs
a) Warm and humid climate b) Hot and humid climate
c) Dry and hot conditions d) Cold and dry climate Ans : d

59. Photoperiodically, gram is a type of plant
a) Long day b) Short day
c) Day neutral d) Intermediate Ans : a

60. The recommended dose of N, P_2O_5 and K_2O kg/ha for deshi gram are
a) 60 – 40 – 40 b) 120 – 40 – 40
c) 30 – 60 – 40 d) 40 – 30 – 40 Ans : c

61. Optimum plant population ha^{-1} for gram has estimated as
a) 50,000 b) 30,0000
c) 80,000 d) 1,00,000 Ans : b

62. The insect pod borer is commonly found on
a) Maize b) Wheat
c) Groundnut d) Gram Ans : d

63. In India Bengal gram is cultivated on
a) Light alluvial soils b) Clay loam soils
c) Black cotton soils d) All these soils Ans: d

64. Bengal gram is grown as a pure as well as a mixed crop in combination with
a) Wheat, barley b) Linseed safflower
c) Mustard d) Any of these crops Ans: d

65. Grain of gram is
a) White colour
b) Yellow colour
c) Brownish red colour, black colour
d) All these colours Ans : d

66. How much seed of gram should be treated with one packet of *Rhizobium* culture
    a) 5 kg b) 10 kg
    c) 15 kg d) 20 kg **Ans : b**

67. Botanical name of brown gram is
    a) *Cicer arietinum* b) *Cicer kabulium*
    c) Both d) None **Ans : a**

68. Botanical name of white gram is
    a) *Cicer arietinum* b) *Cicer kabulium*
    c) Both d) None **Ans : b**

69. Weeds in gram can be controlled by applying
    a) Fluchloralin b) 2,4-D
    c) Both d) None **Ans : a**

70. Gram belongs to family
    a) Liliaceae b) Linaceae
    c) Tiliaceae d) Leguminoceae **Ans : d**

71. The most important crop grown on conserved moisture in our country is
    a) Gram b) Wheat
    c) Mustard d) Barley **Ans : a**

72. Which one of the following operations is associated with the process of nipping in gram
    a) Treating the seed with Rhizobium culture
    b) Typing the branches to avoid lodging
    c) Picking green leaves for vegetable purposes
    d) Plucking the apical buds to promote branches **Ans : d**

73. Awarodhi' variety of gram is resistant to
    a) Wilt b) Blight
    c) Flooding d) Drought **Ans : a**

Greengram

74. The centre of origin of Greengram is
    a) India b) America
    c) S. Africa d) None **Ans : a**

75. In India the productivity of Greengram is highest in the state
    a) Punjab b) Madhya Pradesh
    c) Orissa d) Andhra Pradesh **Ans : a**

76. In India the area of Greengram is highest in the state
a) Punjab b) Maharastra
c) Orissa d) Andhra Pradesh
Ans : b

77. In India ....state is having maximum production of Greengram
a) Punjab b) Maharastra
c) Orissa d) Andhra Pradesh
Ans : b

78. Greengram is also known as
a) Urd b) Moong
c) Khesari d) None
Ans : b

79. Greengram contains about............% protein
a) 60 b) 25
c) 1.3 d) None
Ans : b

80. Scientific name of Greengram is
a) Vigna mungo b) Vigna radiata
c) Phaseolus radiata d) None
Ans : b

81. Greengram belongs to family
a) Gramineae b) Leguminoseae
c) Compositeae d) None
Ans : b

82. Crop grown for food, fodder, manuring and soil conservation is
a) Moong b) Maize
c) Cowpea d) None
Ans : a

83. Greengram originated from ....a wild plant
a) Vigna radiata sublobata b) Phaseolus radiata
c) Vigna mungo d) None
Ans : a

84. The inflorescence in Greengram is
a) Axillary receme b) Panicle
c) Spikelets d) Ear
Ans : a

85. The recommended seed rate for Greengram is _______ kg/ha
a) 8-10 b) 12-15
c) 25-30 d) 20-25
Ans : b

86. Greengram crop planted at a spacing of 30 x 10 cm will have ....lakh plants/ha
a) 1.3 b) 2.3
c) 3.3 d) 4.3
Ans : c

87. Which of the following is the leading state in Greengram production
    a) Bihar b) Uttar Pradesh
    c) Maharastra d) Gujarat Ans : c

88. The place of origin of Greengram is
    a) Burma b) India
    c) Bangladesh d) Ceylon Ans : b

89. For Kharif season Greengram, the recommended optimum seed rate is kg/ha
    a) 8-10 b) 12-15
    c) 15-20 d) 20-30 Ans : b

90. For summer season Greengram, the recommended optimum seed rate is kg/ha
    a) 8-10 b) 12-15
    c) 10-12 d) 18-20 Ans : d

91. Khariff Greengram should be seeded at a row to row spacing of ....cm
    a) 30-45 b) 15-20
    c) 20-25 d) 25-30 Ans : a

92. Summer Greengram should be seeded at a row to row spacing of ....cm
    a) 30-45 b) 15-20
    c) 30-35 d) 25-30 Ans : d

93. The fruit of Greengram is known as ___________
    a) Pod b) Caryopsis
    c) Siliquae d) Dutum Ans : a

94. Greengram was planted at a plant to plant spacing of 10 cm with a population of 2,50,000 plants/ha, its row spacing will be
    a) 30 cm b) 50 cm
    c) 40 cm d) 60 cm Ans : c

95. The protein content in Greengram seeds in about _________ per cent
    a) 40 b) 30
    c) 25 d) 10 Ans : c

96. Most commonly followed method of planting Greengram in India is
    a) Direct seeding in rows
    b) Transplanting on raised seed beds
    c) Direct seeding in puddled beds
    d) Transplanting in standing water in puddled fields Ans : a

97. The type of germination in moongbean is known as

a) Epigeal b) Hypogeal
c) Hypoepigeal d) Epihypogeal **Ans : d**

98. Which one of the following is not correctly matched

Crop		*Seed rate*
a) Wheat	-	100
b) Moong	-	50
c) Maize	-	25
d) Mustard	-	10

Ans : b

99. Greengram is originated in

a) India b) Tropical America
c) China d) Indonesia **Ans : a**

100. Which one *of* the following pulse crops is used as a pulse, a fodder and a green manure crop

a) Groundnut b) Moong
c) Soybean d) Pea **Ans : b**

101. Greengram needs a..............climate

a) Hot and humid b) Cool and humid
c) Hot and dry d) None **Ans : c**

102. It is recommended to apply NPK @................kg/ha to moong

a) 100:60:40 b) 20:40:40
c) 60:40:40 d) None **Ans : b**

103. It is recommended to apply................ @ 1kg a.i. /ha to moong

a) Stomp b) Basalin
c) Isoproturon d) None **Ans : b**

Lentil

104. The centre of origin of lentil is

a) Mediterranean region b) America
c) S. Africa d) None **Ans : a**

105. The inflorescence in lentil is

a) Receme b) Panicle
c) Spikelets d) Ear **Ans : a**

106. Optimum temperatures for growth of lentil ranges between

a) 21-37°C b) 18-20°C
c) 20-25°C d) None **Ans : b**

107. The recommended seed rate for lentil is _______ kg/ha
a) 8-10 b) 75-100
c) 15-18 d) 30-40 Ans : d

108. The leading producer of lentil is
a) Burma b) India
c) Bangladesh d) Ceylon Ans : b

109. The country having maximum acreage of lentil is
a) Burma b) India
c) Bangladesh d) Ceylon Ans : b

110. Whole lentil pulse is commonly known as
a) Masoor b) Matar
c) Channa d) Malka masoor Ans : d

111. The recommended seed rate for late sown lentil is _______ kg/ha
a) 30-40 b) 50-60
c) 15-18 d) 20-25 Ans : b

112. The largest producer of Lentil in the world is _________
a) India b) China
c) Nepal d) Pakistan Ans : a

113. Wherever one irrigation is available, the lentil crop should irrigated at
a) Grain filling b) Tillering stage
c) Pre flowering stage d) Milking stage Ans : c

114. Lentil crop needs
a) Warm and humid climate b) Hot and humid climate
c) Dry and hot conditions d) Cool and dry climate Ans : d

115. The recommended dose of N, P_2O_5 and K_2O kg/ha for lentil are
a) 60 – 40 – 40 b) 120 – 40 – 40
c) 25 – 50-60 – 30-40 d) 40 – 30 – 40 Ans : c

116. The row to row spacing by pora method of sowing of lentil is……cm
a) 50 b) 30
c) 20 d) 10 Ans : b

117. The row to row spacing for late sowing of lentil is……cm
a) 50 b) 30
c) 20 d) 10 Ans : c

118. The insect pod borer is commonly found on

a) Maize b) Wheat
c) Groundnut d) Lentil **Ans : d**

119. Grain of lentil is

a) White colour b) Yellow colour
c) Brownish colour d) All these colours **Ans : c**

120. Botanical name of brown lentil is

a) *Cicer arietinum* b) *Cicer kabulium*
c) *Lens culinaris* d) None **Ans : c**

121. Weeds in lentil can be controlled by applying

a) Fluchloralin b) 2,4-D
c) Both d) None **Ans : a**

122. Lentil contains about.......% protein

a) 10 b) 15
c) 20 d) 25 **Ans : d**

123. Leading lentil producing state in the country

a) Uttar Pradesh b) Uttar Pradesh
c) Bihar d) Rajasthan **Ans : b**

124. Number of chromosomes in lentil

a) 2n=14 b) 2n=24
c) 2n=34 d) 2n=44 **Ans : a**

125. Bold seeded lentil belongs to sub species

a) *Culinaris* b) *Micro sperma*
c) *Macrosperma* d) None **Ans : c**

126. Small seeded lentil belongs to sub species

a) *Culinaris* b) *Micro sperma*
c) *Macrosperma* d) None **Ans : b**

127. Lentil belongs to family

a) *Liliaceae* b) *Linaceae*
c) *Tiliaceae* d) *Leguminoceae* Ans : d

Pea

128. The centre of origin of Pea is

a) Mediterranean b) America
c) S. Africa d) None **Ans : a**

129. The inflorescence in Pea is

a) Axilary receme b) Panicle
c) Spikelets d) Ear **Ans : a**

130. Maximum area under Pea cultivation in India is in

a) M.P. b) U.P.
c) Maharastra d) Bihar Ans : b

131. Optimum temperatures towards flowering and fruiting of Pea ranges between

a) 21-37°C b) 25-30°C
c) 20-25°C d) None **Ans : b**

132. The recommended seed rate for Pea is _______ kg/ha

a) 60-80 b) 75-100
c) 15-18 d) 25-25 **Ans : a**

133. Higher yield of Pea could be achieved by:

a) Use of higher dose of phosphate
b) Adequate amount of N
c) No nitrogen application
d) Closer planting **Ans : a**

134. Pea is commonly known as

a) Arhar b) Channa
c) Matar d) None **Ans : c**

135. What is the ideal temperature for germination of Pea

a) 22°C to 25°C b) 15°C to 20.C
c) 10°C to 15°C d) 25.C to 30.C **Ans : a**

136. The optimum plant population for early Pea is around

a) 2 lacs b) 3 lacs
c) 4 lacs d) 5 lacs **Ans : d**

137. Wherever one irrigation is available, the Pea crop should irrigated at

a) Pod formation stage b) Branching stage
c) Pre flowering stage d) Grain filling stage **Ans : c**

138. Pea crop needs

a) Warm and humid climate b) Hot and humid climate
c) Dry and hot conditions d) Cold and dry climate **Ans : d**

139. The recommended dose of N, P_2O_5 and K_2O kg/ha for garden Pea are

a) 60 – 40 – 40 b) 120 – 40 – 40
c) 20 – 60 – 40 d) 40 – 30 – 40 **Ans : c**

140. The insect pod borer is commonly found on
a) Maize b) Wheat
c) Groundnut d) Pea Ans : d

141. Grain of field Pea is
a) White colour
b) Yellow colour
c) Brownish red colour, black colour
d) All these colours Ans: b

142. How much seed of Pea should be treated with one packet of *Rhizobium* culture
a) 5 kg b) 10 kg
c) 15 kg d) 20 kg Ans : b

143. Botanical name of garden Pea is
a) *Pisum sativum var hortens* b) *Pisum sativum var arvense*
c) *Pisum sativum var gardens* d) None Ans : a

144. Botanical name of field Pea is
a) *Pisum sativum var hortens* b) *Pisum sativum var arvense*
c) *Pisum sativum var gardens* d) None Ans : b

145. Weeds in Pea can be controlled by applying
a) Fluchloralin b) Tribunil
c) 2,4-D d) Both a & b Ans : d

146. Arkel and Asauji are varieties of
a) *Pisum sativum var hortens* b) *Pisum sativum var arvense*
c) *Pisum sativum var gardens* d) None Ans : a

147. Rachna and Harbhajan are varieties of
a) *Pisum sativum var hortens* b) *Pisum sativum var arvense*
c) *Pisum sativum var gardens* d) None Ans : b

148. Pea should be treated with rhizobium inoculation of
a) R. Japonicum b) R. glycine
c) R. leguminosarum d) None Ans : c

149. Wilt disease of pea is caused by
a) *Fusarium oxysporum* b) *Rhizoctonia solani*
c) *Erysiphe polygoni* d) Uromyces fabae Ans : a

150. Root rot disease of pea is caused by
a) *Fusarium oxysporum* b) *Rhizoctonia solani*
c) *Erysiphe polygoni* d) Uromyces fabae Ans : b

151. Powdery mildew disease of pea is caused by

a) *Fusarium oxysporum* b) *Rhizoctonia solani*
c) *Erysiphe polygoni* d) Uromyces fabae **Ans : c**

152. Rust disease of pea is caused by

a) *Fusarium oxysporum* b) *Rhizoctonia solani*
c) *Erysiphe polygoni* d) Uromyces fabae **Ans : d**

153. Pea leaf minor is

a) *Phytomyza atricornis* b) *Etiella zinckenella*
c) *Erysiphe polygoni* d) Uromyces fabae **Ans : a**

154. Pea pod borer is

a) *Phytomyza atricornis* b) *Etiella zinckenella*
c) *Erysiphe polygoni* d) Uromyces fabae **Ans : b**

155. Pea belongs to family

a) *Liliaceae* b) *Linaceae*
c) *Tiliaceae* d) Leguminoceae **Ans : d**

156. Garden pea is cultivated for its

a) Beautiful/colourful flowers b) Green pods used as vegetable
c) Pasturage for hogs d) Dry seed used as pulse – grain **Ans : b**

157. Sweet pea is cultivated exclusively, for

a) Its delicious green pods picked up for vegetable and table purposes.
b) Its dry seeds for pulse grain
c) Ornamental purposes alone excellent pasturage for hogs
d) None **Ans : b**

158. Field pea is cultivated widely in plains of Uttar Pradesh, Bihar, Orissa etc. for

a) Green manuring the poor fields
b) Obtaining food grains to be .used as pulse
c) Getting peas for canning' and freezing purposes
d) Ornamental flowers and decoration purposes in kitchen gardens. **Ans : b**

Pigeonpea

159. The centre of origin of pigeonpea is

a) India and Africa b) America
c) S. Asia d) None **Ans : a**

160. Pigeonpea is commonly known as

a) Arhar b) Channa dal
c) Moong d) Masoor **Ans : a**

161. In India the highest acerage of pigeonpea is in the state

a) Punjab b) Haryana
c) Maharastra d) Mizoram **Ans : c**

162. In India the productivity of pigeonpea is highest in the state

a) Punjab b) Haryana
c) Maharastra d) Mizoram Ans : d

163. In India .... state contributes maximum production of pigeonpea

a) U.P b) Haryana
c) Maharastra d) Mizoram **Ans : a**

164. The inflorescence in pigeonpea is

a) Axillary receme b) Panicle
c) Spikelets d) Ear **Ans : a**

165. Pigeonpea is commonly known as

a) Redgram b) Tur
c) Arhar d) All **Ans : d**

166. Area under pigeonpea cultivation in India is approximately

a) 3.7 m ha b) 4.5m ha
c) 5.0m ha d) 2.9m ha **Ans : a**

167. Botanical name of pigeon pea is

a) *Cajanus radiata* b) *Cajanus mungo*
c) *C.Cajan* d) None **Ans : c**

168. In India pigeonpea ranks ....among pulses crops

a) First b) Second
c) Third d) Fourth **Ans : b**

169. In India pigeonpea is the second most important pulse crop after

a) Gram b) Greengram
c) Blackgram d) Lentil **Ans : a**

170. .... group of pigeon pea is commonly known as arhar

a) *C.indicus var flavus* b) *C.indicus var bicolor*
c) *C.indicus var odorata* d) Both a & b **Ans : b**

171. Pigeonpea has been divided into.........groups

a) 1 b) 2
c) 3 d) 4 **Ans : b**

172. The flowering time of arhar is

a) September-November b) December-February
c) February-April d) May-June **Ans : b**

173. .... is/are groups of pigeonpea

a) *C.indicus var bicolor* b) *C.indicus var flavus*
c) *C.indicus var odorata* d) Both a & b **Ans : d**

174. .... group of pigeonpea is having perennial types, late maturing, talll and bushy habit

a) *C.indicus var flavus* b) *C.indicus var bicolor*
c) *C. indicus var odorata* d) Both a & b **Ans : b**

175. .... group of pigeonpea is having annual types, early maturing, and smaller plants

a) *C.indicus var flavus* b) *C.indicus var bicolor*
c) *C. indicus var odorata* d) Both a & b **Ans : a**

176. .... group of pigeonpea is commonly known as tur

a) *C.indicus var flavus* b) *C.indicus var bicolor*
c) *C. indicus var odorata* d) Both a and b **Ans : a**

177. The flowering time of tur is

a) September-November b) December-February
c) February-April d) May-June **Ans : a**

178. ....needs picking for harvesting

a) Tur b) Gram
c) Both d) None **Ans : a**

179. Branching in pigeonpea begins from...........nodes

a) 3-5 b) 6-10
c) 10-15 d) None **Ans : b**

180. Pigeonpea prefers ....climate

a) Cool and dry
b) Warm tropical and sub tropical climate
c) Temperate humid
d) None **Ans : b**

181. The recommended seed rate for pigeonpea is _______ kg/ha

a) 8-10 b) 12-15
c) 15-18 d) 20-25 **Ans : b**

182. The pigeonpea seed contains...........% protein

a) 18 b) 28
c) 38 d) 48 **Ans : b**

183. The most commonly practiced method of growing pigeonpea is

a) Direct seeding in rows b) Broadcasting
c) Transplanting d) Sowing pre-sprouted seeds **Ans : a**

184. Test weight of pigeonpea is

a) 200-250 gram b) 250-300 gram
c) 300-350 gram d) 100-150 gram **Ans : d**

185. Required plant population of redgram may be obtained by using a seed rate of (kg/ha)

a) 50-75 b) 75-100
c) 90-110 d) 12-15 **Ans : d**

186. Which one of the following varieties of pigeonpea is an extra early maturing

a) Pusa Ageti b) Type I
c) Mukta d) Prabhat **Ans : d**

187. The optimum seed rate for pigeonpea crop is.......... kg/ha

a) 10-12 b) 12-15
c) 15-17 d) 20-25

188. The chromosome number of pigeon pea is 2n=

a) 14 b) 18
c) 22 d) 26 **Ans : c**

189. The optimum row spacing for pigeonpea crop is.......... cm

a) 20-30 b) 30-45
c) 45-60 d) 60-75 **Ans : d**

190. The optimum plant to plant spacing for pigeonpea crop is.......... cm

a) 20 b) 30
c) 45 d) 60 **Ans : a**

191. The optimum plant population for pigeonpea crop is..........

a) 20000-30000 b) 30000-40000
c) 45000-60000 d) 60000-80000 **Ans : d**

192. Medium duration varieties of arhar which mature in 150 days are

a) T-21 b) Pusa Ageti (S-5)
c) BS-5 d) All these **Ans : a**

193. Which of these following arhar varieties are medium varieties maturing in 180-200 days

a) Mukta (R-60), Sharda (S-8), BR-183, Dholi-1234, BR-65, Khargone-2
b) Prabhat, UPAS-120, Pant A-1, Pant A-3
c) T-21, S-5, BS-1
d) NP (WR 15), T-7, T-17, Hyd. 3A, Hyd. 3C Ans. d

194. Which of the following arhar varieties are late varieties maturing in more than 200 days

a) Mukta (R-60), Sharda (S-8), BR-183, dholi-1234
b) NP (WR 25), T-7, T-17
c) Prabhat, UPAS-120, Pant A-1, Pant A-3
d) T-21, S-5, BS-1 **Ans : b**

195. ....soils are best suited for pigeon pea cultivation

a) Well drained alluvial and sandy loam
b) Black cotton soil
c) Sandy soil
d) All the above **Ans : a**

196. ....soils are unfit for pigeon pea cultivation

a) Saline b) Alkaline
c) Waterlogged d) All the above **Ans : d**

197. Pre emergence application of...............@ 3 l/ha for weed control in pigeonpea

a) Basalin b) Lasso
c) Both d) None **Ans : b**

198. Extra early varieties of arhar which muture in 120 days are

a) Prabhat, UPAS-120 b) Pant A-2
c) Pant A-1 d) All the above **Ans. d**

❑❑❑

14 Chapter

Fiber Crop Production

Multiple Choice Questions (MCQ's)

Cotton

1. The centre of origin of cotton is

a) India	b) America
c) S. Africa	d) None

Ans : a

2. In India the productivity of cotton is highest in the state

a) Punjab	b) Haryana
c) W. Bengal	d) None

Ans : a

3. Maximum area under cotton cultivation in India is in

a) Punjab	b) Haryana
c) W. Bengal	d) Maharastra

Ans : d

4. In India the production of cotton is highest in the state

a) Punjab	b) Haryana
c) Gujarat	d) Maharastra

Ans : c

5. The inflorescence in cotton is

a) Axillary	b) Panicle
c) Spikelets	d) Ear

Ans : a

6. Area under cotton cultivation in India is approximately

a) 19 m ha	b) 8m ha
c) 10m ha	d) 29m ha

Ans : b

7. Area under cotton cultivation all over the world is approximately
   a) 19 m ha b) 28m ha
   c) 33 m ha d) 39m ha **Ans : c**

8. India ranks ....in area under cotton cultivation
   a) First b) Second
   c) Third d) Fourth **Ans : a**

9. India ranks ....in production of cotton
   a) First b) Second
   c) Third d) Fourth **Ans : d**

10. Minimum temperature towards germination of cotton is
   a) 18°C b) 20°C
   c) 16°C d) 25 **Ans : c**

11. Minimum temperature towards growth of cotton ranges between
   a) 21-27°C b) 25-30°C
   c) 30-35°C d) None **Ans : a**

12. The recommended seed rate for deshi cotton is _______ kg/ha
   a) 8-10 b) 15-25
   c) 10-18 d) 30-35 **Ans : c**

13. The recommended seed rate for American cotton is _______ kg/ha
   a) 8-10 b) 15-25
   c) 10-18 d) 30-35 **Ans : b**

14. The chromosome number in deshi cotton is _______
   a) n=13 b) n=26
   c) n=52 d) n=20 **Ans : a**

15. The chromosome number in American cotton is _______
   a) n=13 b) n=26
   c) n=52 d) n=20 **Ans : b**

16. Cotton belongs to family
   a) *Solonaceae* b) *Chenopodiaceae*
   c) *Malvaceae* d) *Gossipium* **Ans : c**

17. Genus gossipium includes............species
   a) 20 b) 15
   c) 10 d) 30 **Ans : a**

18. First interspecific hybrid cotton was released by:
    a) Tamil Nadu b) Gujarat
    c) Punjab d) Karnataka **Ans : b**

19. Reproductive structures in succession in cotton:
    a) Flowers-bolls-squares b) Squares-bolls-flowers
    c) Square-flowers-bolls d) Square-bolls-squares **Ans : c**

20. Percentage of lint in seed cotton is generally
    a) 80% b) 60%
    c) 30% d) 10% **Ans : c**

21. Sea Island cotton or Egyptian cotton is
    a) Gossypium arboreum b) Gossypium hirsutum
    c) Gossypium barbadense d) Gossypium herbaceum **Ans : c**

22. Number of species included under genus gossipium
    a) 23 b) 21
    c) 22 d) 20 **Ans : d**

23. Cotton varieties cultivated in India belongs to species
    a) Arboreum b) Herbaceum
    c) Hirsutum d) Barbadanse
    e) All **Ans : e**

24. The seed rate of cotton would be _______kg/ha, if the planting geometry is 75 x 50 cm, and test weight, germination and purity percentage are 100, 90, 90, respectively.
    a) 3.29 b) 6.58
    c) 4.29 d) 5.58 **Ans : a**

25. "Boll shedding" in cotton is due to
    a) Plant is over loaded with bolls b) Inhibitory effect of ABA (Abscisic acid)
    c) Heavy dressing of urea d) Frequent irrigation **Ans : a**

26. Cotton belongs to the_________ family
    a) Compositeae b) Malvaceae
    c) Cruciferеae d) Lineaceae **Ans : b**

27. Delinting of cottonseeds help in ______
    a) Grading the seed
    b) Killing of hibernating boll worm larval and disease pathogens
    c) Quicker germination
    d) All of these **Ans : d**

28. Delinting of cotton seed may be done with ________
    a) Sulphuric acid b) Nitric acid
    c) Citric acid d) Hydrochloric acid **Ans : a**

29. The optimum population of hybrid cotton varieties is around ________ plants/ha

a) 10,000 b) 20,000
c) 30, 000 d) 50,000

Ans : d

30. Gujrat-67 and American nectariless are parents of cotton hybrid ________

a) Hybrid - 4 b) Varalaxmi
b) Hybrid - 6 d) Vikaram

Ans : a

31. Which of the following statements refers to the concept of critical stage of irrigation:

a) When irrigation is not given reduction of yield is maximum
b) When water absorption rate is maximum
c) When water loss by the plant is maximum
d) All of the above

Ans : a

32. Most commonly followed method of planting cotton in India is

a) Direct seeding in rows
b) Transplanting on raised seed beds
c) Direct seeding in puddled beds
d) Transplanting in standing water in puddled fields

Ans : a

33. The first inter-specific hybrid cotton developed' in India' in the year 1972 was

a) Godavari b) Savitri
c) Varalaxmi d) Hybrid-4

Ans : c

34. Parantage of first inter-specific hybrid cotton developed' in India

a) Female - hirsutum X male – barbadense
b) Female - hirsutum X male – arboreum
c) Female - barbadense X male – hirsutum
d) None

Ans : a

35. Sujatha, the first commercial extra long G. barbadense cotton was released in the year

a) 1969 b) 1979
c) 1989 d) 1999

Ans : a

36. The first commercial Hybrid cotton H-4 was released in the year

a) 1970 b) 1980
c) 1969 d) 1999

Ans : a

37. The first commercial Hybrid cotton H-4 was developed by
a) C.T. Patel b) H. Katarki
c) J.C. Mehta d) None
Ans : a

38. The first interspecific hybrid cotton Varalakshmi was developed by
a) C.T. Patel b) H. Katarki
c) J.C. Mehta d) None
Ans : b

39. DCH-32, an extra long staple hybrid cotton was released in the year
a) 1969 b) 1971
c) 1981 d) 1991
Ans : c

40. DCH-32, an extra long staple hybrid cotton was developed through
a) Heterosis breeding b) Interspecific hybridization
c) Mutation d) Selection
Ans : a

41. The first hybrid cotton in the world was developed at
a) Surat b) Varanasi
c) Ahemdabad d) Coiambtore
Ans : a

42. The first hybrid cotton in the world developed to exploit hybrid vigour, which is known as intraspecific hybrids is
a) Godavari b) Savitri
c) Varalaxmi d) Hybrid-4
Ans : c

43. In case of cotton the flower are formed on
a) Monopodial branches b) Sympodial branches
c) All the branches d) None
Ans : b

44. In case of cotton the flower are formed on
a) Lower branches b) Upper branches
c) All the branches d) None
Ans : b

45. Which of the following fiber crop has the highest cultivated area in the world
a) Cotton b) Jute
c) Sunheamp d) None
Ans: a

46. Cotton is originated in
a) SW Asia b) Europe
c) South America d) India
Ans: d

47. India has the largest area in cotton cultivation in the world, which is about

a) 26 m ha b) 36 m ha
c) 8 m ha d) 50 m ha
Ans: c

48. Productivity of cotton is highest in

a) Haryana b) Punjab
c) West Bengal d) Uttar Pradesh
Ans: b

49. The optimum pH of soil for cotton cultivation is varies in range

a) 4 – 6 b) 5.5 – 8.5
c) 6 – 7 d) 7 – 9
Ans: b

50. The recommended dose of N, P_2O_5 and K_2O kg/ha for cotton are

a) 80 – 40 – 40 b) 120 – 40 – 80
c) 30 – 20 – 30 d) 40 – 30 – 40
Ans: a

51. The recommended dose of N, P_2O_5 and K_2O kg/ha for hybrid cotton are

a) 120-150 – 60-75 –40 b) 120 – 40 – 40
c) 80-100 –40- 60 – 40 d) 40 – 30 –40
Ans: a

52. Consider the following pairs of crops and varieties and select the pair, which is correctly matched

a) Barley: Kalyan sona b) Cotton: Sujata
c) Cowpea: Pusa Phalguni d) Green gram: Amber
Ans: b

53. Delinting of cotton seed may be done with

a) Sulphuric acid b) Citric acid
c) Nitric acid d) Hydrochloric acid
Ans: a

54. Optimum plant population ha^{-1} for cotton has estimated as

a) 25,000 – 50,000 b) 30,000 – 60,000
c) 50,000 – 80,000 d) 80,000 - 1,00,000
Ans: c

55. The insect bollworm is commonly found on

a) Maize b) Wheat
c) Cotton d) Castor
Ans: c

56. The recently developed terminator technology has been used in

a) Soybean b) Cotton
c) Tobacco d) Wheat
Ans: b

57. Which one of the following crops has the highest consumption of pesticide

a) Paddy b) Cotton
c) Oilseeds d) Vegetables **Ans: b**

58. Cotton is originated in

a) India b) Tropical America
c) China d) Indonesia **Ans: a**

59. The oil content in the cotton seed ranges from

a) 10-15% b) 15-25%
c) 25-35% d) 40-42% **Ans: b**

60. Which of the following cultivated species of cotton grown in India covers half of the cotton area of the country

a) Gossypium hirsutum b) Gossypium barbadense
c) Gossypium arboreum d) Gossypium herbaceum **Ans : a**

61. Cotton is cultivated in India on

a) 4.0 million hectares b) 2.9 million hectares
c) 8.0 million hectares d) 7.0 million hectares **Ans : c**

62. Sea Island cotton belong to

a) *Gossypium arboreum* b) *Gossypium herbaceum*
c) *Gossypium hirsutum* d) *Gossypium barbadense* **Ans: c**

63. Which proportion of cotton area in India is rainfed

a) 75% b) 25%
c) 40% d) 80% **Ans : a**

64. In India cotton is cultivated on

a) Alluvial soils b) Black cotton soils and lateritic soils
c) Red sandy loam to loam soils d) All these soil types **Ans : d**

65. In case of rainfed cotton cultivated in India, the minimum rainfall requirement is

a) 50 cm b) 100 cm
c) 70-120 cm d) 35-45 cm **Ans : a**

66. Hampi, Khandwa-2, Lakshami, mahalakshami, Lohit and Reba –B50 are the recommended improved varieties of

a) Sunhemp b) Arhar
c) Coconut d) Cotton **Ans : d**

67. Varalaxmi is a inter specific variety of cotton evolved at

a) Udaipur b) Nagpur
c) Dharwad d) Surat **Ans: b**

68. The flower bud in cotton is known as
a) Boll b) Square
c) Capsule d) None
Ans: b

69. The flower opens in ....days after emergence of square in cotton
a) 10-12 b) 12-15
c) 18-24 d) 25-30
Ans: c

70. The fiber in cotton is elongation of .... of seed coat
a) Epidermal cell b) Endodermal cell
c) Capsule d) Cuticle
Ans: a

71. The long fiber of cotton is known as
a) Lint b) Fuzz
c) Thread d) None
Ans: a

72. The short fiber of cotton is known as
a) Lint b) Fuzz
c) Thread d) None
Ans: b

73. ....is a variety of arboreum cotton
a) Shamli b) Hampi
c) Digvijaya d) Sujata
Ans: a

74. ....is a variety of barbadense cotton
a) Shamli b) Hampi
c) Digvijaya d) Sujata
Ans: d

75. ....is a variety of herbaceum cotton
a) Shamli b) Hampi
c) Digvijaya d) Sujata
Ans: c

76. ....is a variety of hirsutum cotton
a) Shamli b) Hampi
c) Digvijaya d) Sujata
Ans: b

77. Planting geometry for irrigated cotton crop is
a) 75x30cm b) 50x30cm
c) 30x30cm d) 50x50cm
Ans: a

78. Planting geometry for rainfed cotton crop is
a) 75x30cm b) 50x30cm
c) 30x30cm d) 50x50cm
Ans: b

79. Critical stages for irrigation to cotton crop is
a) Flowering b) Boll formation
c) Both d) None
Ans: c

80. The herbicide suitable for weed control in cotton crop is
a) Diuron b) Trifluralin
c) Fluchloralin d) All **Ans: d**

81. Topping in cotton crop should be done at..........day stage
a) 80-90 b) 40-50
c) 30-40 d) 50-60 **Ans: a**

82. Topping in cotton is done to
a) Enhance flowering
b) Enhance sympodial branching
c) Enhance monopodial branching
d) Enhance terminal growth **Ans: b**

83. Topping in cotton is done at ....m height
a) 1-1.2 b) 0.5-1.0
c) 1.2-1.5 d) 2 **Ans: a**

84. The fruit in cotton is known as
a) Boll b) Square
c) Capsule d) None **Ans: a**

85. Bad opening of bolls in cotton is known as
a) Mareck b) Tirak
c) Dead boll d) None **Ans: b**

86. ....is favorable for tirak
a) Hot and dry winds at fruiting b) Rains at fruiting
c) Cold winds at fruiting d) Rains at flowering **Ans: a**

87. Tirak is associated with
a) Light sandy soils with N deficiency
b) Sandy loam soils with salinity
c) Both
d) None **Ans: c**

88. Tirak is ....
a) Fungal disease b) Physiological disorder
c) Bacterial disease d) None **Ans: b**

89. The oil content in the cottonseed ranges from
a) 10-15% b) 15-25%
c) 25-35% d) 40-42% **Ans: b**

90. Sea Island cotton belong to
a) *Gossypium arboreum* b) *Gossypium herbaceum*
c) *Gossypium hirsutum* d) *Gossypium barbadense* **Ans: c**

91. Varalaxmi is a inter specific variety of cotton evolved at

a) Udaipur b) Nagpur
c) Dharwad d) Surat **Ans: b**

92. Origin of *Corchorus capsularis* (White jute) is believed to be in the

a) Africa b) Indo-Burma region
c) China d) USA **Ans: a**

93. Weight of one cotton bale is equal to

a) 160 kg b) 170 kg
c) 180 kg d) 178 kg **Ans: b**

94. The seed rate of cotton would be _______kg/ha, if the planting geometry is 75 x 50 cm, and test weight, germination and purity percentage are 100, 90, 90, respectively.

a) 3.29 b) 6.58
c) 4.29 d) 5.58 **Ans : a**

95. "Boll shedding" in cotton is due to

a) Plant is over loaded with bolls b) Inhibitory effect of ABA (Abscisic acid)
c) Heavy dressing of urea d) Frequent irrigation **Ans : b**

96. Cotton belongs to the_________ family

a) *Compositeae* b) *Malvaceae*
c) *Crucifereae* d) *Lineaceae* **Ans : b**

97. Delinting of cotton seeds help in ______

a) Grading the seed
b) Killing of hibernating boll worm larval and disease pathogens
c) Quicker germination
d) All of these **Ans : d**

98. Delinting of cotton seed may be done with ________

a) Sulphuric acid b) Nitric acid
c) Citric acid d) Hydrochloric acid **Ans : a**

99. The optimum population of hybrid cotton varieties is around _________ plants/ha

a) 10,000 b) 20,000
c) 30, 000 d) 40,000 **Ans : a**

100. Gujrat-67 and American nectariless are parents of cotton hybrid ____________

a) Hybrid - 4 b) Varalaxmi
c) Hybrid - 6 d) Vikaram **Ans : a**

101. Postponing of first irrigation to 40-45 days after sowing is always preferable for cotton crop to
a) Promote sympodial branching
b) Prevent excessive vegetative growth
c) Promote early flowering
d) Promote monopodial branches
Ans: c

102. Heavy shedding of buds and bolls occurs in cotton due to
a) Deficiency of nitrogen in the soil
b) Deficiency of phosphorus in the soil
c) Deficiency of magnesium in the soil
d) Water stress at bud formation stage
e) Competitive relationship
Ans: b

Jute

103. The scientific name of jute is
a) *Corchorus sliqua* b) *Corchorus capsularis*
c) *Corchorus hirsutum* d) All
Ans : b

104. The centre of origin of Corchorus capsularis is
a) India and Burma b) America
c) Africa d) None
Ans : a

105. The centre of origin of Corchorus olitorius is
a) India and Burma b) America
c) Africa d) None
Ans : c

106. In India the productivity of jute is highest in the state
a) Assam b) Haryana
c) W. Bengal d) None
Ans : a

107. In India the maximum area and production of jute is in the state
a) Assam b) Haryana
c) W. Bengal d) None
Ans : c

108. The inflorescence in jute is
a) Siliqua b) Panicle
c) Cymose d) Ear
Ans : c

109. Jute belongs to family
a) *Tiliaceae* b) *Liliaceae*
c) *Malvaceae* d) None
Ans : a

110. Area under jute cultivation in India is approximately
a) 0.8 m ha b) 4.5m ha
c) 8m ha d) 2.9m ha
Ans : a

111. Optimum temperatures towards for cultivation of jute is
a) 37°C b) 30°C
c) 34°C d) None
Ans : c

112. Nearly 70% of area under jute cultivation is occupied by
a) Corchorus olitorius b) Corchorus capsularis
c) Corchorus hirsutum d) All
Ans : b

113. White jute is
a) Corchorus olitorius b) Corchorus capsularis
c) Corchorus hirsutum d) All
Ans : b

114. Tossa jute is
a) Corchorus olitorius b) Corchorus capsularis
c) Corchorus hirsutum d) All
Ans : a

115. .... jute is raised on well drained uplands
a) Tossa b) White
c) Both d) None
Ans : a

116. .... jute is raised on uplands and low lands and can withstand water logging
a) Tossa b) White
c) Both d) None

117. ....is an early variety of white jute suitable for double cropping
a) Sonali b) Sabuj sona
c) Naveen d) None
Ans : a

118. ....is a variety of tossa jute
a) Sonali b) Sabuj sona
c) Naveen d) None
Ans : c

119. The recommended seed rate for C. capsularis is _______ kg/ha
a) 8-10 b) 4-6
c) 15-18 d) 20-25
Ans : a

120. The recommended seed rate for C. olitorius is _______ kg/ha
a) 8-10 b) 4-6
c) 15-18 d) 20-25
Ans : b

121. The suitable temperatures for jute ranges between
a) 21-30°C b) 24-37°C
c) 20-25°C d) None
Ans : b

122. The usual plant to plant spacing recommended for jute is ________

a) 8-10 cm
b) 15-20 cm
c) 20-25 cm
d) 25-30 cm

Ans : a

123. The recommended planting geometry for white jute is _______ cm

a) 20 x 10
b) 30 x 10
c) 45 x 15
d) 50 x 20

Ans : b

124. The recommended planting geometry for tossa jute is _______ cm

a) 20 x 10
b) 30 x 10
c) 45 x 15
d) 50 x 20

Ans : a

125. Jute belongs to the family _______

a) Malvaceae
b) Compositeae
c) Lineaceae
d) Tiliaceae

Ans : d

126. The best observed pH for jute cultivation is _______

a) 5.0.5.5
b) 6.0-7.5
c) 7.5 - 8.5
d) 8.5-9.5

Ans : b

127. The optimum time of sowing for capsularis Jute is _________

a) March - April
b) February - March
c) April - May
d) May – June

Ans : a

128. The optimum time of sowing for olitorius Jute is _________

a) March - April
b) February - March
c) April - May
d) May – June

Ans : c

129. Tossa Jute is to be sown at inter row spacing of _________

a) 30
b) 25
c) 10
d) 15

Ans : c

130. For fibre purposes, Jute, may be harvested at any time.

a) After pod formation
b) After flowering
c) Before pod formation
d) Before flowering

Ans : d

131. Ratting is a process by which _________

a) The fibers in the bark get loosened and separated from woody stalk
b) The poor fibre stalk is sorted out
c) The good quality fibre in separated
d) Fibers are classified into various retted groups

Ans : a

132. The observed best temperature for ratting is _________

a) 18^{0}C
b) 25^{0}C
c) 34^{0}C
d) 40^{0}C

Ans : c

133. Which of the following is a parenchymatus fibre
a) Cotton b) Jute
c) Silk d) All of the above **Ans : b**

134. The recommended dose of N, P_2O_5 and K_2O kg/ha for olitorius jute are
a) 60 - 40 - 40 b) 120 - 40 - 40
c) 30 - 20 - 30 d) 40 - 30 - 20 **Ans: a**

135. The recommended dose of N, P_2O_5 and K_2O kg/ha for capsularis jute are
a) 100 - 50 -40 b) 120 - 40 - 40
c) 80 - 40 - 20 d) 40 - 30 -40 **Ans : a**

136. Which of the following soil types are unsuitable for cultivation of jute
a) Deep alluvial soils receiving silt from annual floods
b) Sandy loam soils
c) Clay loam soils
d) Sandy soils, heavy clay soils and soil with low pH **Ans : d**

137. Which of the following crop rotations are not practiced with jute crop under irrigated conditions
a) Jute-paddy-mustard, jute-paddy-pulses, jute-paddy
b) Jute-paddy-potato, jute-paddy-wheat
c) Jute-moong-paddy-potato
d) Cowpea, jute-potato, jute-paddy-barseem **Ans : a**

138. Which of the following crop rotations are practiced with jute crop under rainfed conditions
a) Jute-paddy-potato, jute-paddy-wheat,
b) Jute- wheat
c) Jute-moong-paddy-potato
d) Cowpea-jute-potato, jute-paddy-barseem **Ans : b**

139. Retting of jute is a
a) Chemical process b) Physical process
c) Mechanical process d) Microbiological process **Ans : d**

140. JRC-212, JRC-7447 are varieties of
a) Corchorus olitorius b) Corchorus capsularis
c) Corchorus hirsutum d) All **Ans : b**

141. JRO-7835, JRO-620 are varieties of
a) Corchorus olitorius b) Corchorus capsularis
c) Corchorus hirsutum d) All **Ans : a**

142. The optimum period of sowing for *Capsularis* types (white jute) is

a) March-April
b) April-May
c) July-August
d) November-December

Ans: a

143. The ideal stage of harvest of jute for fibre purpose is before

a) One month old
b) Flowering
c) Small pod stage
d) Mature pod stage

Ans: b

144. Jute belongs to the family ______

a) Malvaceae
b) Compositeae
c) Lineaceae
d) Tiliaceae

Ans : d

145. The best observed pH for jute cultivation is nm _______

a) 5.0.5.5
b) 6.0-7.5
c) 7.5 - 8.5
d) 8.5-9.5

Ans : b

146. The optimum time of sowing for capsularis Jute is _________

a) March - April
b) February - March
c) April - May
d) May – June

Ans : a

147. Tossa Jute is to be sown at inter row spacing of _________

a) 30
b) 25
c) 10
d) 15

Ans : c

148. For fibre purposes, Jute, may be harvested at any time.

a) After pod formation
b) After flowering
c) Before pod formation
d) Before flowering

Ans : d

149. Ratting is a process by which _________

a) The fibers in the bark get loosened and separated from woody stalk
b) The poor fibre stalk is sorted out
c) The good quality fibre in separated
d) Fibers are classified into various retted groups

Ans : a

150. The observed best temperature for ratting is __________

a) 18^0C
b) 25^0C
c) 34^0C
d) 40^0C

Ans : d

Sunnhemp

151. The scientific name of sunnhemp is

a) *Corchorus sliqua*
b) *Corchorus capsularis*
c) *Corchorus hirsutum*
d) *Crotolaria juncea*

Ans : d

152. The centre of origin of sunnhemp is

a) India and Burma
b) America
c) S. Africa
d) None

Ans : a

153. In India the productivity of sunnhemp is highest in the state
a) Assam b) Orissa
c) W. Bengal d) None **Ans : b**

154. In India the maximum area of sunnhemp is in the state
a) U.P b) Orissa
c) W. Bengal d) None **Ans : a**

155. In India the maximum production of sunnhemp is in the state
a) Assam b) Orissa
c) U.P d) M.P. **Ans : b**

156. The inflorescence in sunnhemp is
a) Receme b) Panicle
c) Cymose d) Ear **Ans : a**

157. Sunnhemp belongs to family
a) *Tiliaceae* b) *Liliaceae*
c) *Malvaceae* d) *Leguminoseae* **Ans : d**

158. ....is an early variety of sunnhemp moderately resistant to shoot borer
a) M-19 b) Sabuj sona
c) Naveen d) Bellary **Ans : a**

159. The recommended seed rate for line sowing of sunnhemp is _______ kg/ha
a) 8-10 b) 4-6
c) 50-60 d) 20-25 **Ans : d**

160. The recommended seed rate for sowing by broadcasting of sunnhemp is _______ kg/ha
a) 8-10 b) 4-6
c) 50-60 d) 20-25 **Ans : c**

161. The usual plant to plant spacing recommended for sunnhemp is _________
a) 8-10 cm b) 15-20 cm
c) 20-25 cm d) 25-30 cm **Ans : a**

162. The recommended planting geometry for sunnhemp is _______ cm
a) 20 x 10 b) 30 x 10
c) 45 x 15 d) 50 x 20 **Ans : b**

163. Sunnhemp belongs to the family _______
a) *Malvaceae* b) *Compositeae*
c) *Lineaceae* d) *Leguminoseae* **Ans : d**

164. The optimum time of sowing for seed crop of sunnhemp is __________

a) April b) March
c) May d) August Ans : d

165. Sunnhemp is to be sown at inter row spacing of __________

a) 30 b) 25
c) 10 d) 15 Ans : c

166. For fibre purposes, sunnhemp, may be harvested at

a) Pod formation b) After flowering
c) Before pod formation d) Before flowering Ans : a

167. Ratting is a process by which __________

a) The fibers in the bark get loosened and separated from woody stalk
b) The poor fibre stalk is sorted out
c) The good quality fibre in separated
d) Fibers are classified into various retted groups Ans : a

168. The observed best temperature for ratting of sunnheamp is __________

a) 18-20^{0}C b) 21-27^{0}C
c) 34-36^{0}C d) 36-40^{0}C Ans : b

169. The recommended dose of N, P_2O_5 and K_2O kg/ha for sunnhemp are

a) 100 – 50 –40 b) 120 – 40 – 40
c) 80 – 40 – 20 d) 20 – 50 –30 Ans: d

170. Which of the following soil types are unsuitable for cultivation of sunnhemp

a) Well drained alluvial soils with sandy loam to loamy texture
b) Sandy loam soils
c) Clay loam soils
d) Sandy soils, heavy clay soils and soil with low pH Ans : a

171. Retting of sunnhemp takes .... days in December

a) 3-7 b) 7-12
c) 12-15 d) 21 Ans: c

172. Retting of sunnhemp takes .... days in september

a) 3-7 b) 7-12
c) 12-15 d) 21 Ans: a

173. Retting of sunnhemp is a

a) Chemical process b) Physical process
c) Mechanical process d) Microbiological process Ans : d

174. Green manuring of sannhemp adds approximately _______ kg N/ha to the soils on decomposition.

a) 20-40 b) 40-60
c) 60-80 d) 100-120 **Ans : b**

175. Sunnhemp belongs to the family ________

a) *Leguminoseae* b) *Gramineae*
c) *Compositeae* d) *Tiliaceae* **Ans : a**

176. The items of sunnhemp is ________ in colour

a) Red b) Yellow
c) Blue d) None **Ans : b**

❑❑❑

15 Chapter

Tuber Crop Production

Multiple Choice Questions (MCQ's)

Potato

1. Potato is a ....plant
   a) Annual b) Biennial
   c) Perennial d) None **Ans : c**

2. Potato as a crop is treated as
   a) Annual b) Biennial
   c) Perennial d) None **Ans : a**

3. .... is a recommended variety of potato for hills
   a) Kufri alankar b) Kufri sinduri
   c) Kufri jyoti d) None **Ans : c**

4. .... is an extra early variety of potato
   a) Kufri alankar b) Kufri sinduri
   c) Kufri red d) None **Ans : a**

5. .... is a late variety of potato for hills
   a) Kufri shakti b) Kufri sinduri
   c) Kufri jyoti d) None **Ans : b**

6. .... is a resistant variety of potato against late blight and wart
   a) Kufri alankar b) Kufri sinduri
   c) Kufri jeevan d) None **Ans : c**

7. The most cultivated crop of the world is
a) Potato b) Wheat
c) Maize d) Jowar
Ans : a

8. The centre of origin of potato is
a) Peru and Bolivia b) Russia
c) S. Africa d) None
Ans : a

9. In India the productivity of potato is highest in the state
a) Punjab b) Tamilnadu
c) W. Bengal d) None
Ans : b

10. Maximum acerage under potato is in the state
a) Punjab b) Tamilnadu
c) W. Bengal d) U.P
Ans : d

11. Minimum acerage under potato is in the state
a) Punjab b) Tamilnadu
c) W. Bengal d) Delhi
Ans : d

12. The inflorescence in potato is
a) Siliqua b) Panicle
c) Spikelets d) Ear
Ans : b

13. Area under potato cultivation in India is approximately
a) 39 m ha b) 1.1m ha
c) 50m ha d) 29m ha
Ans : b

14. Number of cultivated species included under genus Solonum
a) 23 b) 21
c) 22 d) 7
Ans : d

15. Number of wild species included under genus Solonum
a) 123 b) 121
c) 154 d) 7
Ans : c

16. Potato varieties cultivated in India belongs to species
a) Andigenum b) Tuberosum
c) Lemmisum d) Indica
Ans : b

17. Optimum temperatures for growth of potato is
a) 18°C b) 24°C
c) 30°C d) 10
Ans : b

18. Optimum temperatures fortuberization of potato is
a) 21-37°C b) 25-30°C
c) 17- 20°C d) None
Ans : c

19. The weight of potato tuber (for sowing) most economical for higher yields is ____
   a) 10-25 grams b) 30-50 grams
   c) 50-80 grams d) 80 - 100 grams **Ans : b**

20. A potato tuber represents
   a) Enlarged underground stem
   b) Enlarged underground root
   c) Enlarged underground nodules
   d) None of these **Ans : a**

21. Potato tuber yield will be ______ q/ha, if crop was spaced at 50 x 20 c,; and average number of tubers/plant and average weight of tubers are 10 and 50 g, respectively.
   a) 250 b) 500
   c) 750 d) 1000 **Ans : b**

22. Potato belongs to the family ____
   a) Crucifereae b) Solanaceae
   c) Compositeae d) None of these **Ans : b**

23. The time of planting of main crop of potato in northern plains is
   a) First fortnight of October b) Second fortnight of October
   c) First fortnight of November d) Second fortnight of November **Ans : b**

24. The variety of potato which possesses the resistance to frost conditions is
   a) Kufri chandramukhi b) Kufri alankar
   c) Kufri sheetman d) Kufri chamatkar **Ans : c**

25. The recommended seed rate for potato in northern plains is ______ q/ha
   a) 10-15 b) 15-20
   c) 20-25 d) 25-30 **Ans : b**

26. Potato belong to the family
   a) Compositeae b) Gramineae
   c) Solanaceae d) Leguminoseae **Ans : c**

27. Which growth stage of potato is critical for irrigation
   a) Germination b) Stolon formation
   c) Tuber bulking d) All **Ans: d**

28. Most critical weed competition period is up to ....days after planting in potato
   a) 45-50 b) 25-30
   c) 30-40 d) None **Ans: a**

29. Most common herbicides used for weed control in potato is/ are

a) Simazine b) ToK E-25
c) Alachlor d) All of these Ans: d

30. Most serious disease of potato is

a) Early blight b) Late blight
c) Wilt d) Wart Ans : a

31. Botanical name of potato is

a) *Solanum tuberosum* b) *Solanum melongena*
c) Both a & b d) None of these Ans: a

32. Which is the leading state in production of potato

a) U.P. b) Rajasthan
c) Haryana d) Gujarat Ans: a

33. Optimum plant population ha^{-1} for potato has estimated as

a) 25,000 – 50,000 b) 30,000 – 60,000
c) 50,000 – 80,000 d) 80,000 - 1,00,000 Ans: d

34. Maximum yield of potato is obtained at a plant geometry of

a) 50 x 20 cm b) 60 x 30 cm
c) 90 x 30 cm d) 105 x 30 cm Ans: a

35. An example of companion cropping is

a) Sugarcane + potato b) Potato + mustard
c) Potato + radish d) Wheat + mustard Ans: a

36. Intercropping of mustard with potato is recommended in

a) Replacement series
b) Additive series
c) Replacement cum additive series
d) None of these Ans: a

37. Which one of the following is TPS variety of potato

a) JH 222 b) HPS 1/113
c) PJ 376 d) Jr 5857 Ans: d

38. The ideal temperature for tuberization in potato is

a) 14°C b) 18°C
c) 21°C d) 34°C Ans: b

39. The main crop of potato is planted from

a) 25th September to 10th October b) 15th October to 25th October
c) 25th October to 25th November d) March – April Ans: c

40. Which of the following are old traditional potato varieties still cultivated in India
   a) Satha, Gola, Phulwa, up-to-date, Darjeeling red round Craig's Defence, Great Scot, President
   b) Kufri red-kufri Kuber, Kufri Kisan, Kufri Kundan
   c) Kufri Safed, Kufri Kumar, Kufri Neela, Kufri Sinduri
   d) Kufri Chandramukhi, Kufri Khasi-Garo, Kufri Naveen **Ans : a**

41. Which of the following are not improved modern varieties of potato
   a) Kufri Red, Kufri Kuber, kufri Kisan, Kufri Kundan, Kufri Chamatkar, Kufri Neelmani
   b) Satha, Gola, Phulwa, Up-to-date Darjeeling Red Round, Craig's Defence, Great Scot, President
   c) Kufri Safed, Kufri Kumar, Kufri Neela, Kufri Sinduri, Kufri Sheetman, Kufri Jyoti
   d) Kufri chandramukhi, Kufri Khasi-garo, Kufri Naveen, Kufri alankar, Kufri Jeevan **Ans : b**

42. Depending upon size of the seed tuber, the seed rate of potato varies from
   a) 50-150 kg/hectare b) 1000-1800 kg/hectare
   c) 800-1500 kg/hectare d) 1500-2000 kg/hectare **Ans : d**

43. Thirty tones of farmyard manure, 120 kg of nitrogen, 100 kg of phosphorus and 70 kg of potassium is a recommended manure-cum-fertilizer dose for a hectare of
   a) Sugarcane crop b) Potato crop
   c) Wheat crop d) Barley crop **Ans : b**

44. Potato contains................% of carbohydrates
   a) 21 b) 2
   c) 0.3 d) 5-6 **Ans: a**

45. Potato contains................% of fat
   a) 21 b) 2
   c) 0.3 d) 5-6 **Ans: c**

46. Inflorescence in potato is
   a) Recemose b) Cymose
   c) Panicle d) None **Ans: a**

47. Late blight of potato is caused by
   a) Alternaria solani b) Phytophthora infestens
   c) Pseudomonassolanacearum d) Synchutrium endobioticum **Ans : b**

48. Brown rot of potato is caused by
a) Alternaria solani
b) Phytophthora infestens
c) Pseudomonassolanacearum
d) Synchutrium endobioticum
Ans : c

49. .... is a common disease in acidic soils in Potato
a) Early blight
b) Late blight
c) Wilt
d) Scab
Ans : d

50. .... tilth is preferred for tuberization in Potato
a) Fine
b) Rough
c) Cloddy
d) None
Ans : a

51. Potato contains................% of protein
a) 21
b) 2
c) 0.3
d) 5-6
Ans: b

52. Best soil for potato cultivation is
a) Sandy loam
b) Clay
c) Clay loam
d) Sandy
Ans: a

53. The weight of potato tuber (for sowing) most economical for higher yields is ____
a) 10-25 grams
b) 30-50 grams
c) 50-80 grams
d) 80 - 100 grams
Ans : b

54. A potato tuber represents
a) Enlarged underground stem
b) Enlarged underground root
c) Enlarged underground nodules
d) None of these
Ans : a

55. Potato tuber yield will be ______ q/ha, if crop was spaced at 50 x 20 c,; and average number of tubers/plant and average weight of tubers are 10 and 50 g, respectively.
a) 250
b) 500
c) 750
d) 1000
Ans : b

56. Potato belongs to the family ____
a) Crucifereae
b) Solanaceae
c) Compositeae
d) None of these
Ans : b

57. The time of planting of main crop of potato in northern plains is
a) First fortnight of October
b) Second fortnight of October
c) First fortnight of November
d) Second fortnight of November
Ans : b

58. The variety of potato which possesses the resistance to frost conditions is

a) Kufri chandramukhi b) Kufri alankar
c) Kufri sheetman d) Kufri chamatkar. **Ans : c**

59. Early blight of potato is caused by

a) Alternaria solani b) Phytophthora infestens
c) Pseudomonassolanacearum d) Synchutrium endobioticum **Ans : a**

60. Wart of potato is caused by

a) Alternaria solani b) Phytophthora infestens
c) Pseudomonassolanacearum d) Synchutrium endobioticum **Ans : d**

61. The recommended seed rate for potato in northern plains is _______ q/ha

a) 10-15 b) 15-20
c) 20-25 d) 25-30 **Ans : b**

❑❑❑

16 Chapter

Plantation Crop Production

Multiple Choice Questions (MCQ's)

Coffee

1. Coffee belongs to family
   a) *Theacae* b) *Rubiaceae*
   c) *Poaceae* d) *Euphorbiaceae* **Ans : b**

2. The centre of origin of coffee camphora is
   a) Uganda b) Ethopia
   c) Liberia d) India **Ans :a**

3. The centre of origin of coffee liberia is
   a) Uganda b) Ethopia
   c) Liberia d) Srilanka **Ans :c**

4. The centre of origin of coffee arabica is
   a) Uganda b) Ethopia
   c) Liberia d) None **Ans :b**

5. The coffee is a............plant
   a) annual b) binneal
   c) perennial d) None **Ans :b**

6. The scientific name of coffee is
   a) *Camellia sinensis* b) Commelina communis
   c) *Coffea arabica* d) None **Ans : a**

7. The scientific name of coffee is
   a) Coffee arabica b) Coffee camphora
   c) Coffea liberia d) None of the above **Ans :d**

8. The coffee is made of
   a) Tender leaves b) Unopened buds
   c) Flowers d) Seeds **Ans :d**

9. .... and..........are principal economic species of coffee cultivated in India
   a) Coffee arabica and Coffee camphora
   b) Coffee Liberia and Coffee camphora
   c) Coffee arabica and Coffee liberia
   d) All **Ans : a**

10. Although coffee can be grown on soils of almost all physical types the ideal soil for its cultivation are
   a) Well drained, deep and friable loams
   b) Forest soils rich in organic matter
   c) Red and lateritic
   d) None of these **Ans : c**

11. The scientific name of robusta coffee is
   a) Coffee arabica b) Coffee camphora
   c) Coffea liberia d) All **Ans :b**

12. The scientific name of arabica coffee is
   a) Coffee arabica b) Coffee camphora
   c) Coffea liberia d) All **Ans : a**

13. Coffee is propagated by
   a) Seed b) Cuttings
   c) Leaf buds d) All these means **Ans : a**

14. Which of the following operations are involved in manufacture of black coffee
   a) Harvested coffee is withered and rolled or distorted by rolling in coffee rolling machines to expose the leaf juice to air for oxidation (fermentation)
   b) Stopping fermentation by passing current of hot air after optimum chemical change is reached
   c) Grading by passing sieves
   d) All these operations **Ans : d**

15. Coffee prefers....................climate
   a) Hot and humid b) Cool and dry
   c) Warm and dry d) Cool and humid **Ans :a**

16. Ideal pH for coffee cultivation....................

a) 4.0 b) 5.0
c) 6.0 d) 7.0 **Ans : c**

17. ....is planted to protect coffee from high speed winds

a) Grewia robusta b) Tectona grandis
c) Delbergia sissoo d) None **Ans : a**

18. ....years old coffee nursery seedlings are used for transplanting in the field

a) Less than 1 b) 1.5
c) 2-3 d) More than 3 **Ans : b**

19. The normal spacing for arabica coffee is

a) 2x2 m b) 1x1m
c) 3x3m d) 4x4m **Ans : a**

20. The normal spacing for robusta coffee is

a) 2x2 m b) 1x1m
c) 3x3m d) 4x4m **Ans : c**

21. The normal plant population/ha for coffee varies from

a) 10000-20000 b) 750-3000
c) 2000-5000 d) 5000 -10000 **Ans : b**

22. Directed spray of ....herbicide is recommended for weed control in coffee

a) Glyphosate b) Alachlor
c) 2,4-D d) Simazine **Ans : a**

23. ....is the most important aspect for quality in coffee

a) Picking b) Roasting
c) Drying d) None **Ans : a**

24. ....seeds produces foxy coffee

a) Under ripe b) Over ripe
c) Over roasted d) Under roasted **Ans : b**

25. The removal of green fruits after harvesting is known as

a) Nipping b) Dipping
c) Stripping d) Desuckering **Ans : c**

26. ....planting is adopted on horizontal slopes

a) Vertical b) Contour
c) Line d) None **Ans : b**

27. The over and under riped coffee seeds are used for making............coffee
a) Chichory b) Cherry
c) Instant d) None
Ans : b

Tea

28. The centre of origin of tea is
a) Burma and China b) America
c) S. Africa d) None
Ans : a

29. In India the productivity of tea is highest in the state
a) Punjab b) Assam
c) W. Bengal d) None
Ans : b

30. The tea is a............plant
a) Annual b) Evergreen
c) Deciduous d) None
Ans : b

31. The scientific name of tea is
a) *Camellia sinensis* b) *Commelina communis*
c) *Coffea arabica* d) None
Ans : a

32. The tea is made of
a) Tender leaves b) Unopened buds
c) Both d) Seeds
Ans : c

33. The lowest leaf of tea is smallest and..........
a) True b) Pseudo
c) Narrow d) Broad
Ans : b

34. The serrated leaves of tea with no tips are called
a) Gol pat b) Janam
c) Pseudo leaf d) None
Ans : a

35. The oval leaves of tea with tips not serrated are called
a) Gol pat b) Janam
c) Pseudo leaf d) None
Ans : b

36. The phylloxy of tea is known as.........system
a) 2-5 b) 5-7
c) 3-6 d) None
Ans : a

37. Although tea can be grown on soils of almost all physical types the ideal soil for its cultivation are
a) Well drained, deep and friable loams
b) Forest soils rich in organic matter
c) Both of these
d) None of these
Ans: c

38. Tea is propagated by

a) Seed and doves   b) Root cuttings
c) Leaf buds   d) All these means   **Ans : a**

39. A formative prune is given to the tea bushes cutting the main branches at about 35-40 cm from the ground level after about

a) 3-4 years from planting   b) 4-5 years from planting
c) 6-7 years from planting   d) 7 years from planting   **Ans : b**

40. New shoots developing on tea bushes after the formative pruning are tipped leaving a growth of

a) 40-50 cm above pruning cut   b) 35-40 cm above pruning cut
c) 20-30 cm above pruning cut   d) 10-15 cm above pruning cut   Ans. c

41. Tea belongs to family

a) Theacae   b) Rubiaceae
c) Poaceae   d) Euphorbiaceae   **Ans : a**

42. Economic life of a tea bush in northern India is

a) 80 years   b) 50 years
c) 40-50 years   d) 40 years   **Ans : c**

43. Economic life of tea bush in southern India is

a) 40 years   b) 50 years
c) 40-50 years   d) 80 or more years   **Ans : d**

44. When plucking in tea is restricted to the terminal bud and two expanded succulent leaves, it is known as

a) Fine plucking   b) Succulent plucking
c) Bud plucking   d) Medium plucking   **Ans : a**

45. When plucking of tea includes terminal bud and 3 expanded succulent leaves, it is known as

a) Fine plucking   b) Medium plucking
c) Coarse plucking   d) Green plucking   **Ans : b**

46. Tea bushes are plucked in northern India from April to December at

a) Monthly intervals   b) Fortnightly intervals
c) Weekly intervals   d) Interval of 3 days   **Ans : c**

47. In northern India harvesting of tea is suspended from December to March because during this period the bushes are

a) Treated with pesticides   b) In bearing
c) Being trained   d) Dormant   **Ans : d**

48. In southern India tea is harvested throughout the year, but harvesting intervals are longer during

a) April-May b) December-March
c) June-August d) September-October Ans : b

49. Which of the following operations are involved in manufacture of black tea

a) Harvested tea is withered and rolled or distorted by rolling in tea rolling machines to expose the leaf juice to air for oxidation (fermentation)
b) Stopping fermentation by passing current of hot air after optimum chemical change is reached
c) Grading by passing sieves
d) All these operations Ans : d

50. What is that special operation which is done during tea manufacture so that the end product is green tea

a) Steaming the shoots soon after plucking to destroy leaf oxidase, the fermentation enzyme
b) Slowing down rolling process
c) Alternate rolling and jerking
d) All the operations Ans : a

51. Tea prefers....................climate

a) Hot and dry b) Cool and dry
c) Warm and humid d) Cool and humid Ans :c

52. Ideal pH for tea cultivation...................

a) 4.5 b) 5.5
c) 6.5 d) 7.5 Ans :b

53. ....is planted to protect tea from high speed winds

a) Grewia robusta b) Tectona grandis
c) Delbergia sissoo d) None Ans :a

54. ....is done to protect the soil from erosion in tea gardens

a) Contour terracing b) Contour lining
c) Bench terracing d) None Ans :c

55. Tea plucking is performed whole year in

a) NE India b) South India
c) Both d) None Ans :b

56. A special..............NPK soluble fertilizer mixture is used in tea nursery

a) 1:1:1 b) 4:3:3
c) 2:1:1 d) 3:2:1 **Ans :a**

57. A ....NPK fertilizer mixture is used for first three years in tea

a) 1:1:1 b) 4:3:3
c) 2:1:1 d) 3:2:1 **Ans :b**

58. ....years old tea nursery seedlings are used for transplanting in the field

a) Less than 1 b) 1.5
c) 2-3 d) More than 3 **Ans :b**

59. ....plantins is found best method of transplanting for tea

a) Square b) Hedge
c) Triangular d) Double hedge **Ans :c**

60. The normal spacing for tea is

a) 120x120 cm b) 120x150 cm
c) 150x100 cm d) 120x75 cm **Ans :d**

61. The normal plant population/ha for tea is

a) 100000 b) 1000
c) 20000 d) 10000 **Ans :d**

62. Directed spray of ....herbicide is recommended for weed control in tea

a) Glyphosate b) Alachlor
c) 2,4-D d) Simazine **Ans :a**

63. ....is the most important insect causing about 75% loss in tea

a) Tea mosquito b) Red spider mite
c) Tea bag worm d) Termite **Ans :b**

Tobacco

64. The centre of origin of tobacco is

a) India and Burma b) America
c) S. Africa d) None **Ans :b**

65. In India the productivity of tobacco is highest in the state

a) Punjab b) Haryana
c) Andhra Pradesh d) Uttar Pradesh **Ans :d**

66. Maximum acerage under tobacco cultivation is in the state

a) Punjab
b) Haryana
c) Andhra pradesh
d) U.P

Ans :d

67. In India the production of tobacco is highest in the state

a) Punjab
b) Haryana
c) Andhra pradesh
d) U.P

Ans :c

68. The inflorescence in tobacco is

a) Receme
b) Panicle
c) Spikelets
d) Ear

Ans :a

69. Optimum temperature for germination of tobacco is

a) 10°C
b) 15°C
c) 28°C
d) 20

Ans :c

70. The recommended seed rate for tobacco is _______ kg/ha

a) 8-10
b) 2-3
c) 5-8
d) 10-12

Ans : b

71. Higher yield and quality of tobacco could be achieved by:

a) Use of higher dose of phosphate
b) Adequate amount of N
c) No nitrogen application
d) Liberal amount of K

Ans :d

72. Tobacco yield is largely reduced mainly due to:

a) Cold wind at the flowering stage
b) Attack of aphids
c) Lack of adequate fertilizer application
d) Failure to apply irrigation in flowering stage

Ans :b

73. The centre of origin of tobacco is

a) India and Burma
b) America
c) S. Africa
d) None

Ans :b

74. The area need to raise tobacco seedlings by for transplanting one hectare area is-

a) 100 -150sqm
b) 50-60 sqm
c) 250-300 sq m
d) 400-500 sqm.

Ans : a

75. Transplanting of seedlings is commonly practiced in

a) Tobacco
b) Maize
c) Sorghum
d) Pearl millet

Ans : a

76. Nicotine content of the leaves of Nicotiana rustica ranges between ______ per cent

a) 2.0-5.5
b) 3.0-4.5
c) 3.5-8.0
d) 8-13

Ans : c

77. Nicotine content of the leaves of Nicotiana tobacum ranges between ______ per cent

a) 0.5-5.5
b) 3.0-4.5
c) 3.5-8.0
d) 8-13

Ans : c

78. Rainfall at the time of maturity of tobacco is not desired as it causes ______

a) Washing of gums and resins from leaf
b) Lodging
c) Rotting of leaves
d) Shattering of leaves

Ans : a

79. Tobacco belong to the family

a) Compositeae
b) Gramineae
c) Solanaceae
d) Leguminoseae

Ans : c

80. The optimum time for sowing nursery of tobacco is ______

a) First fortnight of September
b) Second fortnight of September
c) First fortnight of August
d) Second fortnight of August

Ans : c

81. The recommended seed rate of tobacco to raise seedlings for one hectare area is from

a) 2.0 - 3.0 kg
b) 4.0 - 5.0 kg
c) 0.2 - 0.3 kg
d) 0.4 - 0.5 kg

Ans : a

82. Topping in tobacco means a process of ______

a) Removal of all buds present in the axil of leaves
b) Removal of leaves
c) Removal of terminal bud
d) Buming of leaves

Ans : c

83. The basic aim of topping and desuckering in tobacco is

a) To reduce the plant height
b) To enlarge branching
c) To divert energy and nutrients from flower head to leaves
d) To protect the plants against lodging

Ans : c

84. Deshi type tobacco is

a) Nicotiana tobacum
b) Nicotiana rustica
c) Both
d) None

Ans : a

85. Vilayati type tobacco is
a) Nicotiana tobacum b) Nicotiana rustica
c) Both d) None
Ans : b

86. Which of the following is a stage of curing
a) Yellowing b) Fixing colour
c) Drying d) All
Ans :d

87. The optimum stage of tobacco seedling for transplanting under irrigated condition during kharif season is
a) 2 leaf stage b) 4 leaf stage
c) 6 leaf stage d) 8 leaf stage
Ans : c

88. The fruit of tobacco is known as
a) Capsule b) Pod
c) Siliqua d) Caryopsis
Ans : c

89. .... Is the deciding factor for quality, curing and yield
a) Temperature b) Sunlight
c) Humidity d) All
Ans : c

90. ....ions esults in poor burning quality of leaves
a) Chloride b) Phosphorus
c) Sulphur d) Calcium
Ans : a

91. ....soil reaction is better for production of superior quality leaves of tobacco
a) Mild Alkaline b) Strongly alkaline
c) Mild acidic d) Strongly acidic
Ans : c

92. ....soil is suitable for bidi tobacco cultivation
a) Medium loams b) Sandy soils
c) Heavy soils d) Sandy to silt loams
Ans : a

93. ....soil is suitable for binder (cigar) tobacco cultivation
a) Medium loams b) Sandy soils
c) Heavy soils d) Sandy to silt loams
Ans :c

94. ....soil is suitable for cigarette tobacco cultivation
a) Medium loams b) Light sandy soils
c) Heavy soils d) Sandy to silt loams
Ans : b

95. ....soil is suitable for hookah tobacco cultivation
a) Medium loams b) Sandy soils
c) Heavy soils d) Sandy to silt loams
Ans : d

96. ....soil pH is most suitable for tobacco cultivation
   a) 6-7 b) 7-8
   c) 5-6 d) 4-5 **Ans : c**

97. Optimum age of seedlings for transplanting of N. tobaccum tobacco is........... weeks
   a) 6-7 b) 7-8
   c) 5-6 d) 4-5 **Ans : c**

98. Optimum age of seedlings for transplanting of N. rustica tobacco is........... weeks
   a) 6-7 b) 7-9
   c) 5-6 d) 4-5 **Ans : b**

99. Heavy nitrogen application..................................... of smoking type tobacco
   a) Increases yield and decreases quality
   b) Increases quality and decreases yield
   c) Increases both quality and yield
   d) Decreases both quality and yield **Ans :a**

100. Heavy nitrogen application..................................... of chewing, bidi type tobacco
   a) Increases yield and decreases quality
   b) Increases quality and decreases yield
   c) Increases both quality and yield
   d) Decreases both quality and yield **Ans :c**

101. Better quality of flue cured tobacco is obtained from
   a) Partially N starved plants b) N sufficient plants
   c) Excess N supplied plants d) All **Ans :a**

102. Potash should never be supplied through ....to tobacco
   a) Muriate of potash b) Sulphate of potash
   c) Both d) None **Ans : a**

103. Chloride application ....in tobacco
   a) Decreases quality b) Increases quality
   c) No effect on quality d) None **Ans : a**

104. Thick leaves and brostening of leaves in tobacco is due to
   a) Chlorine b) Calcium
   c) Magnesium d) Boron **Ans :a**

105. Thin leaves and increased yield in tobacco is due to
   a) Chlorine b) Calcium
   c) Magnesium d) Boron **Ans :b**

106. Sand drown in tobacco is due to

a) Chlorine b) Calcium
c) Magnesium d) Boron **Ans :c**

107. Thickening of lower leaves in tobacco is due to

a) Chlorine b) Calcium
c) Magnesium d) Boron **Ans :d**

108. Transplanting of winter crop of tobacco is done in

a) September-October b) October-November
c) November -December d) March-April **Ans :b**

109. Transplanting of summe crop of tobacco is done in

a) May-June b) February-March
c) November -December d) March-April **Ans :d**

110. ....is a total root parasite in tobacco

a) Striga b) Orobanche
c) Cascuta d) All **Ans :b**

111. Pre planting application of EPTC @6kg/ha is recommended in tobacco for control of

a) Striga b) Orobanche
c) Cascuta d) All **Ans :b**

112. The process of removal of flower heads along with few upper leaves in tobacco is

a) Desuckering b) Topping
c) Deheading d) None **Ans :b**

113. Topping results in improved

a) Quality b) Size
c) Texture d) All **Ans :d**

114. Topping results in diversion of energy and nutrients to

a) Leaves b) Stem
c) Fruits d) Flowers **Ans :a**

115. The effect of topping is more pronounced in

a) Younger leaves b) Middle leaves
c) Older leaves d) All **Ans :a**

116. Topping of inflorescence is known as

a) Ttopping high b) Topping low
c) Ttopping full d) None **Ans :a**

117. Topping of inflorescence along with few leaves is known as
a) Topping high b) Topping low
c) Topping full d) None
Ans :b

118. Branching in tobacco is due to
a) Topping b) Desuckering
c) Both d) None
Ans :a

119. Removal of suckers in tobacco is known as
a) Topping b) Desuckering
c) Suckering d) Detoppping
Ans :b

120. To obtain the benefit of topping in tobacco .... is necessary
a) Detopping b) Desuckering
c) Suckering d) All
Ans :b

121. Desuckering is not required in................type tobacco
a) Bidi b) Wrapper
c) Ciggerate d) Hookah
Ans :b

122. Use of ....on top 6 axils suppresses emergence of suckers
a) NAA b) MH
c) Coconut oil d) None
Ans :c

123. Application of ....@2% suppresses emergence of suckers in cheroot tobacco
a) NAA b) MH
c) Coconut oil d) None
Ans :a

124. Flue cured tobacco is raised with low...........level
a) N b) P
c) K d) Cl
Ans :a

125. Flue cured tobacco is harvested by.............method
a) Priming b) Stock cut
c) Both d) None
Ans :a

126. Which of the following is not a stage of curing
a) Yellowing b) Fixing colour
c) Hardening d) Drying
Ans :c

127. Which of the following is air cured tobacco
a) Wrapper b) Chewing
c) Bidi d) Ciggarate
Ans :a

128. Which of the following is fire cured tobacco
a) Wrapper b) Chewing
c) Bidi d) Ciggarate
Ans :b

129. Which of the following tobacco is treated with salt water or jaggery
   a) Wrapper b) Chewing
   c) Bidi d) Ciggarate Ans :b

130. Which of the following is not a method of curing
   a) Fire b) Sun
   c) Chemcal d) Air Ans :c

131. Nicotine content of the leaves of Nicotiana rustica ranges between ______ per cent
   a) 2.0-5.5 b) 3.0-4.5
   c) 3.5-8.0 d) 8-13 Ans : c

132. Rainfall at the time of maturity of tobacco is not desired as it causes _______
   a) Washing of gums and resins from leaf
   b) Lodging
   c) Rotting of leaves
   d) Shattering of leaves Ans : a

133. Tobacco belong to the family
   a) Compositeae b) Gramineae
   c) Solanaceae d) Leguminoseae Ans : c

134. The optimum time for sowing nursery of tobacco is ________
   a First fortnight of September b) Second fortnight of September
   c) First fortnight of August d) Second fortnight of August Ans : c

135. The recommended seed rate of tobacco to raise seedlings for one hectare area is from
   a) 2.0 - 3.0 kg b) 4.0 - 5.0 kg
   c) 0.2 - 0.3 kg d) 0.4 - 0.5 kg Ans : a

136. Topping in tobacco means a process of __________
   a) Removal of all buds present in the axil of leaves
   b) Removal of leaves
   c) Removal of terminal bud
   d) Buming of leaves Ans : c

137. The basic aim of topping and desuckering in tobacco is
   a) To reduce the plant height
   b) To enlarge branching
   c) To divert energy and nutrients from flower head to leaves
   d) To protect the plants against lodging Ans : c

❑❑❑

17 Chapter

Sugar Crop Production

Multiple Choice Questions (MCQ's)

Sugarcane

1. In India the productivity of sugarcane is highest in the state

   a) Punjab　　b) Karnataka
   c) Tamilnadu　　d) None　　**Ans : d**

2. The inflorescence in sugarcane is

   a) Siliqua　　b) Open panicle
   c) Spikelets　　d) Ear　　**Ans : b**

3. The inflorescence in sugarcane is known as

   a) Arrow　　b) Head
   c) Spikelets　　d) Ear　　**Ans : a**

4. Green tops of sugarcane is known as

   a) Arrow　　b) Head
   c) Spikelets　　d) Angola　　**Ans : a**

5. Optimum temperatures for growth and development of sugarcane ranges between

   a) 21-37°C　　b) 28-32°C
   c) 20-25°C　　d) None　　**Ans : b**

6. The most cultivated sugar crop of the world is
   a) Sugarcane b) Sugar beet
   c) Maize d) Jowar
   Ans : a

7. The centre of origin of sugarcane is
   a) India b) America
   c) S. Africa d) None
   Ans : a

8. In India the productivity of sugarcane is highest in
   a) Northern states b) South
   c) Central d) None
   Ans : b

9. Sugarcane belongs to family.
   a) Gramineae b) Chenopodiaceae
   c) Leguminous d) None
   Ans : a

10. Sugarcane is a__________ plant.
    a) C_3 b) C_4
    c) Leguminous d) None
    Ans : b

11. Sugarcane is a__________ plant.
    a) Annual b) Biennial
    c) Perennial d) Triennial
    Ans : c

12. The inflorescence of sugarcane is called a
    a) Spike b) Arrow
    c) Tassel d) Panicle
    Ans : b

13. The average yield of sugarcane in Northern India is low compared to Southern India is because of
    a) Varietal variability b) Soil fertility status
    c) Level of management d) Climatic conditions
    Ans : d

14. The Number of three budded sets needed for planting one hectare of sugarcane is
    a) 10-15 thousand b) 15-25 thousand
    c) 35-40 thousand d) 50-60 thousand
    Ans : c

15. The planting of sugarcane by trench method ________
    a) Improved tillering b) Reduces lodging
    c) Improves germination d) Increases the water use efficiency
    Ans : b

16. Hot water treatment" of sugarcane sets is done because
    a) It increases germination
    b) It controls the soil borne diseases
    c) It breaks dormancy
    d) It promotes the conversion of sucrose into glucose
    Ans : d

17. "Arrowing" refers to
    a) Flower primodia initiation in wheat
    b) Flowering of cotton
    c) Flowering of maize
    d) Flowering of sugarcane. **Ans : d**

18. Which portion of sugarcane stalk is preferred for sowing purposes
    a) Top one-third to half portion
    b) Top one-fourth portion
    c) Bottom one-fourth portion
    d) Bottom one third to half portion **Ans : a**

19. Sugarcane seed sets should essentially have __________
    a) Four buds b) Three buds
    c) Two buds d) One bud **Ans : b**

20. For harvest of sugarcane the brix ratio of the lower and above portion of the stem should be
    a) 0.2 - 0.3 b) 0.5 - 0.6
    c) 0.7 - 0.8 d) 0.9 - 1.0 **Ans : d**

21. The autumn planting of sugarcane is most successful when planted in the month of
    a) September b) October
    c) November d) February **Ans : b**

22. The spring planting of sugarcane is most successful when planted in the month of
    a) September b) October
    c) November d) February **Ans : d**

23. Sugarcane crop yielding 800 quintals of cane per hectare removes N, P_2O_5 and K_2O kg/ha, respectively
    a) 100 – 120, 180 – 280 and 60 – 150
    b) 90 – 100, 60- 70 and 100 – 110
    c) 120 – 40 – 80
    d) 80 – 30/40 – 20 **Ans: a**

24. The recommended dose of N, P_2O_5 and K_2O kg/ha for growing sugarcane in northern India are
    a) 150-200 – 80 –60 b) 120 – 40 – 40
    c) 30 – 20 – 30 d) 40 – 30 – 40 **Ans: a**

25. The recommended dose of N, P_2O_5 and K_2O kg/ha for growing sugarcane in southern India are
    a) 250-300 – 100-120 –60-80 b) 120 – 40 – 40
    c) 30 – 20 – 30 d) 40 – 30 – 40 **Ans: a**

26. The scientific name of noble cane is
    a) Saccharum officinarum b) Saccharum spontaneum
    c) S. barberi d) S. sinense **Ans: a**

27. *Saccharum offinarum* is native of
    a) New Guinea b) India
    c) Indonesia d) China **Ans: b**

28. Which state has highest productivity of sugarcane
    a) U.P. b) W.B.
    c) Karnataka d) Tamil Nadu **Ans: d**

29. Which state has the largest acreage of sugarcane in country
    a) Tamil Nadu b) Karnataka
    c) U.P. d) Bihar **Ans : c**

30. Which state has highest production of sugarcane in country
    a) Tamil Nadu b) Karnataka
    c) U.P. d) Bihar **Ans : c**

31. Which of the following roots are of permanent type on sugarcane sets
    a) Sett roots b) Shoot roots
    c) Prop roots d) All of these **Ans: b**

32. The inflorescence of sugarcane is known as
    a) Arrow b) Panicle
    c) Capitulum's d) Racemose **Ans: a**

33. The best suited temperature for growth of sugarcane lies between
    a) 15-20^{0}C b) 28-32^{0}C
    c) 20-22^{0}C d) 32-35^{0}C **Ans: b**

34. Adsali sugarcane is planted in
    a) July-August b) January-February
    c) February-March d) October-November **Ans: a**

35. Eksali sugarcane is planted in
    a) July-August b) January-February
    c) February-March d) October-November **Ans : b**

36. The top portion of sugarcane should be selected for seed purposes because bud tissues are rich in
    a) Sucrose b) Glucose
    c) Galactose d) Maltose **Ans : b**

37. Which element is essential for sugar translocation in sugarcane
a) P b) K
c) B d) Mo Ans : a

38. Which growth stage of sugarcane is critical for irrigation
a) Germination b) Grand growth phase
c) Formative stage d) Ripening stage Ans: b

39. Most critical weed competition period is up to ....months after planting in sugarcane
a) 2 b) 4
c) 5 d) 6 Ans : b

40. Most common herbicides used for weed control in sugarcane is/are
a) Simazine b) Atrazine
c) Alachlor d) All of these Ans: d

41. Most serious disease of sugarcane is
a) Red stripe b) Red rot
c) Wilt d) Smut Ans: b

42. How many sets are needed to plant one hectare of sugarcane
a) 30,000-35,000 b) 35,000-40,000
c) 40,000-45,000 d) 50,000-55,000 Ans: b

43. Trench method of sugarcane planting is used in
a) North India b) Western India
c) Coastal areas d) Waterlogged area Ans: c

44. With too much increase in nitrogen application, sugar content in juice is
a) Decreased b) Increased
c) Remain constant d) None of the above Ans: a

45. Epricania melanoleuca is a parasitoid effective against
a) Sugarcane scale b) Sugarcane mealy bug
c) Sugarcane pyrilla d) Sugarcane leaf hopper Ans: c

46. Crop logging is used in
a) Sugarcane b) Sugarbeet
c) Maize d) Tea Ans: a

47. Flat-planting furrow-planting and trench planting are the methods of planting
    a) Sugarcane   b) Potato
    c) Dahlia   d) Pepper vires   **Ans. a**

48. Sugarcane is a................ plant
    a) Tropical   b) Temperate
    c) Sub tropical   d) None   **Ans. a**

49. Nobel canes is known as .... in north India
    a) Paunda   b) Ganna
    c) Both   d) None   **Ans. a**

50. The arrangement of spikelets in sugarcane is
    a) Cymose   b) Recemose
    c) Umbel   d) None   **Ans. b**

51. In sugarcane inflorescence the oldest flowers are at
    a) Top   b) Bottom
    c) Middle   d) None   **Ans. b**

52. The sugarcane flowers at the age of ....months
    a) 8-10   b) 10-12
    c) 16-18   d) 6-8   **Ans : b**

53. The soils rich in .... are good for juice quality and lodging resistance
    a) N & P   b) P & Ca
    c) N & K   d) P & K   **Ans : b**

54. The soils rich in .... are good for sugar recovery
    a) N   b) Ca
    c) Fe   d) P &K   **Ans : c**

55. Earthing is done in which of the following crops
    a) Sugarcane   b) Rice
    c) Wheat   d) None   **Ans. a**

56. Tying and wrapping in sugarcane is performed to prevent
    a) Insect infestation   b) Lodging
    c) Disease   d) All   **Ans. b**

57. Tying and wrapping in sugarcane is performed in the month of
    a) August   b) September
    c) October   d) March   **Ans : a**

58. For ratoon crop of sugarcane it is recommended to grow .... varieties
    a) Long duration b) Short duration
    c) Medium d) All
    Ans : b

59. % juice recovery of sugarcane by ordinary crusher is
    a) 55-65 b) 60-70
    c) 70-75 d) All
    Ans : a

60. % juice recovery of sugarcane by power crusher is
    a) 55-65 b) 60-70
    c) 70-75 d) All
    Ans : b

61. % juice recovery of sugarcane by vaccume crusher is
    a) 55-65 b) 60-70
    c) 70-75 d) All
    Ans : c

62. % jaggery recovery from sugarcane
    a) 15-25 b) 9-10
    c) 6-10 d) All
    Ans : b

63. % sugar recovery from sugarcane juice is
    a) 15-25 b) 9-10
    c) 6-10 d) All
    Ans : c

64. % sucrose content in sugarcane
    a) 13-24 b) 9-10
    c) 6-10 d) All
    Ans : a

65. The important by products of sugarcane industry are
    a) Bagasse b) Molasses
    c) Both d) None
    Ans : c

66. Bagasse is used as
    a) Fuel b) Fodder
    c) Fertilizer d) All
    Ans : a

67. Molasses is used for .... preparation
    a) Fuel b) Fodder
    c) Fertilizer d) Alcohol
    Ans : d

68. For producing 70 tonnes of normal healthy sugarcane, the crop removes from soil
    a) 85-110 kg of nitrogen
    b) 180-330 of phosphorus
    c) 60-180 kg of potash and 70-80 kg of calcium
    d) All these nutrients
    Ans : d

69. Sugarcane is a__________ plant.

a) Annual b) Biennial
c) Perennial d) Triennial Ans : c

70. The young inflorescence of sugarcane is called a

a) Spike b) Arrow
c) Tassel d) Panicle Ans : b

71. The average yield of sugarcane in Northern India is low compared to Southern India is because of

a) Varietal variability b) Soil fertility status
c) Level of management c) Climatic conditions Ans : d

72. The Number of three budded sets needed for planting one hectare of sugarcane is

a) 10-15 thousand b) 15-25 thousand
c) 35-40 thousand d) 50-60 thousand Ans : c

73. The lodging of sugarcane is reduced by planting in ________

a) Flat method b) Ridge & furrow method
c) Trench method d) All Ans : c

74. Hot water treatment" of sugarcane sets is done because

a) It increases germination
b) It controls the chloroic streak disease
c) It promotes the conversion of sucrose into glucose
c) Both b & c Ans : b

75. "Arrowing" refers to

a) Rooting in sugarcane
b) Stem elongation in sugarcane
c) Sugar accumulation in sugarcane
d) Flowering of sugarcane. Ans : d

76. Which portion of sugarcane stalk is not preferred for sowing purposes

a) Top one-third to half portion b) Top one-fourth portion
c) Bottom one-fourth portion d) Bottom one third to half portion Ans : c

77. For optimum population Sugarcane seed sets should essentially have __________

a) Four buds b) Three buds
c) Two buds d) Six bud Ans : b

78. Sugarcane should not be harvested if the brix ratio of the lower and above portion of the stem is less than

a) 0.2.0.3 b) 0.5-0.6
c) 0.7-0.8 d) 0.9-1.0 Ans : d

79. For Best autumn crop sugarcane is planted in the month of
a) September b) October
c) November d) February Ans : b

80. Arrowing in sugarcane is
a) Good, for it increases sugar in the cane
b) Good, for it increases cross fertilization and seed setting
c) Considered rather undesirable as it reduces sucrose percent in the cane.
d) Considered good as it leads to extra yields of the cane Ans : c

81. The Botanical name of sugarcane is
a) *Saccharum officinarum* b) *Saccharum spontaneum*
c) *S. barberi* d) *S. sinense* Ans : a

82. Sugarcane is native of
a) New Guinea b) India
c) Indonesia d) China Ans : b

83. Which state has highest productivity of sugarcane
a) U.P. b) W.B.
c) Karnataka d) Tamil Nadu Ans: d

84. Which state has the largest acreage and highest production of sugarcane in country
a) Tamil Nadu b) Karnataka
c) U.P. d) Bihar Ans : c

85. Which of the following roots are of permanent type on sugarcane sets
a) Sett roots b) Shoot roots
c) Prop roots d) All of these Ans: b

86. The emergence of inflorescence in sugarcane is known as
a) Arrowing b) Furrowing
c) Rising d) Nipping Ans : a

87. The best suited temperature for growth of sugarcane lies between
a) 15-20^{0}C b) 28-32^{0}C
c) 26-32^{0}C d) 32-35^{0}C Ans : b

88. Adsali sugarcane is planted in
a) July-August b) January-February
c) February-March d) October-November Ans: a

89. Eksali sugarcane is planted in

a) July-August
b) January-February
c) February-March
d) October-November

Ans: b

90. The top portion of sugarcane should be selected for seed purposes because bud tissues are rich in

a) Sucrose
b) Glucose
c) Galactose
d) Maltose

Ans: b

91. Which element is essential for sugar translocation in sugarcane

a) P
b) K
c) B
d) Mo

Ans: a

92. Which growth stage of sugarcane is critical for irrigation

a) Germination
b) Grand growth phase
c) Formative stage
d) Ripening stage

Ans: b

93. Most critical weed competition period is up to ....months after transplanting exist in sugarcane

a) 2
b) 4
c) 5
d) 6

Ans: b

94. Most common herbicides used for weed control in sugarcane is/are

a) Simazine
b) Atrazine
c) Alachlor
d) All of these

Ans: c

95. Most serious disease of sugarcane is

a) Red stripe
b) Red rot
c) Wilt
d) Smut

Ans : b

96. Under high density planting how many sets are needed to plant one hectare of sugarcane

a) 30,000-35,000
b) 35,000-40,000
c) 40,000-45,000
d) 50,000-55,000

Ans: d

97. Trench method of sugarcane planting is used in

a) North India
b) Western India
c) Coastal areas
d) Waterlogged area

Ans: b

98. With too much increase in nitrogen application, sugar content in juice is

a) Decreased
b) Increased
c) Remain constant
d) None of the above

Ans: a

99. Epricania melanoleuca is a parasitoid effective against

a) Sugarcane scale
b) Blackgram mealy bug
c) Sugarcane pyrilla
d) Blackgram leaf hopper

Ans: c

100. For sugarcane 'N' fertilizer application has to be completed within .....days after planting for better sugar content in juice

a) 20 b) 40
c) 60 d) 80 Ans : c

Sugarbeet

101. The inflorescence in sugarbeat is

a) Siliqua b) Open Panicle
c) Spikelets d) Ear Ans : b

102. Optimum temperatures towards germination of sugarbeat ranges between

a) 21-37°C b) 25-30°C
c) 20-25°C d) None Ans : d

103. Optimum temperatures for growth and sugar accumulation in sugarbeat ranges between

a) 21-37°C b) 25-30°C
c) 20-22°C d) None Ans : c

104. The recommended seed rate for sugarbeat is _______ kg/ha

a) 8-10 b) 12-15
c) 15-18 d) 20-25 Ans : a

105. Higher yield of sugarbeat could be achieved by:

a) Use of higher dose of phosphate
b) Adequate amount of N
c) No nitrogen application
d) Closer planting Ans : b

106. Problem in sugar extraction from roots is due to

a) Use of higher dose of phosphate
b) Use of excess dose of N
c) No nitrogen application
d) Closer planting Ans : b

107. The first man to establish sugar factory by processing of beet roots

a) F.C.Achard b) Mac-Mohan
c) N.E.Borlouge d) K.C.Mehta Ans : a

108. Sugarbeet is a

a) Salad crop b) Vegetable crop
c) Sugar crop d) Tuber crop Ans : c

109. Sugarbeet belong to the family________

a) Chenopodiaceae b) Cruciferae
c) Gramineae d) Liliaceae **Ans : a**

110. In sugarbeet the sucrose in found in ....

a) Stem b) Roots
c) Leaves d) Flower **Ans : b**

111. Sugarbeet is grown in .... season

a) Kharif b) Rabi
c) Zaid d) All of them **Ans : b**

112. In sugarbeet the sugar content is ....%

a) 15-16 b) 10-12
c) 6-8 d) 16-20 **Ans : a**

113. In sugarbeet the sugar recovery is ....%

a) 15-16 b) 10-12
c) 6-8 d) 16-20 **Ans : b**

114. A good crop of sugarbeet yields ....t/ha

a) 15-16 b) 10-12
c) 35-50 d) 16-20 **Ans : c**

115. The optimum row to row spacing for sugarbeet is ....cm

a) 15-20 b) 10-15
c) 40-60 d) 30-40 **Ans : c**

116. The optimum plant to plant spacing for sugarbeet is ....cm

a) 15-20 b) 20-25
c) 40-60 d) 30-40 **Ans : b**

117. The optimum plant population for sugarbeet is .... /ha

a) 50,000 b) 60,000
c) 70,000 d) 1,00,000 **Ans : d**

118. Optimum fertilizer requirement for sugarbeat is

a) 120-80-80 b) 120-60-40
c) 120-60-80 d) None **Ans : a**

119. The critical period for crop weed competition in sugarbeet is...........days

a) 35-45 b) 50-60
c) 20-30 d) 45-55 **Ans : a**

120. Weeds in sugarbeet can effectively be controlled by

a) Pyramin b) Sirmet
c) Betanol d) All
Ans : a

121. Weeds in sugarbeet can effectively be controlled by post emergence application of

a) Pyramin b) Sirmet
c) Betanol d) All
Ans : c

122. Weeds in sugarbeet can effectively be controlled by pre emergence application of

a) Pyramin b) Simazine
c) Betanol d) All
Ans : a

123. Weeds in sugarbeet can effectively be controlled by pre emergence application of

a) Pyramin b) Sirmet
c) Both d) None
Ans : c

124. The seed rate required for sugarbeet is............kg/ha

a) 8-10 b) 10-12
c) 12-15 d) 15-20
Ans : a

125. Sugarbeet *(Beta vulgaris)* originated in Mediterranean region belongs to

a) Gramineae b) Convolvulaceae
c) Chenopodiaceae d) Liliaceae
Ans: c

126. Sugarbeet matures in

a) April/May b) June/July
c) September/October d) January/February
Ans: d

127. For ideal plant population, the seed rate of sugarbeet should be (kg/ha)

a) 4 - 8 b) 8 -10
c) 10 - 15 d) 25 – 30
Ans: b

128. The ideal temperature for tuberization in potato is

a) 14°C b) 18°C
c) 21°C d) 34^0C
Ans: b

129. Sugarbeet *(Beta vulgaris)* originated in Mediterranean region belongs to

a) Gramineae b) Convolvulaceae
c) Chenopodiaceae d) Liliaceae
Ans: c

130. Sugarbeet matures in
a) April/May
b) June/July
c) September/October
d) January/February
Ans: d

131. For ideal plant population, the seed rate of sugarbeet should be (kg/ha)
a) 4 - 8
b) 8 -10
c) 10 - 15
d) 25 – 30
Ans: b

132. Sugarbeet is a
a) Salad crop
b) Vegetable crop
c) Sugar crop
d) Tuber crop
Ans : c

133. Sugarbeet belong to the family__________
a) Phenopodiaceae
b) Cruciferaeae
c) Gramineae
d) Liliaceae
Ans : a

134. In sugarbeet the sucrose in found in ....
a) Stem
b) Roots
c) Leaves
d) Flower
Ans : b

135. Sugarbeet is grown in ... season
a) Kharif
b) Rabi
c) Zaid
d) All of them
Ans : b

136. The seed rate required for sugarbeet is kg/ha
a) 8-10
b) 10-12
c) 12-15
d) 15-20
Ans : a

❑❑❑

18 Chapter

Model Question Papers

Multiple Choice Questions (MCQ's)

Model Paper-I

1. Chemical name of herbicide 'Machete' is
   a) 2, 4-D b) Fluchoralin
   c) Alachlor d) Butachlor

2. Instrument used for measuring wind velocity is known as:
   a) Hydrometer b) Pyranometer
   c) Altimeter d) Anemometer

3. Rice plant belongs to
   a) C_3 b) C_4
   c) CAM d) None of the above

4. Stem nodulating green manure plants is
   a) Sum hemp b) *Susbania aculiata*
   c) *Sesbania rostrate* d) *Tephrosea* sp.

5. Growth of plants towards light is called
   a) Photoperiodism b) Phototropism
   c) Photorespiration d) Photolysis

6. A fodder and pasture legume
   a) Glyricidia b) Stylosanthus
   c) Cowpea d) Sesbania

7. Schoefield is associated with
   a) Neutron moisture metre
   b) Electrical moisture box
   c) Pressure plate apparatus
   d) Longarithm of soil moisture

8. The dry farming research station was started in the year 1936 is
   a) Bijapur
   b) Sholapur
   c) Manjri
   d) Rohtak

9. Law of minimum was proposed by
   a) Theodore de Saussure
   b) Justice Van Liebig
   c) Laes and Gilbert
   d) Jethro Tull

10. The first Agriculture University was established in
    a) Ludhiana
    b) New Delhi
    c) Pantnagar
    d) Coimbatore

11. PUFA content in highest in
    a) Coconut oil
    b) Safflower oil
    c) Groundnut oil
    d) Mustard oil

12. Orobanche is a parasite on
    a) Sorghum
    b) Sugarcane
    c) Safflower
    d) Tobacco

13. Chemical for inhibiting the process of nitrification
    a) 2 4-D
    b) N-serve
    c) CON_2H_4
    d) Paraquat

14. Journal published by Indian Council of Agricultural Research
    a) Indian Journal of Agronomy
    b) Indian Journal of Agricultural Sciences
    c) Agricultural Research
    d) Research

15. Soil particles get deflocculated due to
    a) Calcium
    b) Sodium
    c) Magnesium
    d) Sulphur

16. Dominant from of nitrogen absorbed by plants
    a) NH_4
    b) NO_3
    c) NO_2
    d) NH_3

17. First inter-specific hybrid cotton was released by
    a) Tamil Nadu
    b) Gujarat
    c) Punjab
    d) Karnataka

18. Single superphosphate contains phosphorus in
    a) Citrate soluble form
    b) Insoluble form
    c) Water soluble form
    d) None of these

19. Percentage of lint in seed cotton is normally
    a) 80%
    b) 60%
    c) 30%
    d) 10%

20. Textural class of the soil depends on
    a) Organic matter content
    b) Soil water
    c) Particle size distribution
    d) Soil productivity

21. Sea Island cotton or Egyptian cottons belong to
    a) *Gossypium arboreum*
    b) *Gossypium hirutum*
    c) *Gossypium barbadense*
    d) *Gossypium herbacceum*

22. Plants tolerant to high soil salinity conditions
    a) Heliophytes
    b) Halophytes
    c) Heliotropic
    d) None of these

23. The semi arid climate will have moisture deficit index of
    a) 33.0
    b) -33.3 to 66.6
    c) 0 to 33.3
    d) Less than -66.6

24. In alternate land use system of dry lands, which one of the following fruits is most suitable
    a) Sapota
    b) Guava
    c) Ber
    d) Mango

25. The energy status of the water is called
    a) Pressure
    b) Potential
    c) Tension
    d) Content

26. Duty of water is
    a) Controlled and uniform application of water
    b) Irrigation along the contour
    c) Ratio between actual and potential ET
    d) None of the above

27. The ratio of water stored in the root zone depth of soil to the water delivered to the field from farm supply source
    a) Field water balance
    b) Field application efficiency
    c) Farm irrigation efficiency
    d) Downward flux at the bottom layer

28. Infiltration is
    a) Downward movement of water
    b) Lateral movement of water
    c) Entry of water at the soil surface
    d) Downward flux at the bottom layer

29. Available soil moisture is the moisture content between
    a) Field capacity and wilting point
    b) Field capacity and Hygroscopic co-efficient
    c) Field capacity and permanent wilting point
    d) Wilting point and permanent wilting point

30. The ability of rainfall to cause soil erosion is called
    a) Erosion Rill
    b) Erosivity
    c) Erosion sheet
    d) Erosion splash

31. The most important problem associated with direct sown wet seeded rice is
    a) Weeds
    b) Water
    c) Land management
    d) Fertility

32. The index which expresses the reduction of yield due to the presence of weeds in comparison with weed free situation is
    a) Weed control efficiency
    b) Weed index
    c) Weed control index
    d) None of the above

33. Avena fatua is a problematic weed in
    a) Pulses
    b) Maize
    c) Cotton
    d) Wheat

34. Find out the quantity of simazine WP to be sprayed in one-hectare area (if the rate of application per hectare = 3.00 kg al/ha & active ingredients 80%)
    a) 3.75
    b) 3.5
    c) 3.25
    d) 3.00

35. The term variance is used to denote the
    a) Standard deviation
    b) Square of the standard deviation
    c) Square root of the standard deviation
    d) None of the above

36. Angular transformation refers to
    a) Changing data of binomial distribution to normal distribution
    b) Taking percentage for data
    c) Taking square root for data
    d) None of the above

37. Trend in rainfall can be analysed with
    a) Multiple regression technique
    b) Correlation technique
    c) Moving averages
    d) None of the above

38. The growth pattern of a cereal plant falls in
    a) Linear curve b) Sigmoidal curve
    c) Cubic curve d) Binomial curve

39. Bolting refers to
    a) Stem elongation b) Fruiting
    c) Flowering d) Flower dropping

40. Leaf area index refers to
    a) Leaf area to dry weight of plant
    b) Leaf area to wet weight of plant
    c) Leaf area to land area
    d) Leaf area to dry weight of leaves

41. Linseed is grown in
    a) Kharif b) Rabi
    c) Summer d) All seasons

42. The fraction of photosynthetically active radiation absorbed by leaves is approximately
    a) 0.10 b) 0.30
    c) 0.60 d) 0.80

43. For sowing one hectare of sesamum seed required in kg is
    a) 10 b) 8
    c) 4 d) 1

44. A pasture covered with cultivated plants and used for grazing is known as
    a) Supplemental pasture b) Temporary pasture
    c) Tame pasture d) Rotation pasture

45. Khaira disease of rice is due to the deficiency of
    a) Zinc b) Molybdenum
    c) Boron d) Iron

46. Rock phosphate is useful in the soils which are
    a) Alkaline b) Acidic
    c) Saline d) Neutral

47. Nitrobacter converts
    a) Nitrate to nitrite b) Nitrite to nitrate
    c) Ammonical N to nitrite d) Nitrite to nitrous oxide

48. Magnesium is a constituent of
    a) Cell wall b) Nucleic acid
    c) Enzyme system d) Chlorophyll

49. Tetrazolium chloride is used for
    a) Testing the viability of the seed
    b) Breaking the dormancy
    c) Scarification of seeds
    d) All the above

50. CCC has been found to
    a) Slow down the vegetative growth
    b) Enhance the root growth
    c) Interfere with the biosynthesis of gibberllic acid
    d) All the above

51. The average cropping intensity of India in the 0.5 to 1.0 ha size holding is
    a) 141% b) 128%
    c) 110% d) 106%

52. The exchange of heat between a growing crop and hot air where by air over the crop is cooled
    a) Oasis effect b) Net radiation
    c) Micro environment d) Monochromatic effect

53. The chemical substances released by one species may inhibit species of plants other than the one releasing. It is called
    a) True allelopathy b) Functional allelopathy
    c) Alloinhibition d) Autoinhibition

54. A well-documented function of molybdenum in plant is as a part of enzyme
    a) Proteinase b) Transferase
    c) Lipases d) Nitrate reductase

55. In a field experiment the net plot size was 2.5 x 4 m and the yields were recorded in kg. The conversion factor for presenting yield in q/ha, should be
    a) 10 b) 25
    c) 50 d) 100

56. Most appropriate rhizome sp. for seed inoculation of soybean crop
    a) *R. trifoli* b) *R. leguminoseum*
    c) *R. japonicum* d) *R. Phaseoli*

57. Mycorrhizal fungi helps to increase the availability of plant nutrient
    a) Nitrogen b) Phosphorus
    c) Potassium d) Sulphur

58. The value of the variate occurring most frequently in normal distribution is called
    a) Mean b) Mode
    c) Median d) S.D.

59. Sulphur is used as an amendment to reclaim
    a) Saline soil b) Alkaline soil
    c) Acidic soil d) Sodic soil

60. Most effective herbicide for controlling nut grass in fallow field is
    a) Isoproluron b) Pendimethalin
    c) Glyphosate d) Atrazine

61. The sulphur content of single super phosphate is
    a) 6% b) 12%
    c) 18% d) 24%

62. Correlation coefficient is
    a) Zero b) More than + 1
    c) Less than - 1 d) Lies between - 1 and + 1

63. It has been observed that propanil herbicide interfered with ATP generation and electron transport of mitochondria in
    a) Wheat b) Barley
    c) Sorghum d) Green gram

64. Nitrate Reductase in plants is located in
    a) Chloroplasts b) Cytoplasm
    c) Mitochondria d) Ribosomes

65. The highest area and production of soybean in India is in the following state
    a) U.P. b) M.P.
    c) Rajasthan d) Maharashtra

66. Plant roots absorb sulphur mainly in form of
    a) SO_4 b) SO_2
    c) H_2S d) S

67. The percent of incoming radiation that is reflected is called
    a) Geotemperature b) Albedo
    c) Solarization d) None of these

68. The practice in which runoff is increased and infiltration is reduced in certain areas which can serve as a source of water supply for other areas is called
    a) Moisture conservation b) Water harvesting
    c) Moisture retention d) None of these

69. Pendimethalin is following kind of herbicide
   a) Post emergence   b) Pre-emergence
   c) Soil-incorporated   d) None of these

70. Following nutrient is described as immobile within plants
   a) Carbon   b) Nitrogen
   c) Calcium   d) Magnesium

71. The concept of harvest index was given by
   a) Pendelton   b) Donald
   c) Arnon   d) Bloodward

72. Following is called man-made crop
   a) Triticum aestivum   b) Secale cereale
   c) Oryza sativa   d) None of these

73. For measuring the efficiency cereal legume intercropping system ATER denotes for
   a) Area time equal ratio   b) Area time equivalent ratio
   c) All time equivalent ratio   d) None of these

74. Lathyrus contains a toxin called
   a) Aflatoxin   b) Neurotoxin
   c) Flavotoxin   d) None of these

75. In the hydrolysis of urea by urease the first product form is
   a) Ammonium nitrate   b) Ammonium carbamate
   c) Ammonium carbonate   d) None of these

76. Soils having high content of a swelling clays and develop cracks when dry belong to soil order
   a) Mollisols   b) Aridosols
   c) Entisols   d) Vertisols

77. Relative growth rate of crop is equal to
   a) Net assimilation rate x leaf area ratio
   b) Net assimilation rate x leaf area duration
   c) Net assimilation rate x leaf area index
   d) None of these

78. For conversion of P_2O_5 to P the following factor is used
   a) 0.734   b) 0.347
   c) 0.437   d) None of these

79. Following herbicide belongs to substituted urea group
   a) Alachlor   b) Atrazine
   c) Isoproturon   d) 2, 4-D

80. Humus colloids are composed basically of
  a) Carbon b) Nitrogen
  c) Helium d) Ozone

81. The organic matter of soils is ordinarily obtained by multiplying the organic carbon content by
  a) 1.52 b) 1.62
  c) 1.72 d) 1.82

82. The scientist who developed an equation that was related to growth of plant and supply of plant nutrients
  a) V.J. Kilmer b) R.B. Mikercher
  c) J.W. Friedrich d) E.A. Mistcherlich

83. Soil organic matter has many remarkable effects in soil and one of them is given below, identify that
  a) Increased plasticity b) Encourage granulation
  c) Enhanced cohesion d) Lowered cation exchange

84. Permissible biuret level (%) in urea fertilizer is
  a) 5 b) 4
  c) 3 d) 2

85. Once a phenoxy herbicide penetrates the cuticle, it moves inside the plant
  a) Apoplastically b) Symplastically
  c) By both mechanisms d) None of these

86. Mustard crop planted at a spacing of 30 x 10 cm will have a plants stand of (lakhs/ha)
  a) 1.1 b) 2.2
  c) 3.3 d) 4.4

87. The quantity of urea needed for supplying 69 kg N to sorghum will be
  a) 100 kg b) 125 kg
  c) 150 kg d) 175 kg

88. Retting is process by which
  a) The poor fibre stalk is sorted out
  b) The fibres in the bark get loosened and separated from woody stalk
  c) The good quality fibres are sorted out
  d) Fibres are differentiated into various retted groups.

89. Excess of chlorides in tobacco results in
  a) Inhibition of leaf burn and reduction in storage capacity
  b) Reduction in leaf and aroma
  c) Increase in leaf storage capacity and leaf burn
  d) Increase in leaf size and aroma

90. X^2 test is applied to
    a) Quantitative observations
    b) Qualitative observations
    c) Both of these
    d) None of these

91. Reflectant type of antitranspirant is
    a) PMA
    b) Kaolin
    c) Atrazine
    d) Engine oil

92. An example of sugarcane ripener is
    a) Glyphosine
    b) Glyphosate
    c) Sodium silicate
    d) NAA

93. The salt used in dynamic cloud seedling is
    a) Pentinum chloride
    b) Silver iodide
    c) Silver oxide
    d) Sodium sulphate

94. The law describing light penetration into crop canopy is
    a) Leibig's law
    b) Beer's law
    c) Bray's law
    d) Darcy's law

95. One ha cm of water weighs
    a) 10,000 tonnes
    b) 10,000 kg
    c) 100 tonnes
    d) 10^6 kg

96. The instrument used to measure relative humidity in crop canopy is
    a) Drosometer
    b) Anemometer
    c) Assman psychromter
    d) Wet and dry bulb thermometer

97. Free living association nitrogen fixing bacteria in the soil is
    a) Azospirillum
    b) Azatobacter
    c) Acetobactor
    d) Rhizobium

98. The phenomenon of development of fruit without fertilization is called
    a) Apomixes
    b) Parthenocarpy
    c) Solarization
    d) Tissue culture

99. The man-made gas that is considered responsible for the depletion of zone layer in the atmosphere is
    a) CO_3
    b) CFC
    c) SO_2
    d) N_2O

100. The harvest index of pigeon pea is about
    a) 0.35
    b) 0.50
    c) 0.60
    d) 0.15

101. The International Food Policy Research Institute is situated at

a) Rome b) Geneva
c) Bangkok d) Washington DC

102. The oil content in soybean is

a) 40% b) 20%
c) 33% d) 50%

103. An Agroforestry non-leguminous tree capable of fixing nitrogen in root nodules is

a) Casuarina b) Eucalyptus
c) Acacia d) Neem

104. The base temperature for calculating the growing degree days in most of the cereals is

a) 25^0C b) 30^0C
c) 0^0C d) 10^0C

105. The crop that has highest area in world is

a) Rice b) Wheat
c) Sorghum d) Millets

106. The crop that occupies highest area in India is

a) Rice b) Wheat
c) Sugarcane d) Maize

107. Rothamsted Agricultural Experiment Station is located in

a) U.K. b) U.S.A.
c) France d) Germany

108. The most vital controlling factor of soil temperature is

a) Organic matter b) Soil moisture
c) Soil air d) Soil depth

109. Which is the most common P-supplying fertilizer in India

a) Diammonium phosphate b) Single super phosphate
c) Rock phosphate d) Triple super phosphate

110. In terms of cost (Rs. per kg nutrient) which nutrient element is cheapest in India

a) N b) P
c) K d) Zn

111. Among the following crops, S uptake is highest by

a) Rice b) Mustard
c) Soybean d) Chick pea

112. In acid soils phosphorus is generally fixed as
   a) Calcium phosphate
   b) Zinc phosphate
   c) Silver phosphate
   d) Aluminium phosphate

113. Consumptive use of water is very close to
   a) Water requirement
   b) Evapotranspiration
   c) Transpiration ratio
   d) Irrigation requirement

114. Fe chlorosis is seen on
   a) Upper leaves
   b) Middle leaves
   c) Lower leaves
   d) None of these

115. Soil N is not lost due to
   a) Nitrification
   b) Denitrification
   c) Volatilization
   d) Leaching

116. Growing of plants in water solution of essential nutrients is called
   a) Hydroponics
   b) Aqua culture
   c) Water culture
   d) None of these

117. Lag phase on the response curve refers to
   a) The period of initial slow growth
   b) The period of accelerated fast growth
   c) The period when the yield curve tends to plateau off
   d) The period when the yield curve tends to dip due to decline in response

118. Cardinal temperature refers to
   a) The maximum, minimum & optimum temperature limits of growth
   b) Survival temperature limits of plant
   c) The best temperature for best growth phase
   d) All the above three

119. The most predominant type of barley cultivated in India is
   a) Hull-less six-rowed barley
   b) Hull-less two-rowed barley
   c) Six-row hulled barley
   d) Two-row hulled barley

120. Geographically cultivation of cigarette tobacco is limited to
   a) Central India, laterite soil region
   b) Southern India Black soils region
   c) Eastern Uttar Pradesh & Bihar state an alluvial loams
   d) Southern India laterite & red soil very good drainage

121. Groundnut the chief source of vegetable oil in India, originated in
   a) China
   b) West Africa
   c) Brazil
   d) United States of America

122. The best time of application of herbicide in transplanted flooded paddy fields would be
   a) To apply herbicide 1-2 weeks after transplanting
   b) To apply herbicide 1-2 days after transplanting
   c) To apply herbicide just before puddling
   d) To apply herbicide soon after puddling but before the start of transplanting operation

123. The most commonly used herbicide in sugarcane is
   a) 2- 4-D ester and lasso
   b) Gromoxone & Butachlor
   c) Sencore isoproturon
   d) Atrataf and Tafazine

124. Roguing refers to
   a) Removal of weeds by hand picking
   b) Removal of weeds by use of herbicides
   c) Removal of unwanted plants of same species and weeds
   d) Off-type plants with distinct morphological or phenotypic variations.

125. Harmonal herbicides are
   a) The one which when applied at recommended rates also promote crop growth besides killing the weeds.
   b) The ones which when applied at very low concentrations induce various responses in plants
   c) Total herb-killers and kill all type of growing plants
   d) None of these

126. Amongst the 3 commercial preparation of 2, 4-D available in the market the most volatile and likely to cause extensive damage by drift to sensitive crop like cotton in nearby fields are
   a) Sodium salts of 2, 4-D
   b) Amine salts of 2, 4-D
   c) Esters of 2, 4-D
   d) All are equal

127. Amongst the various herbicides available in the market the most commonly used herbicide in transplanted rice is
   a) Isoproturon
   b) Butachlor
   c) Fluchlorin
   d) Atrataf

128. The most commonly used synthetic organic fertilizer in India is
   a) Annydrons Ammonia
   b) Ammonium sulphate
   c) Urea
   d) DAP

129. In chelated form micro-nutrients are
   a) Adsorbed on clay complex of the soil to be released slowly later on
   b) Held by organic compounds which though soluble in water, do not ionize in soil solution
   c) The compounds of micro nutrients with high solubility in soil
   d) The complex micro nutrient salts fused with special type of glass

130. Fertilizer grade refers to
   a) Quality of fertilizer
   b) The ratio of N: P_2O_5: K_2O present in a fertilizer mixture
   c) The order in which the various fertilizers are to be applied in field before seeding
   d) The gradation done relative to the concentration of active ingredient present in fertilizer.

131. Critical crop growth stage in relation to crop water management view point is the one at which
   a) Water use rate is the highest
   b) Water stress effect is the most acute
   c) Crop can tolerate water deficiency with least adverse effect as yield
   d) Water stress if felt the least by a crop.

132. First irrigation to wheat is recommended at 3-4 weeks stage because
   a) Crop water need is the highest at this stage
   b) Crop requires moist root-zone for development of its adventitious roots
   c) Crop requires moist root zone for seminal root initiation
   d) It requires high moisture to withstand effect of falling low temperatures

133. 1m^3 of water is equal to
   a) 10,000 litres
   b) 100,000 litres
   c) 1000,000 litres
   d) 1,000 litres

134. Aquifer is
   a) The water bearing formation found above-ground on high hills/ mountain ranges
   b) The water bearing formation found under-ground
   c) The perched ground water zone found at shallow depths
   d) The excess water zone found underneath the plough sole layer requiring drainage

135. The depth of seeding in wheat is best related with
   a) Length of radicle
   b) Length of coleoptile
   c) Length of mesocotyl
   d) The depth of placement of embryo with in seed cotyledon

136. The inflorescence of rice is called
   a) Ear
   b) Rachis
   c) Panicle
   d) Rosette

137. Pudding in rice is done primarily to
   a) Provide a soft seed-bed in easy transplanting of seedlings
   b) To kill weeds
   c) To reduce deep percolation losses of water into soil
   d) To accomplish all the above 3 objectives

138. Practice of taking a second crop from stubbles of previous crop is known as
   a) Stubble mulch farming
   b) Double cropping
   c) Relay cropping
   d) Ratooning

139. Which crop is used for producing both the vegetable oil and fibre
   a) Castor
   b) Mesta
   c) Linseed
   d) Safflower

140. Mixed cropping is the concept of
   a) Growing two crops in alternate rows but in the same field together at one time
   b) Is raising of crops and animals together
   c) Mixing the seed and growing them together
   d) Growing one crop within standing crop of other before it I harvested and removed

141. A soil is well flocculated when it contains more of
   a) Potassium
   b) Calcium
   c) Sodium
   d) Sulphur

142. Which of the following contains more of secondary than primary nutrient
   a) Ammonium sulphate nitrate
   b) Ammonium sulphate
   c) Calcium ammonium nitrate
   d) None of the above

143. One nanometer is equal to
   a) 10^{-3} cm
   b) 10^{-5} cm
   c) 10^{-7} cm
   d) 10^{-9} cm

144. G herbaccum is a species of
   a) Jute
   b) Tobacco
   c) Mentha
   d) Cotton

145. 1000-grains weight of basmati rice generally ranges between
   a) 8-10g
   b) 15-20g
   c) 25-30g
   d) 35-40g

146. Pyrite predominantly contains
   a) Fe, S
   b) Fe, Zn
   c) Ca, S
   d) Fe, Al

147. Recommended concentration of zinc sulphate solution for spray in rice crop is
   a) 0.05 percent
   b) 0.50 percent
   c) 1.00 percent
   d) 2.00 percent

148. Cyperus rotundus is a
   a) Halophyte
   b) Sedge
   c) Root parasite
   d) Grassy weed

149. First volume of the Advances in Agronomy was published in the year
   a) 1929
   b) 1939
   c) 1949
   d) 1950

150. A crop grown in pure stand at optimum population is
   a) Sole crop
   b) Mono crop
   c) Main crop
   d) Lease crop

151. A non selective herbicides is
   a) Atrazine
   b) Fluchloralin
   c) Glyphosate
   d) 2, 4-D

152. A perennial fodder legume is
   a) Glyricidia
   b) Stylosanthus
   c) Cowpea
   d) Susbania sp.

153. Sorghum poisoning in cattle is due to
   a) Sodium
   b) Iron
   c) HCN
   d) Nitrate

154. The ratio of the volume of CO_2 released to the volume of O_2 absorbed in the respiratory process is
   a) R.Q.
   b) P.Q.
   c) Q. 10
   d) None

155. Low temperature treatment of seeds and seedlings is known as
   a) Solarisantion
   b) Vernalization
   c) Hardening
   d) Freezing injury

156. Medium range weather forecasting is valid up to
   a) 10 days
   b) 3 days
   c) 30 days
   d) 7 days

157. The typical symptom of phosphorus deficiency in plants is
   a) Chlorosis
   b) Marginal scorching of leaves
   c) Development of purple colour in leaves
   d) All symptoms

158. The percentage of exchangeable Na in non saline alkali soil
a) Less than 15
b) 15
c) More than 15
d) None

159. The following can be used as antitranspirant
a) PMA
b) Urea
c) Machete
d) All three

160. Soil productivity depends on
a) Soil fertility
b) Soil physicxal condition
c) Soil chemical cpndition
d) All the above

161. The process of soaking of the jute plants after harvesting in water to facilitate seperation of the fibre the stalk is called
a) Rotting
b) Retting
c) Soaking
d) Fermentation

162. Which one of the following is a long day plant
a) Barley
b) Pearl millet
c) Cotton
d) Groundnut

163. Single superphosphate is manufactured by reacting
a) H_2SO_4 with gypsum
b) H_2SO_4 with rock phosphate
c) H_2PO_4 with rock phosphate
d) H_3PO_4 with NHO_3

164. Gypsum is applied to reclaim
a) Sodic soils
b) Saline soils
c) Acidic soils
d) None of these

165. An amino acid accumulation observed in plants after the moisture stress and considered as a good indicator of moisture stress is known as
a) Abscicic acid
b) Proline
c) Betane
d) Ethylene

166. When one species yields less and the other more than expected in the intercropping, the competition is known as
a) Mutual inhibition
b) Mutual cooperation
c) Compensation
d) None of the above

167. The biofertilizer used in wetland paddy is
a) Azatobacter
b) Anabaena
c) Azospirillum
d) Rhizobium

168. The nitrogenous fertilizer containing the highest N content is
a) Urea
b) Anhydrous ammonia
c) Ammonium nitrare
d) Ammonium sulphate

169. If gravimetric water content of a soil is 15% and bulk is 1.5 g/cc the volumetric water content is
a) 22.5
b) 10
c) 100
d) None of these

170. Isohytes are lines connecting places of equal
a) Rainfall
b) Humidity
c) Sunshine hours
d) Evaporation

171. In an Randomised Block design, there are 8 treatments and 4 replications, the degree of freedom for error is
a) 32
b) 24
c) 21
d) 31

172. The normal quantity of water given per irrigation in conventional methods is
a) 50 cm
b) 60 cm
c) 20 cm
d) 6 cm

173. The nutrient absorbed both in anionic and cationic form is
a) Calcium
b) Nitrogen
c) Phosphorus
d) Sulphur

174. The normal recovery of sugar from sugarcane is
a) 5%
b) 10%
c) 20%
d) 30%

175. ICRISAT is located in
a) Brazil
b) South Africa
c) USA
d) India

176. The protein content of the pulse blackgram will be about
a) 10%
b) 20%
c) 40%
d) 50%

177. Heavy soils are given this name because
a) Bulk density is high
b) Particle density is high
c) Difficult to till
d) Give more yield

178. Which crop contributes to maximum food grain production in India
a) Wheat
b) Rice
c) Maize
d) Millets

179. The chemical formula of urea is
a) $CO(NH_2)_2$
b) $CONH_2$
c) $CO(NH_2)_4$
d) None of these

180. Inverse yield nitrogen law is propounded by
   a) Spillman
   b) Wilcox
   c) Baule
   d) None of these

181. Example of indeterminate crop
   a) Sorghum
   b) Rice
   c) Setaria
   d) Cotton

182. The most sensitive stage of a crop for alkaline conditions
   a) Germination
   b) Tillering
   c) Flowering
   d) Maturity

183. An idea of total N content of a soil can be had from the soil content of
   a) Total p
   b) Total K
   c) Total Ca
   d) Organic carbon

184. If IW is 60 mm and IW/CPE ratio is 1.2, the CPE at which irrigation shall be scheduled is
   a) 20 mm
   b) 50 mm
   c) 40 mm
   d) None of these

185. Shade loving crops are termed as
   a) Mesophytes
   b) Heliophytes
   c) Sciophytes
   d) Hydrophytes

186. If the nutrient content is 100 ppm, the percent content is
   a) 0.001
   b) 0.01
   c) 0.1
   d) None of these

187. Vertical mulches are useful in
   a) Sandy soils
   b) Red soils
   c) Black soils
   d) Loamy soils

188. Water use efficiency of a crop is calculated using the formula
   a) Y/ET
   b) Y/E
   c) Y/ PET
   d) Y/T

189. P=0.05 indicate probability percentage of
   a) 50%
   b) 5%
   c) 0.5%
   d) 0.05%

190. Approximate seed rate of rice for broadcasting directly in the field is (kg/ha)
   a) 10
   b) 40
   c) 80
   d) 160

191. The content of phosphorus in diammonium phosphate is
   a) 18% b) 64%
   c) 50% d) 46%

192. Luxury consumption is usually referred to the nutrient
   a) Nitrogen b) Phosphorus
   c) Potassium d) Sulphur

193. A green manure, cum fodder fibre crop is
   a) Sunnhemp b) Cowpea
   c) Sesbania d) Mesta

194. The ratio of economic yield to total biological yield is called is
   a) Harvest ratio b) Harvest Index
   c) Net assimilation Ratio d) Land equivalent ratio

195. If a tonne of urea cost Rs. 3680 the unit cost of nitrogen will be (Rs/kg)
   a) 4 b) 6
   c) 8 d) 10

196. Drip method of irrigation can be recommended to
   a) Rice b) Pearl Millet
   c) Groundnut d) Cotton

197. Integrated nutrient management means
   a) Use of fertilizers only b) Use of organic manures only
   c) Use of biofertilizers only d) All sources of nutrients

198. Groundnut requires application of Gypsum for
   a) Lowering soil pH b) Supply of calcium
   c) Supply of sulphur d) Supply of calcium and sulphur

199. Which of the following crop has the highest cultivated area in the world
   a) Rice b) Wheat
   c) Barley d) Bajra

200. The number of replications is equal to number of treatments in the design
   a) Latin square
   b) Randomised Block Design
   c) Completely Randomized Block Design
   d) Split plot

Model Paper-II

1. ….. (1674- 1741 AD) conducted several experiments on plant nutrition and published a book, "Horse Hoeing Husbandry"
   a) Sir Humphry Davy b) Aristotle
   c) Jethro Tull d) Van Helmounst

2. When seedlings are transplanted, they establish by producing
   a) Seminal roots b) Crown roots
   c) Fibrus roots d) Nodal roots

3. If the seeds having 100% germination and test weight of 21g are to be sown at 40 x 10 cm spacing, the net seed rate for one ha will be ……………..
   a) 5.25 kg b) 5.0 kg
   c) 4.25 kg d) 4.0 kg

4. When primary tillage is completely avoided and secondary tillage is restricted to seedbed preparation in the row zone only it is …………
   a) Minimum tillage b) Zero tillage
   c) Row zone tillage d) Advance

5. In case of rice when urea is placed in the reduced zone it reduces ………… losses
   a) Leaching and dentrification b) Leaching and volatilization
   c) Runoff and leaching d) Denitrification and volatilization

6. The weeds that belong to the family ………… are known as sedges
   a) *Gramineae* b) *Convolvulaceae*
   c) *Euphorbiaceae* d) *Cyperaceae*

7. The critical period of weed competition for transplanted rice is ………… days after transplanting
   a) 30-50 b) 15-45
   c) 20-60 d) 25-60

8. Butachlor and propanil belong to ………….. group of herbicide
   a) Ureas b) Carbamates
   c) Aliphatic d) Amides

9. The …………….. under the trade name Collego is normally applied to rice and soybean
   a) Bioherbicide b) Chemical herbicide
   c) Mycoherbicide d) Natural plant origin herbicide

10. If the recommended dose of oxydiazon (0.25% a/i) to rice is 0.5 kg ai/ha, how much amount of commercial product you will apply in the crop.
    a) 1 kg
    b) 2 kg
    c) 1.5 kg
    d) 2.5 kg

11. Moisture stress in soils and plants during crop growth due to imbalance between soil moisture and evapotranspiration of a crop is ....
    a) Agricultural drought
    b) Hydrological drought
    c) Meteorological drought
    d) Hysiological drought

12. The air entry value to tensiometer cup is as high as ....
    a) 0.75 bar
    b) 0.80 bar
    c) 0.85 bar
    d) 0.90 bar

13. If a crop is to be irrigated at 0.75 ID/CPE ratio with 60 mm depth, it should be irrigated after ....CPE
    a) 60 mm
    b) 75 mm
    c) 80 mm
    d) 70 mm

14. Under saline soil conditions ....irrigation could help in obtaining better crop stand and yield
    a) Sprinkler
    b) Subsurface
    c) Check basin
    d) Furrow

15. Use of brackish water for agriculture has become more relevant due to .... irrigation:
    a) Sprinkler
    b) Furrow
    c) Check basin
    d) Micro

16. Drainage lowers underground water table so as to facilitate increased ....
    a) Water holding capacity
    b) Root zone depth
    c) Microbial activity
    d) Porosity

17. In alkali soils crop growth is affected due to ....
    a) Higher salt concentration
    b) Poisonous ion concentration
    c) Poor physical properties
    d) Microorganisms

18. At initial stages of reclamation of sodic soils, the cropping sequence which is more remunerative is
    a) Rice-wheat-dhaincha
    b) Rice-sunflower-green gram
    c) Rice-cotton-soybean
    d) Rapeseed-wheat-dhaincha

19. .... is the most tolerant crop in respect of alkalinity and salts:

a) Alfalfa b) Barely
c) Cotton d) Rice

20. In acidic soils the plant growth is affected due to ....ions:

a) Fe and Cl b) Al and Cu
c) Fe and Al d) Cl and Cu

21. Cotton cultivar with terminator gene was first introduced by ....seed company:

a) Kargil b) Pioneer
c) Nath d) Monsanto

22. Jowar crop cultivated for fodder purpose should harvested ....

a) One month after sowing b) Two months after sowing
c) At flowering d) At maturity

23. Potato was originated in ....

a) South Africa b) North America
c) Europe d) Australia

24. The most important climatic factor affecting wheat yield in India is ....

a) Sunshine hours b) Rainfall
c) Temperature d) Humidity

25. The formula for calculating degree days is :

a) Max. temp + Min. temp/2 + Base temp.
b) Max. temp – Min. temp/2 - Base temp.
c) Max. temp + Min. temp x 2 - Base temp.
d) Max. temp - Min. temp x 2 + Base temp.

26. The spices adopted to low light intensity are ....

a) Cardamom and clove b) Ginger and clove
c) Cumin and cardamom d) Turmeric and clove

27. The climate factor responsible for the quality of beetle vine is ....

a) Temperature b) Humidity
c) Light intensity d) Rainfall

28. The most ground water polluting fertilizers are ......

a) Nitrate fertilizers b) Ammonical fertilizers
c) Potassic fertilizers d) Amide fertilizers

29. In a strip plot design with three replications, six horizontal plots and three vertical plots, the degrees of freedom of error used to test the variation between the interaction are ....

a) 4
b) 10
c) 20
d) 30

30. Homogeneity of experimental material is expected in ....

a) Latin square design
b) Completely randomized design
c) Split plot design
d) Randomized block design

31. Statistical procedures employed in making inferences about treatments hold good only when there is :

a) Randomization
b) Replication
c) Randomization and local control
d) Randomization and replication

32. Precise estimate of experimental error is obtained through ....

a) Replication
b) Randomization
c) Randomization and local control
d) Randomization and replication

33. To test the equality of mean yield of a crop from two independent fields treated alike except the fertilizer, one has to use ....

a) Chi-square test
b) F-test
c) Z-test
d) T-test

34. Which of the following is an example of an agrisilviculture system:

a) Teak + Napier
b) Wheat + Poplar
c) Maize + Cowpea
d) Coffee + Coconut + Pineapple

35. Cultivation of food crops between hedgerows of multipurpose shrubs known as ....

a) Alley farming
b) Silviculture
c) Shifting cultivation
d) Silvi-pastoral system

36. Which of the following is a multipurpose fodder, fuel wood, timber and fast growing tree species:

a) Terminalia bellerica
b) Pongamia pinnata
c) Leucaena leucocephala
d) Acacia coinciana

37. Kharif, Rabi and Ziad words belong to

a) Urdu
b) Persian
c) Arabic
d) French

38. The alternate host of gram caterpillar is one of the following

a) *Amaranthus*
b) *Crotalaria* sp.
c) *Agropyron repens*
d) *Cenchrus* sp.

39. Skin allergy is caused by one of the following

a) Amaranthus
b) Brush weeds
c) Cyperus
d) Parthenium

40. Which one of the following weed comes in the category of sedges

a) *Chloris barbata*
b) *Eclipta alba*
c) *Cyperus rotundus*
d) *Leuscus aspera*

41. Which one of the following weeds belong to the Leguminaecae family

a) *Elesusine indica*
b) *Melilotus alba*
c) *Panicum repens*
d) *Cyperus deformis*

42. Which of the following herbicides is used for chemical tillage

a) 2, 4, D
b) Grammaxone
c) MCPA
d) MCPB

43. Which one of the following weedicides is available in wettable powder form

a) 2, 4-D
b) Pendamethaline
c) Isoproturon
d) MCPB

44. Which one of the following weedicide is recommended for wheat + mustard

a) 2, 4-D
b) Pendamethaline
c) Isoproturon
d) MCPB

45. If the field capacity is 18% and permanent wilting points is 6%, 50% of the available soil moisture will be

a) 8 %
b) 10%
c) 12%
d) 14%

46. Which of the following antitranspirant belongs to stomatal closing group

a) Cycocel
b) PMA
c) Silicone
d) Paraquat

47. Which one of the following chemical is used as antitranspirant in dryland crops:
    a) 2, 4-D   b) MCPA
    c) Cycocel   d) Paraquat

48. The best crop for intercropping with autumn planted sugarcane is :
    a) Potato   b) Mustard
    c) Wheat   d) Toria

49. For transplanting of rice in late sown condition the optimum age of seedling is:
    a) 20 days   b) 25 days
    c) 30 days   d) 35 days

50. After thinning the number of plants per hectare at 50 cm x 20 cm is :
    a) 50000   b) 75000
    c) 100000   d) 125000

51. What sort of aberration in rainfall affects crop growth most:
    a) Lesser rainfall   b) Higher rainfall
    c) Uncertain distribution   d) No rain in reproductive stage

52. Which of these influence most the water requirement of a crop:
    a) Weather   b) Canopy development
    c) Water table   d) Soil condition

53. Which one be called as Dryland?
    a) When rainfall received is less than 1000 mm
    b) When rainfall received is less than 500 mm
    c) When rainfall received is less than 750 mm
    d) Hilly terrain with low rainfall less than 500mm

54. In undulating terrain one should go for ....
    a) Normal crop raising   b) Strip cropping
    c) Pastoral cropping   d) Mixed pastoral and tree cropping

55. Higher yield of rice could be ensured by adequate supply of ....fertilizer at the tillering stages
    a) Nitrogen   b) Potash
    c) Both N & K   d) Phosphorus

56. By delayed planting of rice in the second season the yield decreases most due to ....
    a) Poor tillering   b) Higher temperature at flowering
    c) Heavy insect infestation   d) For want of water

57. For greater precision in field experiments with 2-3 sets of factors it is better to adopt the design...
    a) Randomized block
    b) Confounding
    c) Split plot
    d) Incomplete block

58. Better drainage helps crop growth most due to ....
    a) Better aeration of the soil
    b) Better mineralisation of manures and fertilizers
    c) Better root growth
    d) Better availability of nutrients

59. Tillering in rice is encouraged by ....
    a) Optimum moisture content of the soil
    b) Saturation of the soil
    c) Shallow depth of standing water
    d) Higher depth of standing water

60. Which of the following statement is false
    a) 2, 4-D is specific to kill monocots
    b) MCPA is specific for killing dicots
    c) Aquatic weed "chara chinensis" can be controlled by 2, 4-D
    d) Glyphosate kills woody plants

61. Higher yield of groundnut could be achieved by:
    a) Use of higher dose of phosphate
    b) Adequate amount of N
    c) No nitrogen application
    d) Closer planting

62. Mustard yield is largely reduced mainly due to:
    a) Cold wind at the flowering stage
    b) Attack of aphids
    c) Lack of adequate fertilizer application
    d) Failure to apply irrigation in flowering stage

63. Chemical name of herbicide 'Machete' is:
    a) Isproturon
    b) Fluchloralin
    c) Alachlor
    d) Butachlor

64. Instrument used for measuring wind velocity is known as:
    a) Hydrometer
    b) Pyranometer
    c) Altimeter
    d) Anemometer

65. Air dried fodder is called as:
    a) Silage
    b) Hay
    c) Green fodder
    d) Stove

66. Stem nodulating green manure plant is
   a) Sunnhemp b) Sesbania aculiata
   c) Sesbania rostrata d) Tephrosea sps.

67. The term PET was coined by:
   a) Sarkar and Biswas b) Haigreaves
   c) Thomthwaite d) Mendel

68. Chemical used for the control of transpiration
   a) Kaolin b) Lime
   c) PMA d) ABA

69. A fodder and pasture legume
   a) Glyricidia b) Stylosanthus
   c) Cowpea d) Sesbania

70. Hydrolysis of urea results in the loss in nitrogen in the form of
   a) NH_4 b) NH_3
   c) NH_2 d) NH_4OH

71. One acre inch of water is equivalent to litres of water:
   a) 102800 b) 126800
   c) 140000 d) 180000

72. The first dry farming research station was started in the year 1936 at
   a) Bijapur b) Sholapur
   c) Manjri d) Rohtak

73. Glyphosate is a .... herbicide
   a) Selective b) Pre-emergence
   c) Non-selective d) Preplanting

74. Orobanche is a parasite on
   a) Sorghum b) Sugarcane
   c) Safflower d) Tobacco

75. Famous Rosthamsted Agricultural Experiment Station was established by:
   a) Liebig b) DeSaussure
   c) Jethro Tull d) Lawes and Gilbert

76. Person who established the relationship between plant growth factor:
   a) Liebig b) Van Helmont
   c) Tisdale d) Mistscherlich

77. Soil particles get deflocculated due to
a) Calcium b) Sodium
c) Magnesium d) Sulphur

78. Availability of H_2PO_4 is greatest when soil pH is
a) Low b) High
c) Neutral d) None of these

79. First interspecific hybrid cotton was released by
a) Tamil Nadu b) Gujarat
c) Punjab d) Karnataka

80. Natural farming was advocated by
a) Gilbert b) Fakuoka
c) Lawes d) Norman E. borlaug

81. Reproductive structures in succession in cotton
a) Flowers-bolls-squares b) Squares-bolls-flowers
c) Square-flowers-bolls d) Square-bolls-squares

82. Growth of Parthenium weed can be controlled by weed
a) Cassia serecia b) Cyperus rotundus
c) Cynodon doctylon d) Momosa pudica

83. Watershed Management helps to increase productivity in
a) Irrigated areas b) Rainfed areas
c) Humid areas d) None of these areas

84. Single superphosphate contains phosphorus in:
a) Citrate soluble form b) Insoluble form
c) Water soluble form d) None of these

85. Simarauba glauca has a high potential for
a) Protein b) Sugar
c) Edible oil d) None of these

86. Percentage of lint in seed cotton is generally
a) 80% b) 60%
c) 30% d) 10%

87. Angstrom is a unit of length equal to
a) 10^{-2} cm b) 10^{-4} cm
c) 10^{-6} cm d) 10^{-2} cm

88. The study of the relation of agricultural crops and environment is called as
a) Autoecology b) Agro-ecology
c) Agrology d) Agrometeorology

89. Cultivation of woody plants, particularly those used for decoration and shade is known
   a) Autoecology b) Arboriculture
   c) Silviculture d) Aviculture

90. Sea Island cotton or Egyptian cottons belongs to
   a) Gossypium arboreum b) Gossypium hirsutum
   c) Gossypium barbadense d) Gossypium herbaceum

91. Plants tolerant to high soil salinity conditions
   a) Heliophytes b) Hylophytes
   c) Heliotropic d) None of these

92. Dust mulch is very effective in
   a) Coarse textured soil b) Heavy textured soil
   c) Sandy soil d) None of these

93. The value of albedo in most o the dry land soils ranges from
   a) 0.15 to 0.35 b) 0.00 to 0.10
   c) 0.50 to 0.75 d) 0.76 to 0.80

94. Normally drought is assessed using
   a) Aridity b) Dryness
   c) Aridity anomaly index d) Deficient rainfall

95. A year in which rainfall of a particular place is short by more than ....is known as drought year:
   a) Twice the mean b) Twice the standard deviation
   c) Twice the median d) Twice standard error

96. Which one of the following is considered as rain water recycling
   a) Farm pond b) Wells
   c) Percolation pond d) Compartmental bunding

97. The most appropriate classification scheme which relates climate to vegetation is:
   a) Penman classification b) Thornthwite classification
   c) Koppens classification d) None of the above

98. The plastic much helps to
   a) Reduce soil moisture b) Increase the evaporation
   c) Increase soil temperature d) None of the above

99. A recorded rainfall of ....within a period of 24 hours is considered as rainy day
   a) More than 2.5 mm b) More than 5.0 mm
   c) More than 3.0 mm d) More than 1.0 mm

100. ....is a rabi crop raised under residual soil moisture in dry lands

a) Cotton b) Maize
c) Chickpea d) Black gram

101. Most important problem with Indian vertisols of SAT is

a) Less nutrient b) Water stagnation
c) Iron deficiency d) Erosion

102. At permanent wilting point the approximate value of PF would be

a) 1.6 b) 5.8
c) 4.2 d) 1.0

103. Moisture moves through soil pores at a rate proportional to the

a) Hydraulic gradient b) Hydraulic head
c) Soil porosity d) None of the above

104. The lateral influx of water in leveled freely drained soil is

a) Very active b) Negligible
c) No relationship d) Moderately active

105. Duty of water is

a) Controlled and uniform application of water
b) Irrigation along the contour
c) Ratio between actual and potential ET
d) None of the above

106. Advance and recession curves are used to estimate

a) Channel length b) Channel ha-m
c) Channel depth d) None of the above

107. India's annual water input from all the sources is approximately

a) 350 million ha-m b) 400 million ha-m
c) 100 million ha-m d) 650 million ha-m

108. Infiltration is:

a) Downward movement of water
b) Lateral movement of water
c) Entry of water at the soil surface
d) Downward flux at the bottom layer

109. A minor irrigation project will have a capacity to irrigate

a) <3000 hectare b) <2000 ha
c) < 4000 ha d) 1000 ha

110. Water conveyance across the road is possible with:
   a) Inverted siphons
   b) Culverts
   c) Spill ways
   d) None of the above

111. A method of drainage where the drains are placed in a direction more or less at right angles to the direction of the steepest slope of the land to be drained is known as:
   a) Tile drainage
   b) Transverse drainage
   c) Crisscross drainage
   d) Steep drainage

112. The total storage capacity created in the various river systems in India is:
   a) 40 million ha-m
   b) 22.2 million ha-m
   c) 33.3 mullion ha-m
   d) 44.4 million ha-m

113. Water required to produce one ton of rice is:
   a) 800 tonnes
   b) 1000 tonnes
   c) 2000 tonnes
   d) 3000 tonnes

114. USWB class A plan evaporimeter is having:
   a) Diameter of 120 cm with depth of 25 cm
   b) Diameter of 100 cm with depth of 25 cm
   c) Diameter of 25 cm with depth of 120 cm
   d) Wilting point and permanent wilting point

115. Water requirement is equal to:
   a) IR + ER + S
   b) ER + S
   c) IR
   d) IR – (ER + S)

116. Field water use efficiency is:
   a) Y/WR
   b) Y/ET
   c) Y/E
   d) Rate of photosynthesis/ rate of transpiration

117. The maximum rate of soil moisture used by crop under moderate humid condition is:
   a) 4 mm per day
   b) 6 mm per day
   c) 8 mm per day
   d) 2 mm per day

118. ....is the line of equal precipitation
   a) Isohyet
   b) Isotach
   c) Isobar
   d) Isotherm

119. When the hydraulic conductivity is about ....cm per hour, it is classified as moderate
   a) 0.001 to 0.01
   b) 0.01 to 0.1
   c) 0.1 to 1.0
   d) 1 to 10

120. Vertical mulching is the incorporation of .... mulch

a) Soil b) Vegetative
c) Straw d) Roughage

121. Common name of 2-methyl, 4-chlorophenoxy acetic acid is:

a) 2, 4-D b) MCPB
c) MCPA d) 2,4, 5-T

122. Approximately........% of water taken by the roots is transpired to the atmosphere through stomatal pores in the leave

a) 50 b) 80
c) 90 d) 99

123. A single plant of *Phalaris minor* produces about ....seeds

a) One lakh b) 2000
c) 75,000 d) 1,50,000

124. The index which expresses the reduction in yield due to the presence of weed free situation is:

a) Weed control efficiency b) Weed index
c) Weed control index d) None of the above

125. Diallate and Triallate herbicides come under:

a) Nitrates group b) Phenols group
c) Triazoles group d) Thiocarbamates group

126. ....is an annual stem

a) Dodder b) Broomrape
c) Witchweed d) Lantana camera

127. Weeds Science journal is published in:

a) UK b) India
c) Netherlands d) USA

128. ....is a growth regulating substance:

a) Ethepan b) Atrazine
c) Dalapon d) Divron 2, 4-D

129. For rice plants..........stage have to face greater set back due to presence of weeds in the field

a) Seedling stage b) Vegetative stage
c) Grain formation d) Flowering stage

130. The indicator plants used for atrazine herbicide bio assay:

a) Minor millets b) Oat, mustard
c) Soybean, cucumber d) Sorghum

131. An application of spray or dust to a restricted area at base of plant is called:
   a) Spot application b) Lay by application
   c) Directed d) Band treatment

132. ....are polar compounds having both electrically positive and negative regions
   a) Lipophilic substances b) Solvents
   c) Hydrophilic substances d) Surfactants

133. An example of worst perennial weed:
   a) Paspalum conjugatum b) Eleusine indica
   c) Eichhornia crassipes d) Chenopodium album

134. ....is the pre-emergence herbicide used for controlling weeds in jute:
   a) Benthiocarp b) MSMA
   c) DSMA d) Nitrofen

135. Herbicide which can be applied before planting/sowing:
   a) Glyphosate b) Alachlor
   c) Atrazine d) Fluchloralin

136. Weeds which will grow well on alkali soil:
   a) Barnyard grass b) Nut grass
   c) Quack grass d) *Chenopodium album*

137. One of the main difference between rice and Echinochloa cruspalli (weed) is:
   a) Leaves are rough and hairy (rice) and leaves are smooth (weed)
   b) Stems are erect and produce tillers (rice) and do not produce tillers (weed)
   c) Have auricles and ligules (rice) and no auricles and ligules (weed)
   d) No difference till flowering

138. The .....of a series of n values is the nth root of the product of n values of the series
   a) Mode b) Median
   c) Geometric mean d) None of the above

139. The term variance is used to denote the:
   a) Standard deviation
   b) Square of the standard deviation
   c) Square root of the standard deviation
   d) None of the above

140. Trend in rainfall can be analysed with:
a) Multiple regression technique b) Correlation technique
c) Moving averages d) None of the above

141. Quartiles are useful to separate:
a) Extreme values b) Mean of the values
c) Normalized values d) None of the above

142. The growth pattern of a cereal plant falls in:
a) Linear curve b) Sigmoidal curve
c) Cubic curve d) Binomial curve

143. Dividing experimental units into compact blocks satisfies:
a) Randomization b) Uniformity
c) Local control d) Replication

144. According to ……. "the index numbers are used to measure the changes in some quantity which cannot be observed directly"
a) Bowley b) Blair
c) Karl Pearson d) Goudriaan

145. Historgram represents
a) Cumulative frequency b) Frequency distribution
c) Mean deviation d) None of the above

146. Nursery area requirement for transplanting of 1 ha of field is:
a) 0.2 acre b) 0.1 ha
c) 0.2 ha d) 0.1 acre

147. A pasture covered with cultivated plants and used for grazing is known as:
a) Supplemental pasture b) Temporary pasture
c) Tame pasture d) Rotation

148. Bolting refers to:
a) Stem elongation b) Flowering
c) Fruiting d) Flower dropping

149. Lithosphytes are plants which are usually found on:
a) Soil surface b) Rock surface
c) Water bodies d) None of the above

150. The origin of Toria is:
a) China b) Afghanistan
c) India d) Middle East

151. Javanica a subspecies of O. sativa is grown in:
a) Indonesia
b) India
c) China
d) Thailand

152. Seed rate of rice for broad casting in the field directly is .... Kg/ha
a) 80
b) 60
c) 50
d) 35

153. Botanical name of black gram is:
a) Panicum miliane
b) Phaseolus mungo
c) Cicer arietinum
d) Cajanus cajan

154. For getting higher yield and to control 'scab' the pH requirement for potato crop is:
a) 5-5.2
b) 4-4.2
c) 6-6.2
d) 7-7.2

155. Borax + gypsum is recommended to groundnut crop for ....
a) More flower production
b) More yield
c) Pod development
d) To increase the test weight of the kernel

156. The equivalent acidity (per 100 kg of fertilizer) of urea is:
a) 80 kg
b) 95 kg
c) 110 kg
d) 128 kg

157. Luxury consumption is usually referred to nutrient:
a) Sulphur
b) Nitrogen
c) Phosphorus
d) Potassium

158. Mar is referred to:
a) Pig manure
b) Poultry manure
c) Raw humus
d) Iron

159. Nitrobacter converts:
a) Nitrate to nitrite
b) Nitrite to nitrate
c) Ammonical N to nitrite
d) Nitrite to nitrous oxide

160. ....micro nutrients are required for N fixation
a) Copper and boron
b) Chlorine and iron
c) Cobalt and molybdenum
d) Zinc and sulphur

161. Tetrazobium chloride is used for:
a) Testing the viability of the seed
b) Breaking the dormancy
c) Scarification of seeds
d) All the above

162. Oxisols are mostly found in:
a) All climates
b) Temperate climates
c) Sub-tropical climate
d) Tropical climate

163. ....contains, the antiquality constituents called saponin:
a) Luceme
b) Leucaena
c) Sweet clover
d) Sorghum

164. In cotton water stress during the period of early flowering causes:
a) Shedding of new flower bud
b) Shedding of current flowering
c) Shedding of bolls
d) All of the above

165. Planting......direction may yield more than the other directions:
a) North-south
b) East-west
c) North east-south west
d) None of the above

166. The tree species that can be grown on water logged are:
a) Casuarinas
b) Neem
c) Willos, bamboos
d) Protopis

167. Most of the cultivated groundnuts are grouped under:
a) Nambyquarc
b) Asiatica
c) Gigantia
d) Oleifera

168. The begassee as a by product from the sugar industry is about ......% of the weight of the cane crushed:
a) 10
b) 20
c) 30
d) 40

169. Diversity Index measures the:
a) Multiplicity of farm products in a year
b) Multiplicity of farm products in a season
c) Multiplicity of farm products in two years
d) None of the above

170. One foot candle ....lux:
a) 1.0764
b) 10.764
c) 107.64
d) 1076.4

171. Chemical used for delineating cotton is:
a) KCl
b) HNO_3
c) H_2SO_4
d) Paraquat

172. Sorghum poisoning in cattle is due to:
a) Sodium
b) Iron
c) HCN
d) Nitrate

173. Movement (growth) of plants towards light is called:
a) Photosynthesis
b) Phototropism
c) Photoperiodism
d) Heliotropism

174. Visible spectrum of light ranges from:
a) 390-760 nm
b) 300-700 nm
c) 420-720 nm
d) 500-880 nm

175. Medium range weather forecasting is valid upto ....days
a) 10 days
b) 3 days
c) 30 days
d) 7 days

176. Material used for artificial rain making is:
a) KH_2PO_4
b) Silver lodide
c) $HgCl_2$
d) Common salt

177. The herbicide recommended to control phalaris minor in wheat is:
a) 2, 4-D
b) Isoproturon
c) Alachlor
d) Fluchloralin

178. 'Lag phase' on the response curve:
a) The period of initial slow growth
b) The period of accelerated fast growth
c) The period when the yield curve tends to plateau off
d) The period when the yield curve tends to dip due to decline in response

179. Successful recycling of high C: N ratio rice straw on agricultural lands could be accomplished:
a) By adding nitrogenous fertilizers at incorporation
b) By adding cellulytic micro organisms (fungi)
c) By giving an additional irrigation after incorporation
d) By adopting all above three options together

180. Cardinal temperature refers to:
a) The maximum, minimum and optimum temperature limits of growth
b) Survival temperature limits of plant
c) The best temperatures for best growth phase
d) All the above three

181. Cytokinins are plant growth regulators used:
a) To inhibit the growth of plants
b) To stimulate cell division and delay the senescence
c) To break dormancy in bulbs and tubers
d) To hasten ripening of fruits

182. The most pre-dominant type of barley cultivated in India is:
a) Hull-less six-rowed barley
b) Hull-less two-rowed barley
c) Six-row hulled barley
d) Two-row hulled barley

183. Groundnut the chief source of vegetable oil in India originated in:
a) China
b) West Africa
c) Brazil
d) United States of America

184. Which of the following is most suited for cultivation under extremely dry conditions:
a) Brassica napus
b) Brassica juncea
c) Brassica campestris var. yellow sarso
d) Eruca sativa

185. Which of the following 4 species of cotton cultivated in India accounts for the largest hectarage occupancy:
a) Gossypium arboreum
b) Gossypium hercaum
c) Gossypium barbadense
d) Gossypium hirsutum

186. Mentha cultivation is done primarily in North India to get:
a) Mentha seed for use as condiments
b) Mentha herbs for use in medicines
c) Mentha oil for use in commerce
d) For obtaining a high quality fodder for milch cattle

187. 'Congress grass' is a common weed found:
a) In cultivated irrigated lands
b) In dry lands farming situation
c) In waste lands and along the rails road
d) In high altitudes with high rains

188. Phosphorus is most easily absorbed by plants:
a) HPO_4
b) As H_2PO_4
c) As PO_4
d) With equal ease in all the above

189. Concentration of hydrogen at pH9 would be how much different than at pH7:
a) 10 times greater
b) 100 times greater
c) 10 times lesser
d) 100 times lesser

190. Annidation refers to:
a) Direct or indirect harmful effects that one plant has on another
b) Complementary interactions which occur both in space and time
c) Interaction effect of fertilizer levels and plant population densities
d) Date of planting and fertilizer rates

191. In chelected form micro nutrients are:
   a) Absorbed on clay-complex of the soil to be released slowly, later on
   b) Held by compounds which though soluble in water, do not ionize in soil solution
   c) The compounds of micro nutrients with high solubility in soil
   d) The complex micronutrients salts fused with special type of glass

192. Which of the following nutrient element is the most immobile in plant:
   a) Sulphur
   b) Calcium
   c) Boron
   d) Magnesium

193. The depth of seedling in wheat is best related with:
   a) Length of radicle
   b) Length of coleoptile
   c) Length of mesocotyle
   d) The depth pf placement of embryo with in seed cotyledon

194. Ergot is a serious disease of:
   a) Sorghum
   b) Pearl millet
   c) Maize
   d) Rice

195. Triticale is a:
   a) New legume fodder crop developed for milch cattle
   b) New cereal crop prepared by crossing wheat with rye
   c) New grass developed for growing on low lands
   d) A species of wheat not found in India

196. Boro crop of rice in East India is planted in fields in:
   a) June-July
   b) November-December
   c) February-March
   d) None of these

197. Puddling in rice fields is done, primarily to:
   a) Provide a soft seed-bed to ease transplanting of seedlings
   b) To kill weeds
   c) To reduce deep percolation losses of water into soil
   d) To accomplish all the above 3 objectives

198. Gram for good germination requires:
   a) Fine seed bed of good tilth
   b) Very fine powdry and loose tilth
   c) Cloddy and rough seed bed
   d) Firm and fine tilth

199. 'Tikka disease' is an important disease of:
   a) Potato
   b) Pigeonpea
   c) Groundnut
   d) Maize

200. Which crop is used for producing both the vegetable oil and fibre:
a) Castor
b) Mesta
c) Linseed
d) Safflower

Model Paper-III

1. Growing plants in fixed order in a sequence for a specified duration on a specified area is called
a) Crop sequence
b) Crop intensity
c) Crop rotation
d) Cropping system

2. Cropping history refers to:
a) History of the crop under cultivation
b) History of farm operations carried out during the preceding years
c) Calendar of operations to be followed during the crop season
d) History and origin of crop

3. 'CAP' storage refers to storage of food grains:
a) In well ventilated ware-houses
b) In storage bins
c) On high plinths in the open polythene covers
d) In the fields under farmers conditions

4. 'Lag phase' on the response curve refers to:
a) The period of initial slow growth
b) The period of accelerated fast growth
c) The period when the yield curve tends to plateau off
d) The period when the yield curve tends to dip due to decline in response.

5. Successful recycling of high C N ratio-rice straw on agricultural lands could be accomplished by:
a) By adding nitrogenous fertilizers at incorporation
b) By adding cellulytic micro organisms (fungi) at incorporation
c) By giving an additional irrigation after incorporation
d) By adopting all above three options together

6. Base temperature refers to:
a) Temperature range required for optimum rate of the plant
b) The lowest temperature which is needed for the seed to initiate germination
c) The lowest temperature at which there is virtually no growth
d) The lowest temperature at which the plant dies

7. Cardinal temperature refers to:
a) The maximum, minimum & optimum temperature limits of growth
b) Survival temperature limits of plant
c) The best temperature for best growth phase
d) All the above three

8. Puddling in rice fields is done primarily
   a) To achieve a soft seed bed to ease planting of seedlings of paddy
   b) To kill the weeds
   c) To prevent deep percolation of water in paddy fields
   d) To achieve all these 3 objectives

9. Cytokinnins are plant growth regulators used
   a) To inhibit the growth of plants
   b) To stimulate cell division and delay the senescence
   c) To break dormancy in bulbs and tubers
   d) To hasten ripening of fruits

10. The most predominant type of barley cultivated in India is:
    a) Hull-less six rowed barley b) Hull-less two rowed barley
    c) Six-row hulled barley d) Two-row hulled barley

11. The best rice growing area with consistent highest average yield performance in India is
    a) Southern Indian Plateau region (Tamilnadu, Karnataka, Kerela)
    b) Eastern Indian Plains region (Assam, West Bengal and Orissa)
    c) North-western Indo-Gangetic plains region of Punjab, Haryana, Western Uttar Pradesh
    d) The Kashmir valley

12. Buck wheat is one of extensively growth crop in
    a) Indo-Gangetic irrigated plains of North India
    b) Europe and America continents
    c) African continent
    d) All the above three

13. Which of the following is most suited for cultivation under extremely dry conditions:
    a) *Brassica napus*
    b) *Brassica juncea*
    c) *Brassica campestris var yellow sarson*
    d) *Eurica sativa*

14. Which of the following 4 species of cotton cultivated in India accounts for the largest hectarage occupancy
    a) *Gossypium arboreum* b) *Gossypium herbaceum*
    c) *Gossypium barbadense* d) *Gossypium hirsutum*

15. The heavy weed infestation in wheat can be controlled by
    a) Stomp (Gromoxone) b) Isoproturon
    c) 2, 4-D d) Basalin (Fluchloralin)

16. The best time of application of herbicide in transplanted flooded paddy fields would be
    a) To apply herbicide 1-2 weeks after transplanting
    b) To apply herbicide 1-2 days after transplanting
    c) To apply herbicide just before puddling
    d) To apply herbicide soon after puddling but before the start of transplanting operation

17. 'Congress grass' is a common weed found:
    a) In cultivated irrigated lands
    b) In dry land farming situation
    c) In waste lands and along the rail and road
    d) On high altitudes with high rains

18. Rogueing refers to:
    a) Removal of weeds by hand picking
    b) Removal of weeds by use of herbicides
    c) Removal of plants of the same species and weeds
    d) Off-type plants with distinct morphological or phenotypic variations

19. Hormonal herbicides are
    a) The one which when applied at recommended rates also promote crop growth besides killing the weeds.
    b) Total herb-killers and kill all type of growing plants
    c) None of these

20. Amongst the 3 commercial preparation of 2, 4-D available in the market, the most volatile and likely to cause extensive damage by drift to sensitive crops like cotton in nearby fields is
    a) Sodium salt of 2, 4-D
    b) Amine form of 2, 4-D
    c) Esters of 2, 4-D
    d) All are equal

21. Amongst the various herbicides available in the market the most commonly used herbicide in transplanted rice is
    a) Isoproturon
    b) Butachlor
    c) Fluchlorin
    d) Atrataf

22. Phosphorus is most easily absorbed by plants as
    a) HPO_4
    b) H_2PO_4
    c) PO_4
    d) With equal ease in all the above-three forms

23. The most commonly used synthetic organic fertilizer in India is
    a) Anhydrous Ammonia
    b) Ammonium sulphate
    c) Urea
    d) DAP

24. Concentration of hydrogen ions at pH 9 would be how much different than at pH7
    a) 10 times greater
    b) 100 times greater
    c) 10 times lesser
    d) 100 times lesser

25. Mycorrhiza refers to:
    a) Nitrogen fixing organisms commonly found in association with legume plants
    b) Free floating fresh water fem commonly found in rice fields
    c) Phosphorus solubilizing bacteria and fungi associated with legumes
    d) All the above three.

26. Annidation refers to:
    a) Direct or indirect harmful effects that are plant has on another
    b) Complementary interactions occurs both in space and time
    c) Interaction effect of fertilizer levels and plant population densities
    d) Date of planting and fertilizer rates

27. In Chelated form micro-nutrients are:
    a) Adsorbed on clay-complex of the soil to be released slowly, later on
    b) Held by organic compounds which through soluble in water, do not ionize in soil solution
    c) The compounds of micro nutrients with high solubility in soil
    d) The complex micro nutrients salts fused with special types of glass

28. 'Khaira' disease is associated with
    a) Iron deficiency in rice
    b) Nitrogen toxicity in rice
    c) Zinc deficiency in rice
    d) Bacterial leaf blight disease of potato.

29. Which of the following nutrient element is the most immobile in plant?
    a) Sulphur
    b) Calcium
    c) Boron
    d) Magnesium

30. Liming is adding of
    a) Gypsum to reclaim alkali soils
    b) Calcium oxide to reclaim alkali soils
    c) Calcium oxide and Magnesium oxide obtained through dolo-mite lime stone to reclaim acid soils
    d) Fertilizing the soil with calcium containing fertilizers to correct calcium deficiency alone

31. Fertilizer grade refers to
    a) Quality of fertilizer
    b) The ratio of N: P_2O_5: K_2O present in a fertilizer mixture
    c) The order in which the various fertilizers are to be applied before seeding
    d) The gradation done relative to the concentration of active ingredient present in fertilizer

32. Fertilizer ratio designate
    a) The relative percent proportion of 3 major nutrients in the fertilizer mixture
    b) The relative nutrient ratio of major nutrients present in the fertilizer mixture
    c) The relative proportion of three major nutrients keeping the percentage of nitrogen as one
    d) The percent active nutrient content in a fertilizer

33. First irrigation to wheat is recommended at 3-4 weeks stage because
    a) Crop water need in the highest at this stage
    b) Crop require moist root-zone for development of its adventitious roots
    c) Crop require moist root-zone for seminal root initiation
    d) It requires high moisture to withstand effect of rapidly falling temperature

33. Ponding of water in transplanted rice is considered essential for:
    a) About 2 weeks at transplanting time only
    b) Throughout the crop growth season, except near the harvest
    c) 4-6 weeks stage only
    d) 2 weeks at transplanting and then again 2 weeks at anthesis stage

34. A soil is said to be saline when its soluble salt content is high and it tests an electrical conductivity value of
    a) Less than 8 mmhos/cm at 25^0C
    b) Less than 4 mmhos/cm at 25^0C
    c) More than 8 mmhos/cm at 25^0C
    d) More than 4 mmhos/cm at 25^0C

35. Allelopathy refers to:
    a) Accumulation of salts in soil and their toxic effect on crop
    b) Release of exudates from the plant parts with harmful/beneficial effect on the succeeding crop
    c) Release of exudates from plants with harmful effect only on succeeding crops
    d) Exchange of ions through medium of filaments of fungi

36. Agroforestry refers to:
    a) Growing of trees in solid plantation on vast scale
    b) Growing of trees on farm lands in association with cultivated crops.
    c) Growing of trees on waste lands
    d) Growing of trees and horticultural crops in associations

37. Apparent specific gravity refers to:
    a) Bulk density of soil
    b) Ratio of density of 'soil + air' of given volume to the density of water of same volume
    c) Ratio of density of soil solids to the density of water of the same volume
    d) None of these

38. Available soil moisture regime refers to:
    a) Water available between the field capacity and permanent wilting percent point
    b) A point to which soil water is allowed to deplete within the range of available moisture before application of irrigation
    c) Water available between the field capacity and Hygroscopic moisture point
    d) None of these

39. 1 m^3 of water is equal to
    a) 10.000 litres
    b) 100.000 litres
    c) 1000.000 litres
    d) 1.000 litres

40. One cumec flow over 24 hours would equal to
    a) 364 ha m of water
    b) 6.84 ha m of water
    c) 864 ha m of water
    d) 9.64 ha m of water

41. Time required to irrigate 12 ha of land with 72 cm of water for a 20 litres/sec stream would be
    a) 60 hrs
    b) 130 hrs
    c) 168 hrs
    d) 192 hrs

42. The depth of seeding in wheat is best related with
    a) Length of radicle
    b) Length of coleoptile
    c) Length of mesocotyle
    d) The depth of placement of embryo within seed cotyledon

43. Ergot is a serious disease of
    a) Sorghum
    b) Pear millet
    c) Maize
    d) Rice

44. Triticale is a
   a) New legume fodder crop developed for milch cattle
   b) New cereal crop prepared by crossing wheat with rye
   c) New grass developed for growing on low lands
   d) A species of wheat not found in India

45. Arable farming refers to
   a) Enclosing the natural lands for raising game and wild animal
   b) Growing of filed crops
   c) Management of natural grass lands
   d) Raising and management of naturally growing forest trees

46. In hybrid sunflower cultivation which of the following statement is true
   a) Seed is raised afresh for each sowing
   b) The seed can be sown for a few generations before it is replaced
   c) The seed can be used for two seasons only
   d) The seed can be used indefinitely, year after year without any replacement

47. Gram for good germination requires
   a) Fine seed bed of good tilth
   b) Very fine powdery and loose tilth
   c) Cloddy and rough seed bed
   d) Film and fine tilth

48. Which crop is used for producing both the vegetable oil and fibre
   a) Castor
   b) Mesta
   c) Linseed
   d) Safflower

49. The number of individuals lying on both side of the central value would be the same in case of
   a) Arithmathematical Mean
   b) Mode
   c) Median
   d) In all the above 3 cases

50. Which of the following crop has the highest cultivated area in the world
   a) Rice
   b) Wheat
   c) Barley
   d) Bajra

51. The genus Oryza includes
   a) 14 species
   b) 22 species
   c) 24 species
   d) 30 species

52. The varieties which belong to species Oryza glaberrima are found in
   a) Europe
   b) Asia
   c) America
   d) Africa

53. Rice grown in Indonesia belongs to
a) O. glaberrima
b) O. sativa indica
c) O. sativa japonica
d) O. sativa javanica

54. Rice crop needs
a) Warm and humid climate
b) Hot and humid climate
c) Dry and hot conditions
d) Cold and dry climate

55. For ripening of rice, temperature should be lies between
a) 21-37^{0}C
b) 26.5-29.5^{0}C
c) 20-25^{0}C
d) 15-20^{0}C

56. Which of the following variety is a dwarf mutant
a) IR-8
b) Jaya
c) Pusa-2-21
d) Jagannath

57. The seed rate of rice cultivation in broadcasting and drilling, is sufficient for nursery raising
a) 100, 60 kg/ha
b) 60, 100 kg/ha
c) 45, 100 kg/ha
d) 100,45 kg/ha

58. For transplanting one hectare area of rice how much area is sufficient for nursery raising
a) 100 m^2
b) 500 m^2
c) 1000 m^2
d) 1500 m^2

59. The Kresek occurs in early stage of plant growth of rice in
a) BLB
b) Bacterial leaf streak
c) Tungro virus
d) False Smut

60. *Oryza sativa* is a diploid species chromosome
a) 18
b) 24
c) 28
d) 42

61. Gene responsible for dwarfing characters in rice is
a) Tift 23A
b) Dee-gee-woo-gen
c) Norin-10
d) Opaque-2

62. The scientific name of common bread wheat is
a) *T. dicocum*
b) *Triticum aestivum*
c) *T. spherococcum*
d) *T. spelta*

63. Indian dwarf wheat belongs to
a) *T. durum*
b) *Triticum aestivum*
c) *T. spherococcum*
d) *T. spelta*

64. The inflorescence of wheat is known as
   a) Ear
   b) Raceme
   c) Panicle
   d) Umbel

65. The normal seed rate of wheat is
   a) 50 kg/ha
   b) 75 kg/ha
   c) 100 kg/ha
   d) 125 kg/ha

66. Triticale is a cross between
   a) Wheat x rye
   b) Oat x barley
   c) Wheat x barley
   d) None of these

67. Phalaris minor belongs to the family
   a) Cyperacene
   b) Gramineae
   c) Solanaceae
   d) Malvaceae

68. The nitrogen losses in rice can be reduced by placing NH_4 fertilizer in
   a) Oxidise zone
   b) Reduced zone
   c) Both a & b
   d) None of these

69. Green revolution has been most successful in
   a) 1999-2000
   b) 2000-01
   c) 2001-02
   d) 2004-05

70. At pH 4.0 the predominant ionic form of phosphorus present is
   a) HPO_4
   b) $H_2PO_4^-$
   c) PO_4^{3-}
   d) None of these

71. Directorate of Wheat Research (DWR) is located at
   a) Hyderabad
   b) New Delhi
   c) Karnal
   d) Bikaner

72. *Saccharum officinarum* is native of
   a) New Guinea
   b) India
   c) Indonesia
   d) China

73. Which state has highest productivity of sugarcane
   a) U.P.
   b) W.B.
   c) Karnataka
   d) Tamil Nadu

74. Which state has the largest acreage and highest production of sugarcane in country
   a) Tamil Nadu
   b) Karnataka
   c) U.P.
   d) Bihar

75. The inflorescence of sugarcane is known as

a) Arrow
b) Panicle
c) Capitulum's
d) Racemose

76. AICRP on sugarcane was started in

a) 1959
b) 1960
c) 1970-71
d) 1985-86

77. Which growth stage of sugarcane is critical for irrigation

a) Germination
b) Grand growth phase
c) Formative stage
d) Ripening stage

78. Most critical weed competition period is up to ....months after transplanting exist in sugarcane

a) 2
b) 4
c) 5
d) 6

79. Most common herbicides used for weed control in sugarcane is/are

a) Simazine
b) Atrazine
c) Alachlor
d) All of these

80. Which one of the following types of resistance is present in a crop variety showing high degree of resistance to a pathogen but becoming susceptible to it after large scale cultivation for 4 to 5 years or so

a) Vertical resistance
b) Horizontal resistance
c) Durable resistance
d) General resistance

81. Which of the following food grain is most coarse

a) Panicum miliaceum
b) Echinochloa frumentacea
c) Durable resistance
d) Paspalum scrobiculatum

82. Which of the following pairs of crop and critical stage of irrigation is not correctly matched

a) Bajra: Ear head
b) Cotton: Pre-flowering
c) Wheat: Crown root initiation
d) Groundnut: pod development

83. Contribution of flag leaf in photosynthates is about

a) 52%
b) 40%
c) 35%
d) 20%

84. Which one of the following crops is most sensitive to both excess moisture and drought

a) Direct seeded rice
b) Maize
c) Sunflower
d) Sorghum

85. Atrazine used as an antitranspirant
   a) Reduces the growth of the crop
   b) Does not reflect light from plant leaf surface
   c) Affects the closure and opening of stomata
   d) Forms thin layer on leaf surface

86. Productivity of an intercrop per unit area of ground compared with that expected from sole crop sown in the same proportions is termed as
   a) Land equivalent ratio
   b) Competitive ratio
   c) Land equivalent coefficient
   d) Crop performance ratio

87. Which one of the following fertilizer schedule is recommended to optimize the productivity in intercropping system
   a) Full recommended dose of both the main crop ands intercrop
   b) Full recommended dose of main crop and half of intercrop
   c) Half recommended dose of main crop and intercrop
   d) Full recommended dose of main crop only

88. The capacity of a soil to resist appreciable change in pH value is called
   a) CEC
   b) Buffering capacity
   c) Percentage base saturation
   d) Anion-exchange capacity

89. Consider the following pairs of crops and varieties and select the pair, which is not correctly matched
   a) Barley: Clipper
   b) Cotton: Sujata
   c) Cowpea: Pusa Phalguni
   d) Green gram: UPAS 120

90. Which of the following is the ideal temperature range for tillering stage in wheat
   a) 10-15^{0}C
   b) 16-20^{0}C
   c) 20-23^{0}C
   d) 23-25^{0}C

91. The element which does not enter into permanent organic combination in plants is
   a) Nitrogen
   b) Phosphorus
   c) Magnesium
   d) Potassium

92. The plateau with little growth rate and continuing accumulation of nutrient element in the plants can be defined as
   a) Severe deficiency range
   b) Toxic range
   c) Moderate deficiency
   d) Sufficiency range

93. The major organic cementing agent in soil aggregate formation is
   a) Lipids
   b) Proteins and protein derivatives
   c) Polysaccharides
   d) Organic acids

94. Tetrazoloin test is used to determine
   a) Seed purity
   b) Seed germination
   c) Seed viability
   d) Seed quality

95. How much quantity of true potato seed is required for one hectare planting
   a) 100 g
   b) 200 g
   c) 300 g
   d) 400 g

96. The largest producer of mothbean in India
   a) Rajasthan
   b) Punjab
   c) Madhya Pradesh
   d) Uttar Pradesh

97. The seed rate for Sesamum is
   a) 2-3 kg
   b) 3-4 kg
   c) 4-5 kg
   d) 5-6 kg

98. The highest production of mustard in India is in the state of
   a) Gujarat
   b) U.P.
   c) Punjab
   d) Rajasthan

99. The fruit of rapeseed and mustard is known as
   a) Pod
   b) Grain
   c) Siliqua
   d) Caryopsis

100. The inherent capacity of soil to supply plant nutrients in adequate amount and suitable proportion is called
   a) Fertility index
   b) Soil fertility
   c) Soil productivity
   d) None of these

101. The criteria of essentiality of nutrients in plants was given by
   a) DJ Nicholas
   b) Arnon and stout
   c) JS Kanwar
   d) Rajendra Prasad

102. Potato is a
   a) Modified stem
   b) Modified root
   c) Modified leaf
   d) Modified flower

103. The method of irrigation employed where surface is undulating
   a) Flood irrigation
   b) Spinkler irrigation
   c) Drip irrigation
   d) Subsurface irrigation

104. The depth of sowing of bajra is

a) 1-2 cm b) 3-4 cm
c) 4-5 cm d) 5-6 cm

105. Which is generally used for correcting soil acidity

a) Gypsum b) Lime
c) Iron pyrites d) Both a & b

106. The main objective of growing a catch crop is to

a) Add more residues to the soil
b) Prevent cracking of soil
c) Suppress weeds
d) Get an additional income without further investment

107. Biuret content in urea should not exceed .... .according to fertilizer control order, 1957

a) Furrows b) Ridges/bunds
c) Soil smooth d) All of these

108. The force of attraction binds the molecules of the sane kind is

a) Adhesion b) Matric force
c) Cohesion d) None of these

109. The upper limit of the soil moisture available for the plant growth is

a) PWP (15 bars) b) Field capacity (1/3 bars)
c) Hygroscopic coefficient d) Matric suction

110. A device for measuring percolation and leaching losses from a column of soil under controlled conditions is known as

a) Infiltrometer b) Evaporimeter
c) Psycrometer d) Lysimeter

111. A calibrated device for measuring the flow of water in open conduit is known as

a) V-notch b) Parshall flume
c) Watermeter d) None of these

112. The instrument used for measuring depth of water table is known as

a) Lysimeter b) Odometer
c) Piezometer d) Evaporimeter

113. The seed rate of hybrid maize is

a) 20-25 kg/ha b) 18-20 kg/ha
c) 30-35 kg/ha d) 10-15 kg/ha

114. Botanical name of Ragi is
a) Eleusine coracana
b) Echinochloa frumentacea
c) Panicum miliacium
d) None of these

115. The most serious pest of bengalgram is
a) Pod borer
b) Cut worm
c) Aphid
d) None of these

116. The oil and protein content of groundnut are
a) 20% & 50%
b) 26% & 45%
c) 45% & 26%
d) 50% & 26%

117. Raising of crop with the least tillage operations is called
a) Zero tillage
b) Minimum tillage
c) No tillage
d) Heavy tillage

118. For which type of fertilizer India is fully dependent on imports
a) N-fertilizer
b) P-fertilizer
c) K-fertilizer
d) None of these

119. National Agriculture Policy aims the growth rate of
a) >2.5% pa
b) >3.0% pa
c) >4.0% pa
d) >5.0 pa

120. Which is the leading state in production of groundnut
a) U.P.
b) Rajasthan
c) Haryana
d) Gujarat

121. Remote sensing helps in studying
a) Cropped area
b) Soil characters
c) Underground water
d) All of the above

122. Sunflower act as an indicator plant to diagnose the deficiency of
a) Boron
b) Iron
c) Nitrogen
d) Phosphorus

123. Which of the following elements is not considered as fertilizer nutrient
a) C
b) N
c) S
d) Zn

124. The type of germination in mungbean is known as
a) Epigeal
b) Hypogeal
c) Hypoepigeal
d) Epihypogeal

125. Bacteria responsible for N fixation in soybean is

a) *R. phaseoli* b) *R. glycicum*
c) *R. japonicum* d) *R. leguminoseum*

126. Sesamum belongs to the family

a) *Chenopodiaceae* b) *Papilionaceae*
c) *Leguminosae* d) *Pedaliaceae*

127. Fluchloralin can be in soybean as

a) Pre-emergence b) Post-emergence
c) Pre-plant incorporation d) None of these

128. Topping of safflower plants is beneficial

a) To reduce lodging b) To promote branching and flowering
c) To reduce water loss d) All of these

129. Delinting of cotton seed may be done with

a) Sulphuric cid b) Citric acid
c) Nitric acid d) Hydrochloric acid

130. Optimum plant population ha^{-1} for cotton has estimated as

a) 25,000 - 50,000 b) 30,000 - 60,000
c) 50,000 - 80,000 d) 80,000 - 1,00,000

131. The insect bollworm is commonly found on

a) Maize b) Wheat
c) Cotton d) Rice

132. Stomata closing can be induced by

a) Kaoline b) Linseed oil
c) 2, 4 -D d) PMA

133. The most critical stage of maize from irrigation point of view is

a) Silking stage b) Tasseling stage
c) Boot stage d) Dough stage

134. Relationship between water and fertilizer as production factors in crops is

a) Additive b) Synergetic
c) Antagonistic d) None of these

135. Growing of coconut, black pepper and ginger simultaneously in the same field is called

a) Relay cropping b) Inter cropping
c) Multiple cropping d) Multistoried cropping

136. Stevenson screen is related to

a) Bacteriology b) Biotechnology
c) Agrometeorology d) Remote sensing

137. Botanical name of Sawan grass is

a) Lasiurus sindicus b) Panicum maximum
c) Cynodon dactylon d) Cenchrus ciliaris

138. RMO-40 is a variety of

a) Mungbean b) Urdbean
c) Horsebean d) Monthbean

139. Amongst oil cakes, the highest nitrogen content is in

a) Castor cake b) Neem cake
c) Groundnut cake d) Coconut cake

140. The term "Evergreen Revolution" has been given by

a) Dr. AS Faroda b) Dr. MS Swaminathan
c) Dr. VL Chopra d) Dr. RS Paroda

141. Maximum residual acidity is associate with the continuous soil application of

a) Urea b) Ammonium nitrate
c) Ammonium sulphate d) CAN

142. The quantity of SSP (16% P_2O_5) required for application to one acre of rice-field at a dose of 60 kg P_2O_5 ha^{-1} is

a) 250 kg b) 200 kg
c) 150 kg d) 100 kg

143. The most important potential contaminant of food produced on sewage sludge amended soils is

a) Chromium b) Lead
c) Zinc d) Cadmium

144. Which one of the following nutrients is constituent of cell wall in the plants

a) Phosphorus b) Calcium
c) Sulphur d) Potassium

145. Which of the following phytohormone is involved in nutrient mobilization

a) Gibberellin b) Cytokinin
c) Auxin d) ABA

146. Heavy applications of potassium fertilizers often lead to reduced absorption of

a) Mg and Ca b) B and Mn
c) N and P d) Zn and Cu

147. If a plant shows extensive interveinal chlorosis of the older leaves and late this reaches the younger leaves, the symptoms is due to lack of

a) Magnesium b) Calcium
c) Phosphorus d) Copper

148. Phenylmercuric acetate (PMA) is a chemical used in agricultural crops in order to

a) Increase CO2 uptake b) Reduce respiration
c) Reduce transpiration d) Increase transpiration

149. Water requirement (WR) includes the losses due to

a) Evapotranspiration and application of water
b) Consumptive use of water and application of water
c) Consumptive use, water required for special operations and other economically unavoidable losses of water.
d) ET and water required for special operations.

150. If on a 20-acre farm, crops are raised on 15 acres in *rabi*, 15 acres in *kharif* and 20 acres in *zaid* in a year, what shall be the cropping intensity

a) 300% b) 400%
c) 250% d) 150%

151. Which one of the following pairs is not correctly matched

a) Bajra- Tift 23 D_2A b) Maize- T Cytoplasm
c) Wheat-Rht1 Rht2 d) Rice-Norin 10

152. Which one of the following disease not include floral abnormalities

a) Downey mildew of mustard b) Green ear of bajra
c) White rust of crucifers d) None of these

153. Which one of the following pairs is not correctly matched

a) Paddy-gall fly: Silver leaf b) Rice bug: Repugnant smell
c) Rice stem borer: Black ears d) Rice zig-zag leaf hopper: Orange leaf

154. Which one of the following diseases of pigeon pea is transmitted by Eryophid mite

a) Phytophthora stem rot b) Cercospora leaf spot
c) Yellow mosaic d) Sterility mosaic

155. Which one of the following was introduced in the country to eradicate prickly pear (*Opuntia dilleni*) weed

a) Chrysomella spp.
b) Dactylopius tomentosus
c) Zygogramma bicolarata
d) Neochetina spp.

156. Which one of the following microbial agents is being commercially exploited as biocontrol agent

a) Penicillium notatum
b) Bacillus subtilis
c) Trichoderma viridiae
d) Sclerotium rolfsii

157. Which one of the following causes more wastage of herbicide by drift

a) High volume sprayer
b) Ultra-low volume sprayer
c) Hand sprayer
d) Low volume sprayer

158. Application of organic materials with C: N ratio wider than 20 to 30: 1 leads to .... of soil N

a) Nitrification
b) Fixation
c) Mineralization
d) Immobilization

159. For sugarcane 'N' fertilizer application has to be completed within .....days after planting for better sugar content in juice

a) 20
b) 40
c) 60
d) 80

160. At field capacity the value of *pF* is

a) 0.0
b) 2.5
c) 4.2
d) 6.0

161. At permanent wilting point (PWP) the *pF* value is

a) 0.0
b) 2.5
c) 4.2
d) 6.0

162. The % pore space in a soil having a bulk density of 1.16 mg m^{-3} and particle density of 2.62 mg m^{-3} is

a) 26.2
b) 44.2
c) 55.7
d) 81.7

163. Among the following the set of essential nutrient elements, which are absorbed as anions by plants is

a) Boron, chlorine, copper, iron, manganese
b) Copper, iron, phosphorus, sulphur, zinc
c) Potassium, manganese, molybdenum, phosphorus, zinc
d) Boron, chloride, molybdenum, phosphorus, sulphur

164. A fertilizer which supplies three essential plant nutrients is

a) DAP b) MOP
c) SSP d) SOP

165. The Fe-chelate suitable for application to crops grown on acid soils is

a) Fe-EDTA b) Fe-EDDHA
c) Fe-DTPA d) Fe-HEDTA

166. Which one of the following fertilizers contain water soluble P

a) SSP b) DCP
c) MAP d) Both a & c

167. Stem nodulation occurs in green manure crop of

a) *Sesbania aculeate* b) *Sesbania cannabina*
c) *Crotolaria juncea* d) *Aeschynomene afraspera*

168. Hybrid rice for commercial production was first evolved in

a) India b) China
c) Japan d) USA

169. In tobacco, the basic aim of topping and desuckering is to

a) Reduce the plant height
b) Encourage the branching
c) Divert energy and nutrients from flower heads to leaves
d) Protect the plants against lodging

170. Crop logging is used in

a) Sugarcane b) Sugarbeet
c) Maize d) Tea

171. The neutron scattering method is not useful in estimating the moisture in

a) Acidic soils b) Alluvial soils
c) Lateritic soils d) Organic soils

172. Application of gypsum is required more for

a) Paddy b) Bersecm
c) Lucerne d) Groundnut

173. The rotation intensity of maize-wheat+gram-moong is

a) 200% b) 250%
c) 300% d) 400%

174. An example of companion cropping is

a) Sugarcane + potato b) Potato + mustard
c) Potato + radish d) Wheat + mustard

175. Intercropping of mustard with potato is recommended in
   a) Replacement series
   b) Additive series
   c) Replacement cum additive series
   d) None of these

176. Which clay mineral is rich in potash
   a) Montmorillonite   b) Kaolinite
   c) Illite   d) Chlorite

177. Which of the following plants does not show photorespiration
   a) Pea   b) Wheat
   c) Rice   d) Maize

178. In which of the following, composite and synthetic cultivars are used
   a) Rice   b) Wheat
   c) Maize   d) Cotton

179. The recently developed terminator technology has been used in
   a) Rice   b) Cotton
   c) Tobacco   d) Wheat

180. Which cation has higher aggregation capacity in soil
   a) Al^{3+} b) Ca^{2+}
   c) Mg^{2+} d) K^{+}

181. Which cation has low adsorption capacity on clay,
   a) H   b) Ca
   c) Fe   d) Na

182. Excess uptake of which element is known as luxury consumption
   a) N   b) P
   c) K   d) Zn

183. Which one of the following causes pungency of mustard oil
   a) Phenols   b) Amino acids
   c) Glucosinolates   d) Erucic acid

184. The formula LER-1/LER x value of combined intercrop represents
   a) MAI   b) LER
   c) LAI   d) IER

185. LEISA is related to

a) Organic farming
b) Inorganic farming
c) Natural farming
d) All of these

186. Soil structure is improved by application of

a) Urea
b) Super phosphate
c) Muriate of potash
d) Zinc sulphate

187. The principal way in which P and K ions move from the soil to the root of field crops

a) Root interception
b) Mass flow
c) Diffusion
d) None

188. Immobilization of sulphur take place when the '5' content of organic matter is less than

a) 0.45%
b) 0.60%
c) 0:30%
d) 0.15%

189. Organic carbon is a measure of

a) Available N in soil
b) Available P in soil
c) Available K in soil
d) Available Mg in soil

190. Approximate weight of surface 15 cm soil of one hectare field is

a) 1×10^6 kg
b) 1.5×10^6 kg
c) 2×10^6 kg
d) 2.24×10^6 kg

191. Concentration of which element is highest in soil

a) O_2
b) Fe
c) Al
d) Si

192. If the gravimetric moisture content of soil is 15% and the bulk density is 1.4 g/cc, the volumetric moisture content is

a) 18.0%
b) 21.0%
c) 24.0%
d) 27.0%

193. Which one of the following is not correctly matched Sunflower oil is rated as'good quality oil because it contains

a) Low quantities of saturated fatty acids
b) Low quantities of unsaturated fatty acid
c) High quantities of saturated fatty acid
d) High quantities of unsaturated fatty acid

194. Which crop of the following has the double symbibtic relationship with nitrogen fixing bacteria

a) Phaseolus vulgaris
b) Cajanus cajan
c) Sesbania rostrata
d) Glycine max

195. Zero tillage system was first used successfully in 1950 in pasture renovation in

a) Germany
b) Japan
c) United Kingdom
d) USA

196. The total volume of water on melting of ice

a) Increases
b) Remains constant
c) Decreases
d) None of these

197. The normal form of plant is maintained by

a) Water
b) Gases
c) Solids
d) None

198. The panicle initiation stage in rice plants comes

a) Immediate after transplanting
b) After maximum tillering stage
c) After boot leaf stage
d) At any time after transplanting

199. The first GM potato developed at CPRI for increasing protein content in tubers consists of genes from.

a) Chickpea
b) Pigeonpea
c) Field pea
d) Grain amaranthus

200. CO_2 content in soil air is

a) 0.03%
b) 0.003%
c) 0.25%
d) 0.50%

Model Paper-IV

1. Which of the following crop has the highest cultivated area in the world

a) Rice
b) Wheat
c) Barley
d) Bajra

2. Rice is originated in

a) SW Asia
b) Europe
c) South America
d) India and Burma

3. The sowing time of boro or dalua rice is

a) Nov. - Dec.
b) June-July
c) May-June
d) March - April

4. Dead heart and white head damage to rice is caused by

a) Gall midge
b) Leaf roller
c) Army worm
d) Stem borer

5. Grassy stunt virus disease of rice is transmitted by

a) Green leaf hopper
b) Brown plant hopper
c) White fly
d) Gundhi bug

6. Gene responsible for dwarfing characters in rice is
   a) Tift 23A
   b) Dee-gee-woo-gen
   c) Norin-10
   d) Opaque-2

7. Idea of super rice was given by
   a) GH Shull
   b) GS Khush
   c) VL Chopra
   d) Yoshida

8. Indian dwarf wheat belongs to
   a) T. durum
   b) Triticum aestivum
   c) T. spherococcum
   d) T. spelta

9. Select the pair which is not correctly matched
   a) South America
   b) Central Asia (Turkey)
   c) SE Asia
   d) Europe

10. N, P_2O_5 and K_2O contents of DAP are
    a) 46 - 18 - 0
    b) 18 - 46 - 0
    c) 0 - 18 - 46
    d) 0 - 46 - 18

11. The N fertilizer use efficiency in rice can be increased by using
    a) S-coated urea
    b) Urea super granules
    c) BGA
    d) Both a & b

12. The nitrogen losses in rice can be reduced by placing NH_4 fertilizer in
    a) Oxidise zone
    b) Reduced zone
    c) Both a & b
    d) None of these

13. *Saccharum officinarum* is native of
    a) New Guinea
    b) India
    c) Indonesia
    d) China

14. Which state has highest productivity of sugarcane
    a) Uttar Pradesh
    b) West Bengal
    c) Karnataka
    d) Tamil Nadu

15. Which state has the largest acreage and highest production of sugarcane in country
    a) Tamil Nadu
    b) Karnataka
    c) U.P.
    d) Bihar

16. Which of the following roots are of permanent type on sugarcane sets
    a) Sett roots
    b) Shoot roots
    c) Prop roots
    d) All of these

17. The best suited temperature for growth of sugarcane lies between
    a) 15-20^{0}C   b) 20-25^{0}C
    c) 26-32^{0}C   d) 32-35^{0}C

18. Adsali sugarcane is planted in
    a) July-August   b) January-February
    c) February-March   d) October-November

19. The top portion of sugarcane should be selected for seed purposes because bud tissues are rich in
    a) Sucrose   b) Glucose
    c) Galactose   d) Maltose

20. Which element is essential for sugar translocation in sugarcane
    a) P   b) K
    c) B   d) Mo

21. Which growth stage of sugarcane is critical for irrigation
    a) Germination   b) Grand growth phase
    c) Formative stage   d) Ripening stage

22. Most critical weed competition period is up to ....months after transplanting exist in sugarcane
    a) 2   b) 4
    c) 5   d) 6

23. Most common herbicides used for weed control in sugarcane is/are
    a) Simazine   b) Atrazine
    c) Alachlor   d) All of these

24. Most serious disease of sugarcane is
    a) Red stripe   b) Red rot
    c) Wilt   d) Smut

25. Trench method of sugarcane planting is used in
    a) North India   b) Western India
    c) Coastal areas   d) Waterlogged area

26. With too much increase in nitrogen application, sugar content in juice is
    a) Decreased   b) Increased
    c) Remain constant   d) None of the above

27. Epricania melanoleuca is a parasitoid effective against
    a) Sugarcane scale   b) Rice mealy bug
    c) Sugarcane pyrilla   d) Rice leaf hopper

28. Which one of the following types of resistance is present in a crop variety showing high degree of resistance to a pathogen but becoming susceptible to it after large scale cultivation for 4 to 5 years or so

a) Vertical resistance b) Horizontal resistance
c) Durable resistance d) General resistance

29. Which of the following food grain is most coarse

a) Panicum miliaceum b) Echinochloa frumentacea
c) Durable resistance d) Paspalum scrobiculatum

30. Which of the following pairs of crop and critical stage of irrigation is not correctly matched

a) Bajra: Ear head b) Cotton: Pre-flowering
c) Wheat: Crown root initiation d) Groundnut: pod development

31. Contribution of flag leaf in photosynthates is about

a) 52% b) 40%
c) 35% d) 20%

32. Which one of the following crops is most sensitive to both excess moisture and drought

a) Direct seeded rice b) Maize
c) Sunflower d) Sorghum

33. Which one of the following pairs of weeds and crops is not correctly matched

a) Striga: Sorghum b) Cuscuta: Lucerne
c) Typha: Sugarcane d) Orobanche: Tobacco

34. Which one of the following is mycoheribicide

a) Diquat b) Met sulfuron methyl
c) Collego d) Bromacil

35. Which one of the following crop rotation is the best for maintaining soil fertility

a) Maize-toria-wheat b) Paddy-wheat-cowpea
c) Paddy-potato-green gram d) Soybean-wheat-green gram

36. Productivity of an intercrop per unit area of ground compared with that expected from sole crop sown in the same proportions is termed as

a) Land equivalent ratio b) Competitive ratio
c) Land equivalent coefficient d) Crop performance ratio

37. Which one of the following fertilizer schedule is recommended to optimize the productivity in intercropping system
    a) Full recommended dose of both the main crop ands intercrop
    b) Full recommended dose of main crop and half of intercrop
    c) Half recommended dose of main crop and intercrop
    d) Full recommended dose of main crop only

38. The capacity of a soil to resist appreciable change in pH value is called
    a) CEC
    b) Buffering capacity
    c) Percentage base saturation
    d) Anion-exchange capacity

39. Mechanical analysis of soil makes use of
    a) Darcy's law
    b) Stoke's law
    c) Schofied's law
    d) Ohm's law

40. Which one of the following planting f=geometry is recommended to optimize system productivity in intercropping system in semi-arid and sub-humid ecosystem
    a) Normal planting of intercrops between two rows of normal planted main crop
    b) Skip row planting
    c) Three rows of main crop and one row of intercrops
    d) Four rows of main crop and one row of intercrops

41. Humic acid is a fraction of humus, which is
    a) Soluble in alkali and acid
    b) Soluble in alkali and soluble in acid
    c) Insoluble in alkali and soluble in acid
    d) Insoluble in both in alkali and acid

42. The process of replacement of one atom by another atom of similar size, in a crystal lattice of a soil clay without disrupting or changing the crystal structure of the mineral is termed as
    a) Ion exchange
    b) Isomerism
    c) Isomorphic substitution
    d) Polymorphism

43. Which one of the following sequences is being used in a combination seed treatment
    a) Rhizobium, fungicide and insecticide
    b) Rhizobium, insecticide and fungicide
    c) Fungicide, insecticide and Rhizobium
    d) Insecticide, Rhizobium and fungicide

44. Consider the following pairs of crops and varieties and select the pair, which is not correctly matched
    a) Barley: Clipper
    b) Cotton: Sujata
    c) Cowpea: Pusa Phalguni
    d) Green gram: UPAS 120

45. The plateau with little growth rate and continuing accumulation of nutrient element in the plants can be defined as
    a) Severe deficiency range
    b) Toxic range
    c) Moderate deficiency
    d) Sufficiency range

46. The major organic cementing agent in soil aggregate formation is
    a) Lipids
    b) Proteins and protein derivatives
    c) Polysaccharides
    d) Organic acids

47. Tetrazoloin test is used to determine
    a) Seed purity
    b) Seed germination
    c) Seed viability
    d) Seed quality

48. How much quantity of true potato seed is required for one hectare planting
    a) 100 g
    b) 200 g
    c) 300 g
    d) 400 g

49. The largest producer of mothbean in India
    a) Rajasthan
    b) Punjab
    c) Madhya Pradesh
    d) Uttar Pradesh

50. The seed rate for Sesamum is
    a) 2-3 kg
    b) 3-4 kg
    c) 4-5 kg
    d) 5-6 kg

51. The highest production of mustard in India is in the state of
    a) Gujarat
    b) U.P.
    c) Punjab
    d) Rajasthan

52. The fruit of rapeseed and mustard is known as
    a) Pod
    b) Grain
    c) Siliqua
    d) Caryopsis

53. The inherent capacity of soil to supply plant nutrients in adequate amount and suitable proportion is called
    a) Fertility index
    b) Soil fertility
    c) Soil productivity
    d) None of these

54. The criteria of essentiality of nutrients in plants was given by
   a) DJ Nicholas
   b) Arnon and stout
   c) JS Kanwar
   d) Rajendra Prasad

55. Potato is a
   a) Modified stem
   b) Modified root
   c) Modified leaf
   d) Modified flower

56. The method of irrigation employed where surface is undulating
   a) Flood irrigation
   b) Spinkler irrigation
   c) Drip irrigation
   d) Subsurface irrigation

57. The depth of sowing of bajra is
   a) 1-2 cm
   b) 3-4 cm
   c) 4-5 cm
   d) 5-6 cm

58. Which is generally used for correcting soil acidity
   a) Gypsum
   b) Lime
   c) Iron pyrites
   d) Both a & b

59. The main objective of growing a catch crop is to
   a) Add more residues to the soil
   b) Prevent cracking of soil
   c) Suppress weeds
   d) Get an additional income without further investment

60. Biuret content in urea should not exceed .... .according to fertilizer control order, 1957
   a) 2%
   b) 5%
   c) 1%
   d) 3%

61. The force of attraction binds the molecules of the sane kind is
   a) Adhesion
   b) Matric force
   c) Cohesion
   d) None of these

62. The upper limit of the soil moisture available for the plant growth is
   a) PWP (15 bars)
   b) Field capacity (1/3 bars)
   c) Hygroscopic coefficient
   d) Matric suction

63. A device for measuring percolation and leaching losses from a column of soil under controlled conditions is known as
   a) Infiltrometer
   b) Evaporimeter
   c) Psycrometer
   d) Lysimeter

64. A calibrated device for measuring the flow of water in open conduit is known as

a) V-notch b) Parshall flume
c) Watermeter d) None of these

65. The instrument used for measuring depth of water table is known as

a) Lysimeter b) Odometer
c) Piezometer d) Evaporimeter

66. The seed rate of hybrid maize is

a) 20-25 kg/ha b) 18-20 kg/ha
c) 30-35 kg/ha d) 10-15 kg/ha

67. Botanical name of Ragi is

a) Eleusine coracana b) Echinochloa frumentacea
c) Panicum miliacium d) None of these

68. The most serious pest of bengalgram is

a) Pod borer b) Cut worm
c) Aphid d) None of these

69. The oil and protein content of groundnut are

a) 20% & 50% b) 26% & 45%
c) 45% & 26% d) 50% & 26%

70. Raising of crop with the least tillage operations is called

a) Zero tillage b) Minimum tillage
c) No tillage d) Heavy tillage

71. For which type of fertilizer India is fully dependent on imports

a) N-fertilizer b) P-fertilizer
c) K-fertilizer d) None of these

72. National Agriculture Policy aims the growth rate of

a) >2.5% pa b) >3.0% pa
c) >4.0% pa d) >5.0 pa

73. Which is the leading state in production of groundnut

a) U.P. b) Rajasthan
c) Haryana d) Gujarat

74. Remote sensing helps in studying

a) Cropped area b) Soil characters
c) Underground water d) All of the above

75. Sunflower act as an indicator plant to diagnose the deficiency of

a) Boron b) Iron
c) Nitrogen d) Phosphorus

76. Which of the following elements is not considered as fertilizer nutrient

a) C b) N
c) S d) Zn

77. The type of germination in mungbean is known as

a) Epigeal b) Hypogeal
c) Hypoepigeal d) Epihypogeal

78. Bacteria responsible for N fixation in soybean is

a) R. phaseoli b) R. glycicum
c) R. japonicum d) R. leguminoserum

79. Sesamum belongs to the family

a) Chenopodiaceae b) Papilionaceae
c) Leguminosae d) Pedaliaceae

80. Fluchloralin can be in soybean as

a) Pre-emergence b) Post-emergence
c) Pre-plant incorporation d) None of these

81. Topping of safflower plants is beneficial

a) To reduce lodging b) To promote branching and flowering
c) To reduce water loss d) All of these

82. Delinting of cotton seed may be done with

a) Sulphuric acid b) Citric acid
c) Nitric acid d) Hydrochloric acid

83. Optimum plant population ha^{-1} for cotton has estimated as

a) 25,000 – 50,000 b) 30,000 – 60,000
c) 50,000 – 80,000 d) 80,000 - 1,00,000

84. The insect bollworm is commonly found on

a) Maize b) Wheat
c) Cotton d) Rice

85. Stomata closing can be induced by

a) Kaoline b) Linseed oil
c) 2, 4 –D d) PMA

86. The most critical stage of maize from irrigation point of view is

a) Silking stage b) Tasseling stage
c) Boot stage d) Dough stage

87. Relationship between water and fertilizer as production factors in crops is
    a) Additive b) Synergestic
    c) Antagonistic d) None of these

88. Growing of coconut, black pepper and ginger simultaneously in the same field is called
    a) Relay cropping b) Inter cropping
    c) Multiple cropping d) Multistoried cropping

89. Stevenson screen is related to
    a) Bacteriology b) Biotechnology
    c) Agrometeorology d) Remote sensing

90. Botanical name of Sewan grass is
    a) Lasiurus sindicus b) Panicum maximum
    c) Cynodon dactylon d) Cenchrus ciliaris

91. RMO-40 is a variety of
    a) Mungbean b) Urdbean
    c) Horsebean d) Monthbean

92. Amongst oil cakes, the highest nitrogen content is in
    a) Castor cake b) Neem cake
    c) Groundnut cake d) Coconut cake

93. The term "Evergreen Revolution" has been given by
    a) Dr. AS Faroda b) Dr. MS Swaminathan
    c) Dr. VL Chopra d) Dr. RS Paroda

94. Maximum residual acidity is associate with the continuous soil application of
    a) Urea b) Ammonium nitrate
    c) Ammonium sulphate d) CAN

95. The quantity of SSP (16% P_2O_5) required for application to one acre of rice-field at a dose of 60 kg P_2O_5 ha^{-1} is
    a) 250 kg b) 200 kg
    c) 150 kg d) 100 kg

96. The most important potential contaminant of food produced on sewage sludge amended soils is
    a) Chromium b) Lead
    c) Zinc d) Cadmium

97. Which one of the following nutrients is constituent of cell wall in the plants
    a) Phosphorus b) Calcium
    c) Sulphur d) Potassium

98. Which of the following phytohormone is involved in nutrient mobilization
    a) Gibberellin
    b) Cytokinin
    c) Auxin
    d) ABA

99. Heavy applications of potassium fertilizers often lead to reduced absorption of
    a) Mg and Ca
    b) B and Mn
    c) N and P
    d) Zn and Cu

100. If a plant shows extensive interveinal chlorosis of the older leaves and late this reaches the younger leaves, the symptoms is due to lack of
    a) Magnesium
    b) Calcium
    c) Phosphorus
    d) Copper

101. Phenylmercuric acetate (PMA) is a chemical used in agricultural crops in order to
    a) Increase CO_2 uptake
    b) Reduce respiration
    c) Reduce transpiration
    d) Increase transpiration

102. Water requirement (WR) includes the losses due to
    a) Evapotranspiration and application of water
    b) Consumptive use of water and application of water
    c) Consumptive use, water required for special operations and other economically unavoidable losses of water.
    d) ET and water required for special operations.

103. If on a 20-acre farm, crops are raised on 15 acres in *rabi*, 15 acres in *kharif* and 20 acres in *zaid* in a year, what shall be the cropping intensity
    a) 300%
    b) 400%
    c) 250%
    d) 150%

104. Which one of the following pairs is not correctly matched
    a) Bajra- Tift 23 D_2A
    b) Maize- T Cytoplasm
    c) Wheat-Rht1 Rht2
    d) Rice-Norin 10

105. Which one of the following causes more wastage of herbicide by drift
    a) High volume sprayer
    b) Ultra-low volume sprayer
    c) Hand sprayer
    d) Low volume sprayer

106. Postponing of first irrigation to 40-45 days after sowing is always preferable for cotton crop to
    a) Promote sympodial branching
    b) Prevent excessive vegetative growth
    c) Promote early flowering
    d) Promote monopodial branches

107. In functional allelopathy
   a) Toxic substances are released as such from the plant
   b) A precursor is released which is converted into active substances by some microorganism
   c) There is no question of release of any toxic substance
   d) Release of nitrogen from nodule of legume take place.

108. Which one of the following operations is associated with the process of nipping in gram
   a) Treating the seed with Rhizobium culture
   b) Typing the branches to avoid lodging
   c) Picking green leaves for vegetable purposes
   d) Plucking the apical buds to promote branches

109. Which one of the following is the best example of catch crop
   a) Linseed
   b) Mustard
   c) Toria
   d) Groundnut

110. Temporal complementarity's in intercropping results from
   a) The growth patterns of the component crops differing in time,
   b) The growth patterns of the component crops differing in height,
   c) Differences in yield
   d) Differences in cost of cultivation

111. Stem nodulation occurs in green manure crop of
   a) Sesbania aculeate
   b) Sesbania cannabina
   c) Crotolaria juncea
   d) Aeschynomene afraspera

112. In tobacco, the basic aim of topping and desuckering is to
   a) Reduce the plant height
   b) Encourage the branching
   c) Divert energy and nutrients from flower heads to leaves
   d) Protect the plants against lodging

113. Crop logging is used in
   a) Sugarcane
   b) Sugarbeet
   c) Maize
   d) Tea

114. The neutron scattering method is not useful in estimating the moisture in
   a) Acidic soils
   b) Alluvial soils
   c) Lateritic soils
   d) Organic soils

115. Heavy shedding of buds and bolls occurs in cotton due to
   a) Deficiency of nitrogen in the soil
   b) Deficiency of phosphorus in the soil
   c) Deficiency of magnesium in the soil
   d) Water stress at bud formation stage

116. In furrow method of irrigation
   a) Only 1/4th of the furrow is wetted
   b) Only 4/5th of the furrow wetted
   c) Only 1/2 of the furrow is wetted
   d) Only 3/4th of the furrow is wetted

117. If the rate of application per hectare is 3.00 kg a)i., the quantity of simazine WP (80% a)i.) required to be sprayed in 0.20 hectare area would be
   a) 0.50 kg
   b) 0.75 kg
   c) 1.25 kg
   d) 1.87 kg

118. Yield advantage in an intercropping system occurs due to the development of
   a) Temporal complementarity
   b) Spatial complementarity
   c) Both a& b
   d) Competitive relationship

119. At field capacity the water is held at
   a) 0.033 MPa
   b) 0.30 MPa
   c) 3.00 MPa
   d) 30.00 Mpa

120. 'A' value concept was given by
   a) Sorenson
   b) Beckett
   c) Schofield
   d) Fried and Dean

121. Which one of the following has organic form of sulphur
   a) Purine
   b) Cysteine
   c) RNA
   d) Phytin

122. The process of formation of nitrogen and nitrous oxide gases from ammonical fertilizers in soil is known as
   a) Ammonification
   b) Nitrification
   c) Denitrification
   d) Mineralization

123. *Brassica juncea* has been evolved after hybridization between
   a) B. nigra and B. campestris
   b) B. nigra and B. oleracea
   c) B. campestris and B. oleracea
   d) B.campestris and B. carinata

124. Dolomite is
   a) $CaCO_3$
   b) $MgSO_4$
   c) $Ca(OH)_2$
   d) $MgCO_3CaCO_3$

125. Salts in the xylem ducts of the root are carried upward with
   a) Photosynthesis
   b) Transpiration
   c) Respiration
   d) Guttation

126. The organic residues with wider C:N ratios as compared to narrow C:N ratios are decomposed at
a) Faster rate
b) Slow rate
c) Equal rate
d) Faster rate followed by slow rate

127. Which form of nitrogen is available in urea
a) Ammonical
b) Amide
c) Nitrite
d) Nitrate

128. At what pH value, phosphate availability is the highest in the soil
a) 5.5
b) 6.5
c) 7.5
d) 8.5

129. Bunchy top in sugarcane is caused by
a) Root borer
b) Stock borer
c) Internote borer
d) Top shoot borer

130. Bacterial diseases are controlled by use of chemicals
a) Antibiotics
b) Viricides
c) Fungicides
d) Kelthane

131. Application of potash increases
a) Disease resistance in plants
b) Resistance for water logging
c) Frost resistance in plants
d) None of these

132. The downward movement of surface soil water is known as
a) Infiltration
b) Percolation
c) Leaching
d) Wash out

133. The herbicides containing carbon and hydrogen in their molecule are called
a) Arsenic
b) Acid
c) Organic herbicide
d) Salt

134. 'Alley cropping' means
a) Growing of pastures in between two widely spaced rows of fast growing trees,
b) Growing of field crops in between two widely spaced rows of fast growing trees,
c) Growing of only short duration crops in between two widely spaced rows of fast growing trees,
d) Growing of only fodder crops in between two widely spaced rows of fast growing trees.

135. 'Relay cropping' means
   a) Growing of more than one crop on the same land in a year,
   b) Growing of three crops on the same land in a year,
   c) Growing of four crops on the same land in a year in such a manner that the following crop is sown before the harvest of preceding crop,
   d) None of these.

136. How much seed of legume crop should be treated with one packet of *Rhizobium* culture
   a) 5 kg b) 10 kg
   c) 15 kg d) 20 kg

137. The huskless variety of barley is
   a) Manjula b) KK 141
   c) Dolma d) RS 6

138. Which one of the following is not correctly matched

	Crop		*Variety*
a)	Mustard	-	Basanti
b)	Groundnut	-	Chandra
c)	Lentil	-	Malika
d)	Arhar	-	Neelam

139. The crown roots in wheat appear
   a) Above soil surface
   b) Below soil surface but above seed
   c) Below soil surface and below seed
   d) Below seed

140. A short duration crop in between two main seasonal crops is termed as
   a) Cash crop b) Intercrop
   c) Companion crop d) Catch crop

141. Awarodhi' variety of gram is resistant to
   a) Wilt b) Blight
   c) Flooding d) Drought

142. In which crop, the use of BGA as a biofertilizer, will be most useful
   a) Maize b) Potato
   c) Rice d) Sugarcane

143. Which vegetable oil is good for heart patient
   a) Groundnut oil b) Mustard oil
   c) Soybean oil d) Sunflower oil

144. Nitrogen is applied into transplanted rice in the proportion of
   a) 50% at basal+ nil at tillering+50% at panicle emergence stage,
   b) Nil at basal+50% at tillering+50% at panicle emergence stage
   c) 25% at basal+50% at tillering+25% at panicle emergence stage
   d) 50% at basal+25% at tillering+25% at panicle emergence stage.

145. The optimum plant population per hectare of sorghum is
   a) 50,000 plants
   b) 1,00,000 plants
   c) 1,50,000 plants
   d) 2,00,000 plants

146. The crop grown in U.P. during all the three seasons is
   a) Urd
   b) Sorghum
   c) Moong
   d) Maize

147. Irrigations in gram are recommended at
   a) Early flowering and maturity stage
   b) Late flowering and maturity stage
   c) Early flowering and pod formation stage
   d) Before flowering and pod formation stage

148. UPAS 120 is a variety of
   a) Wheat
   b) Pigeonpea
   c) Barley
   d) Urd

149. Interculture in groundnut is avoided at
   a) Flowering stage
   b) Seedling stage
   c) Pegging stage
   d) None of these

150. The ginning percentage in cotton can be worked out by the formula
   a) Weight of lint/weight of cotton seed x 100,
   b) Weight of lint/weight of seed cotton x 100,
   c) Weight of cotton seed/weight of lint x 100,
   d) Weight of seed cotton/weight of lint x 100

151. Maximum yield of mustard is obtained at a plant geometry of
   a) 45 x 20 cm
   b) 60 x 30 cm
   c) 90 x 30 cm
   d) 105 x 30 cm

152. Application of gypsum is required more for
   a) Paddy
   b) Bersecm
   c) Lucerne
   d) Groundnut

153. The rotation intensity of maize-wheat+gram-moong is
   a) 200%
   b) 250%
   c) 300%
   d) 400%

154. An example of companion cropping is
   a) Sugarcane + potato
   b) Potato + mustard
   c) Potato + radish
   d) Wheat + mustard

155. Intercropping of mustard with potato is recommended in
   a) Replacement series
   b) Additive series
   c) Replacement cum additive series
   d) None of these

156. Which clay mineral is rich in potash
   a) Montmorillonite
   b) Kaolinite
   c) Illite
   d) Chlorite

157. The carbohydrates produced in leaves can not be translocated to different growing parts of plant in the absence of
   a) Manganese
   b) Boron
   c) Zinc
   d) Iron

158. 'Aruna' is a mutant variety of
   a) Pea
   b) Cotton
   c) Castor
   d) Soybean

159. Which soil has the highest cation exchange capacity
   a) Loam
   b) Loamy sand
   c) Sandy loam
   d) Clay loam

160. During germination radical and plumule develop from
   a) Embryo
   b) Endosperm
   c) Hilum
   d) Seed coat

161. In which of the following, composite and synthetic cultivars are used
   a) Rice
   b) Wheat
   c) Maize
   d) Cotton

162. The recently developed terminator technology has been used in
   a) Rice
   b) Cotton
   c) Tobacco
   d) Wheat

163. Which one of the following is not correctly matched

	Crop		*Variety*
a)	Cotton	-	Digvijaya
b)	Soybean	-	Bragg
e)	Groundnut	-	AK 12-24
d)	Sunflower	-	Prabhat

164. Which cation has higher aggregation capacity in soil

a) Al^{3+} b) Ca^{2+}
c) Mg^{2+} d) K^{+}

165. Which cation has low adsorption capacity on clay,

a) H b) Ca
c) Fe d) Na

166. Black cotton soil is rich in

a) Montmorillonite b) Kaolinite
c) Illite d) Chlorite

167. Which form of nitrogen is a,bsorbed by paddy under waterlogged conditions

a) N_2 b) NH_4ion
c) NO_2ion d) Nitrate ion

168. Phosphate containing mineral is

a) Dolomite b) Aptatite
c) Marble d) Pyrites

169. PDM 11 is a variety of

a) Urd b) Arhar
c) Moong d) Lobia

170. Excess uptake of which element is known as luxury consumption

a) N b) P
c) K d) Zn

171. The soil carried in saltation consists of particle size

a) 0.1 to 0.5 mm b) 0.1 to 0.2 mm
c) 0.5 to 1.0 mm d) None of these

172. Which one of the following causes pungency of mustard oil

a) Phenols b) Amino acids
c) Glucosinolates d) Erucic acid

173. Texture of the soil can be changed by

a) Use of fertilizer b) Use of manures
c) Use of tillage practices d) None of these

174. The slow growing species of legume root nodule bacteria are included in bacterial genus

a) Rhizobium b) Brady rhizobium
c) Sino-rhizobium d) Azo-rhizobium

175. Substances responsible for bread making quality of wheat is
   a) Gluten
   b) Globulin
   c) Glycine
   d) Lycine

176. Which one of the following increases phosphate solubility
   a) Clostridium
   b) Pseudomonas
   c) Azotobacter
   d) Nitrosomonas

177. Seed work board is required for
   a) Viability test of seed
   b) Germination test of seed
   c) Purity test of seed
   d) Blending of seed

178. The formula LER-1/LER x value of combined intercrop represents
   a) MAI
   b) LER
   c) LAI
   d) IER

179. LEISA is related to
   a) Organic farming
   b) Inorganic farming
   c) Natural farming
   d) All of these

180. Under drought the sorghum plants synthesize Dhurin in
   a) Roots
   b) Shoots
   c) Leaves
   d) All the plant parts

181. Which one of the following varieties of barley is nematode resistant
   a) Jyoti
   b) Ratna
   c) Karan-19
   d) Raj Kiran (RD-387)

182. All the grain legumes have a high photo respiration because of
   a) C 3 mechanism
   b) More vegetative growth
   c) Indeterminate
   d) Pod position

183. Pollen viability of wheat is related to supply of
   a) Zinc
   b) Molybdenum
   c) Boron
   d) Magnesium

184. Which one of the following is a green house gas
   a) Oxygen
   b) Ammonia
   c) Methane
   d) Chlorine

185. Khaira disease in rice is caused due to
   a) Fungal infection
   b) Zinc deficiency
   c) Excessive application of potassium
   d) Bacterial infection

186. 'Tarai' soils are deficient in

a) N
b) P
c) S
d) Zn

187. Process of loosening for separating fibres from plant is known as

a) Curing
b) Retting
c) Wetting
d) Netting

188. Flow of water in saturated soil is described by

a) Poiseuille's law
b) Darcy's law
c) Fick's'law
d) Both a& b

189. The use of tensiometer is confined up to metric potential of

a) -0.8 bar
b) -0.6 bar
c) -0.4 bar
d) -0.2 bar

190. Each unit change in pH represents a change in activity of H+ or OH ions by

a) One fold
b) Ten fold
c) Hundred fold
d) Thousand fold

191. Relative weeds are

a) Plants of same crops but other variety in the field
b) Plants of other crop in the crop field
c) Seasonal weeds in the crop field
d) All of these

192. How many agroclimatic zones are categorized in India

a) 20
b) 18
c) 14
d) 15

193. The most concentrated fertilizer used for nutrient supply is

a) Urea
b) DAP
c) Anhydrous ammonia
d) SSP

194. Rancidity in sunflower oil is caused by

a) Reduction
b) Oxidation
c) Esterification
d) Nitrification

195. First hybrid maize Ganga-1 was developed in India in

a) 1957
b) 1961
c) 1964
d) 1965

196. Who introduced the technique of production of double cross hybrid maize

a) EM East
b) DF Jones
c) GH Shull
d) Mendel

197. Soil sickness is caused by

a) Maize
b) Jowar
c) Linseed'
d) Both b & c

198. CSH-l was first sorghum hybrid released in

a) 1961
b) 1962
c) 1963

199. Fertilization application in maize should be completed before

a) Tasseling
b) Silking
c) Ripening
d) Knee high stage

200. Under stress condition, which amino acid accumulated in crop plants

a) Methionine
b) Tryptophan
c) Proline
d) Phenyl alanine

Model Paper-V

1. First inter-specific hybrid cotton was released by

a) Tamil Nadu
b) Gujarat
c) Punjab
d) Karnataka

2. Textural class of the soil depends on

a) Organic matter content
b) Soil water
c) Particle size distribution
d) Soil productivity

3. Plants tolerant to high soil salinity conditions

a) Heliophytes
b) Halophytes
c) Heliotropic
d) None of these

4. The semi arid climate will have moisture deficit index of

a) 33.0
b) -33.3 to 66.6
c) 0 to 33.3
d) Less than -66.6

5. A fodder and pasture legume

a) Glyricidia
b) Stylosanthus
c) Cowpea
d) Sesbania

6. Hydrolysis of urea results in the loss in nitrogen in the form of

a) NH_4
b) NH_3
c) NH_2
d) NH_4OH

7. Person who established the relationship between plant growth factor

a) Liebig
b) Van Helmont
c) Tisdale
d) Mistscherlich

8. Soil particles get deflocculated due to:
   a) Calcium b) Sodium
   c) Magnesium d) Sulphur

9. Availability of H_2PO_4 is greatest when soil pH is:
   a) Low b) High
   c) Neutral d) None of these

10. First interspecific hybrid cotton was released by:
    a) Tamil Nadu b) Gujarat
    c) Punjab d) Karnataka

11. Natural farming was advocated by:
    a) Gilbert b) Fakuoka
    c) Lawes d) Norman E. borlaug

12. The energy status of the water is called
    a) Pressure b) Potential
    c) Tension d) Content

13. Avena fatua is a problematic weed in
    a) Pulses b) Maize
    c) Cotton d) Wheat

14. Find out the quantity of atrazine WP to be sprayed in one-hectare area (if the rate of application per hectare = 3.00 kg ai/ha & active ingredients 60%)
    a) 3.75 b) 5.00
    c) 3.25 d) 3.00

15. The term variance is used to denote the
    a) Standard deviation
    b) Square of the standard deviation
    c) Square root of the standard deviation
    d) None of the above

16. Trend in rainfall can be analysed with
    a) Multiple regression technique b) Correlation technique
    c) Moving averages d) None of the above

17. The growth pattern of a cereal plant falls in
    a) Linear curve b) Sigmoidal curve
    c) Cubic curve d) Binomial curve

18. The percent of incoming radiation that is reflected is called
    a) Geotemperature b) Albedo
    c) Solarization d) None of these

19. The practice in which runoff is increased and infiltration is reduced in certain areas which can serve as a source of water supply for other areas is called
    a) Moisture conservation
    b) Water harvesting
    c) Moisture retention
    d) None of these

20. Pendimethalin is following kind of herbicide
    a) Post emergence
    b) Pre-emergence
    c) Soil-incorporated
    d) None of these

21. The organic matter of soils is ordinarily obtained by multiplying the organic carbon content by
    a) 1.52
    b) 1.62
    c) 1.72
    d) 1.82

22. The harvest index of pigeonpea is about
    a) 0.35
    b) 0.50
    c) 0.60
    d) 0.15

23. The most vital controlling factor of soil temperature is
    a) Organic matter
    b) Soil moisture
    c) Soil air
    d) Soil depth

24. Fe chlorosis is seen on
    a) Upper leaves
    b) Middle leaves
    c) Lower leaves
    d) None of these

25. Soil N is not lost due to
    a) Nitrification
    b) Denitrification
    c) Volatilization
    d) Leaching

26. Growing of plants in water solution of essential nutrients is called
    a) Hydroponics
    b) Aquaculture
    c) Water culture
    d) None of these

27. The most predominant type of barley cultivated in India is
    a) Hull-less six-rowed barley
    b) Hull-less two-rowed barley
    c) Six-row hulled barley
    d) Two-row hulled barley

28. Groundnut the chief source of vegetable oil in India, originated in
    a) China
    b) West Africa
    c) Brazil
    d) United States of America

29. The most commonly used herbicide in sugarcane is
   a) 2- 4-D ester and lasso
   b) Gromoxone & Butachlor
   c) Sencore isoproturon
   d) Atrataf and Tafazine

30. Amongst the 3 commercial preparation of 2, 4-D available in the market the most volatile and likely to cause extensive damage by drift to sensitive crop like cotton in nearby fields are
   a) Sodium salts of 2, 4-D
   b) Amine salts of 2, 4-D
   c) Esters of 2, 4-D
   d) All are equal

31. Amongst the various herbicides available in the market the most commonly used herbicide in transplanted rice is
   a) Isoproturon
   b) Butachlor
   c) Fluchlorin
   d) Atrataf

32. The most commonly used synthetic organic fertilizer in India is
   a) Annydrons Ammonia
   b) Ammonium sulphate
   c) Urea
   d) DAP

33. $1m^3$ of water is equal to
   a) 10,000 litres
   b) 100,000 litres
   c) 1000,000 litres
   d) 1,000 litres

34. The depth of seeding in wheat is best related with
   a) Length of radicle
   b) Length of coleoptile
   c) Length of mesocotyl
   d) The depth of placement of embryo with in seed cotyledon

35. The inflorescence of rice is called
   a) Ear
   b) Rachis
   c) Panicle
   d) Rosette

36. Practice of taking a second crop from stubbles of previous crop is known as
   a) Stubble mulch farming
   b) Double cropping
   c) Relay cropping
   d) Ratooning

37. Which crop is used for producing both the vegetable oil and fiber
   a) Castor
   b) Mesta
   c) Linseed
   d) Safflower

38. A soil is well flocculated when it contains more of
   a) Potassium
   b) Calcium
   c) Sodium
   d) Sulphur

39. Which of the following contains more of secondary than primary nutrient
    a) Ammonium sulphate nitrate b) Ammonium sulphate
    c) Calcium ammonium nitrate d) None of the above

40. Pyrite predominantly contains
    a) Fe, S b) Fe, Zn
    c) Ca, S d) Fe, Al

41. Low temperature treatment of seeds & seedlings is known as
    a) Solarisation b) Vernalization
    c) Hardening d) Freezing injury

42. The following can be used as antitranspirant
    a) PMA b) Urea
    c) Machete d) All three

43. Single superphosphate is manufactured by reacting
    a) H_2SO_4 with gypsum b) H_2SO_4 with rock phosphate
    c) H_2PO_4 with rock phosphate d) H_3PO_4 with NHO_3

44. Gypsum is applied to reclaim
    a) Sodic soils b) Saline soils
    c) Acidic soils d) None of these

45. The nitrogenous fertilizer containing the highest N content is
    a) Urea b) Anhydrous ammonia
    c) Ammonium nitrare d) Ammonium sulphate

46. If gravimetric water content of a soil is 12 % and bulk density is 1.5 g/cc the volumetric water content is
    a) 18 b) 8
    c) 100 d) None of these

47. Isohytes are lines connecting places of equal
    a) Rainfall b) Humidity
    c) Sunshine hours d) Evaporation

48. In an Randomised Block design, there are 8 treatments and 4 replications, the degree of freedom for error is
    a) 32 b) 24
    c) 21 d) 31

49. Example of indeterminate crop
    a) Sorghum b) Rice
    c) Setaria d) Cotton

50. Water use efficiency of a crop is calculated using the formula
   a) Y/ET b) Y/E
   c) Y/ PET d) Y/T

51. If the seeds having 80% germination and test weight of 20g are to be sown at 50 x 10 cm spacing, the net seed rate for one ha will be ....
   a) 5.25 kg b) 5.0 kg
   c) 4.25 kg d) 4.0 kg

52. Butachlor and propanil belong to .... group of herbicide
   a) Ureas b) Carbamates
   c) Aliphatic d) Amides

53. If a crop is to be irrigated at 0.5 IW/CPE ratio with 50 mm depth, it should be irrigated after ....CPE
   a) 60 mm b) 75 mm
   c) 25 mm d) 100 mm

54. Under saline soil conditions ....irrigation could help in obtaining better crop stand and yield
   a) Sprinkler b) Subsurface
   c) Check basin d) Furrow

55. Use of brackish water for agriculture has become more relevant due to .... irrigation:
   a) Sprinkler b) Furrow
   c) Check basin d) Micro

56. The most ground water polluting fertilizers are ......
   a) Nitrate fertilizers b) Ammonical fertilizers
   c) Potassic fertilizers d) Amide fertilizers

57. The alternate host of gram caterpillar is one of the following:
   a) Amaranthus b) Crotalaria sp.
   c) Agropyron repens d) Cenchrus sp.

58. Skin allergy is caused by one of the following:
   a) Amaranthus b) Brush weeds
   c) Cyperus d) Parthenium

59. Which one of the following weed comes in the category of sedges:
   a) Chloris barbata b) Eclipta alba
   c) Cyperus rotundus d) Leuscus aspera

60. After thinning the number of plants per hectare at 40 cm x 25 cm is :
    a) 50000 b) 75000
    c) 100000 d) 125000

61. Which one be called as Dryland?
    a) When rainfall received is less than 1000 mm
    b) When rainfall received is less than 500 mm
    c) When rainfall received is less than 750 mm
    d) Hilly terrain with low rainfall less than 500mm

62. In undulating terrain one should go for ....
    a) Normal crop raising b) Strip cropping
    c) Pastoral cropping d) Mixed pastoral and tree cropping

63. Air dried fodder is called as:
    a) Silage b) Hay
    c) Green fodder d) Stove

64. Stem nodulating green manure plant is
    a) Sunnhemp b) Sesbania aculiata
    c) Sesbania rostrata d) Tephrosea sps.

65. The term PET was coined by:
    a) Sarkar and Biswas b) Haigreaves
    c) Thomthwaite d) Mendel

66. Chemical used for the control of transpiration
    a) Kaolin b) Lime
    c) PMA d) ABA

67. The study of the relation of agricultural crops and environment is called as
    a) Autoecology b) Agro-ecology
    c) Agrology d) Agrometeorology

68. Cultivation of woody plants, particularly those used for decoration and shade is known
    a) Autoecology b) Arboriculture
    c) Silviculture d) Aviculture

69. Sea Island cotton or Egyptian cottons belongs to
    a) Gossypium arboreum b) Gossypium hirsutum
    c) Gossypium barbadense d) Gossypium herbaceum

70. Plants tolerant to high soil salinity conditions
    a) Heliophytes b) Hylophytes
    c) Heliotropic d) None of these

71. The value of albedo in most o the dry land soils ranges from
    a) 0.15 to 0.35 b) 0.00 to 0.10
    c) 0.50 to 0.75 d) 0.76 to 0.80

72. Normally drought is assessed using
    a) Aridity b) Dryness
    c) Aridity anomaly index d) Deficient rainfall

73. A year in which rainfall of a particular place is short by more than ....is known as drought year:
    a) Twice the mean b) Twice the standard deviation
    c) Twice the median d) Ttwice standard error

74. Which one of the following is considered as rain water recycling
    a) Farm pond b) Wells
    c) Percolation pond d) Compartmental bunding

75. The lateral influx of water in leveled freely drained soil is
    a) Very active b) Negligible
    c) No relationship d) Moderately active

76. Water requirement is equal to:
    a) IR + ER + S b) ER + S
    c) IR d) IR – (ER + S)

77. Diversity Index measures the:
    a) Multiplicity of farm products in a year
    b) Multiplicity of farm products in a season
    c) Multiplicity of farm products in two years
    d) None of the above

78. One foot candle ....lux:
    a) 1.0764 b) 10.764
    c) 107.64 d) 1076.4

79. Chemical used for delineating cotton is:
    a) KCl b) HNO_3
    c) H_2SO_4 d) Paraquat

80. Material used for artificial rain making is:
    a) KH_2PO_4 b) Silver lodide
    c) $HgCl_2$ d) Common salt

81. The herbicide recommended to control phalaris minor in wheat is:
    a) 2, 4-D b) Isoproturon
    c) Alachlor d) Fluchloralin

82. 'Lag phase' on the response curve:
   a) The period of initial slow growth
   b) The period of accelerated fast growth
   c) The period when the yield curve tends to plateau off
   d) The period when the yield curve tends to dip due to decline in response

83. Successful recycling of high C: N ratio rice straw on agricultural lands could be accomplished:
   a) By adding nitrogenous fertilizers at incorporation
   b) By adding cellulytic micro organisms (fungi)
   c) By giving an additional irrigation after incorporation
   d) By adopting all above three options together

84. Cardinal temperature refers to:
   a) The maximum, minimum and optimum temperature limits of growth
   b) Survival temperature limits of plant
   c) The best temperatures for best growth phase
   d) All the above three

85. Cytokinins are plant growth regulators used:
   a) To inhibit the growth of plants
   b) To stimulate cell division and delay the senescence
   c) To break dormancy in bulbs and tubers
   d) To hasten ripening of fruits

86. The most pre-dominant type of barley cultivated in India is:
   a) Hull-less six-rowed barley
   b) Hull-less two-rowed barley
   c) Six-row hulled barley
   d) Two-row hulled barley

87. Groundnut the chief source of vegetable oil in India originated in:
   a) China
   b) West Africa
   c) Brazil
   d) United States of America

88. Which of the following is most suited for cultivation under extremely dry conditions:
   a) Brassica napus
   b) Brassica juncea
   c) Brassica campestris var. yellow sarso
   d) Eruca sativa

89. Which of the following 4 species of cotton cultivated in India accounts for the largest hectarage occupancy:
   a) Gossypium arboreum
   b) Gossypium hercaum
   c) Gossypium barbadense
   d) Gossypium hirsutum

90. Mentha cultivation is done primarily in North India to get:
a) Mentha seed for use as condiments
b) Mentha herbs for use in medicines
c) Mentha oil for use in commerce
d) For obtaining a high quality fodder for milch cattle

91. 'Congress grass' is a common weed found:
a) In cultivated irrigated lands
b) In dry lands farming situation
c) In waste lands and along the rails road
d) In high altitudes with high rains

92. Phosphorus is most easily absorbed by plants:
a) HPO_4
b) As H_2PO_4
c) As PO_4
d) With equal ease in all the above

93. Concentration of hydrogen at pH9 would be how much different than at pH7:
a) 10 times greater
b) 100 times greater
c) 10 times lesser
d) 100 times lesser

94. Annidation refers to:
a) Direct or indirect harmful effects that one plant has on another
b) Complementary interactions which occur both in space and time
c) Interaction effect of fertilizer levels and plant population densities
d) Date of planting and fertilizer rates

95. Puddling in rice fields is done, primarily to:
a) Provide a soft seed-bed to ease transplanting of seedlings
b) To kill weeds
c) To reduce deep percolation losses of water into soil
d) To accomplish all the above 3 objectives

96. Gram for good germination requires:
a) Fine seed bed of good tilth
b) Very fine powdry and loose tilth
c) Cloddy and rough seed bed
d) Firm and fine tilth

97. 'Tikka disease' is an important disease of:
a) Potato
b) Pigeonpea
c) Groundnut
d) Maize

98. Which crop is used for producing both the vegetable oil and fibre:
a) Castor
b) Mesta
c) Linseed
d) Safflower

99. Integrated nutrient management means

a) Use of fertilizers only
b) Use of organic manures only
c) Use of biofertilizers only
d) All sources of nutrients

100. Groundnut requires application of Gypsum for

a) Lowering soil pH
b) Supply of calcium
c) Supply of sulphur
d) Supply of calcium and sulphur

101. Which of the following crop has the highest cultivated area in the world

a) Rice
b) Wheat
c) Barley
d) Bajra

102. The number of replications is equal to number of treatments in the design

a) Latin square
b) Randomised Block Design
c) Completely Randomized Block Design
d) Split plot

103. Growing plants in fixed order in a sequence for a specified duration on a specified area is called

a) Crop sequence
b) Crop intensity
c) Crop rotation
d) Cropping system

104. Cropping history refers to:

a) History of the crop under cultivation
b) History of farm operations carried out during the preceding years
c) Calendar of operations to be followed during the crop season
d) History and origin of crop

105. Arable farming refers to

a) Enclosing the natural lands for raising game and wild animal
b) Growing of filed crops
c) Management of natural grass lands
d) Raising and management of naturally growing forest trees

106. At pH 4.0 the predominant ionic form of phosphorus present is

a) HPO_4
b) $H_2PO_4^-$
c) PO_4^{3-}
d) None of these

107. The inflorescence of sugarcane is known as

a) Arrow
b) Panicle
c) Capitulum's
d) Racemose

108. Contribution of flag leaf in photosynthates is about
a) 52% b) 40%
c) 35% d) 20%

109. Botanical name of Buck wheat is
a) Fagopyrum esculentum b) Echinochloa frumentacea
c) Panicum miliacium d) None of these

110. Optimum plant population ha^{-1} for Maize has estimated as
a) 25,000 - 50,000 b) 30,000 - 60,000
c) 50,000 - 80,000 d) 80,000 - 1,00,000

111. Water requirement (WR) includes the losses due to
a) Evapotranspiration and application of water
b) Consumptive use of water and application of water
c) Consumptive use, water required for special operations and other economically unavoidable losses of water.
d) ET and water required for special operations.

112. If on a 10-acre farm, crops are raised on 10 acres in *rabi*, 8 acres in *kharif* and 7 acres in *zaid* in a year, what shall be the cropping intensity
a) 300% b) 400%
c) 250% d) 150%

113. At field capacity the value of *pF* is
a) 0.0 b) 2.5
c) 4.2 d) 6.0

114. At permanent wilting point (PWP) the *pF* value is
a) 0.0 b) 2.5
c) 4.2 d) 6.0

115. The % pore space in a soil having a bulk density of 1.25 mg m^{-3} and particle density of 2.5 mg m^{-3} is
a) 20.00 b) 30.75
c) 50.00 d) 81.75

116. Bolting refers to
a) Stem elongation b) Fruiting
c) Flowering d) Flower dropping

117. Leaf area index refers to
a) Leaf area to dry weight of plant
b) Leaf area to wet weight of plant
c) Leaf area to land area
d) Leaf area to dry weight of leaves

118. The fraction of photosynthetically active radiation absorbed by leaves is approximately
   a) 0.10 b) 0.30
   c) 0.60 d) 0.80

119. Nitrate Reductase in plants is located in
   a) Chloroplasts b) Cytoplasm
   c) Mitochondria d) Ribosomes

120. The highest area and production of soybean in India is in the following state
   a) U.P. b) M.P.
   c) Rajasthan d) Maharashtra

121. Plant roots absorb sulphur mainly in form of
   a) SO_4 b) SO_2
   c) H_2S d) S

122. If the gravimetric moisture content of soil is 25% and the bulk density is 1.5 g/cc, the volumetric moisture content is
   a) 16.66% b) 37.5%
   c) 20.0% d) 27.0%

123. The best suited temperature for growth of sugarcane lies between
   a) 15-20°C b) 20-25°C
   c) 26-32°C d) 32-35°C

124. Adsali sugarcane is planted in
   a) July-August b) January-February
   c) February-March d) October-November

125. The capacity of a soil to resist appreciable change in pH value is called
   a) CEC b) Buffering capacity
   c) Percentage base saturation d) Anion-exchange capacity

126. Tetrazoloin test is used to determine
   a) Seed purity b) Seed germination
   c) Seed viability d) Seed quality

127. How much quantity of true potato seed is required for one hectare planting
   a) 100 g b) 200 g
   c) 300 g d) 400 g

128. The seed rate of hybrid maize is
   a) 20-25 kg/ha
   b) 18-20 kg/ha
   c) 30-35 kg/ha
   d) 10-15 kg/ha

129. Botanical name of Ragi is
   a) Eleusine coracana
   b) Echinochloa frumentacea
   c) Panicum miliacium
   d) None of these

130. Stevenson screen is related to
   a) Bacteriology
   b) Biotechnology
   c) Agrometeorology
   d) Remote sensing

131. Heavy applications of potassium fertilizers often lead to reduced absorption of
   a) Mg and Ca
   b) B and Mn
   c) N and P
   d) Zn and Cu

132. If a plant shows extensive interveinal chlorosis of the older leaves and late this reaches the younger leaves, the symptoms is due to lack of
   a) Magnesium
   b) Calcium
   c) Phosphorus
   d) Copper

133. In furrow method of irrigation
   a) Only $1/4^{th}$ of the furrow is wetted
   b) Only $4/5^{th}$ of the furrow wetted
   c) Only 1/2 of the furrow is wetted
   d) Only $3/4^{th}$ of the furrow is wetted

134. The herbicides containing carbon and hydrogen in their molecule are called
   a) Arsenic
   b) Acid
   c) Organic herbicide
   d) Salt

135. Irrigations in gram are recommended at
   a) Early flowering and maturity stage
   b) Late flowering and maturity stage
   c) Early flowering and pod formation stage
   d) Before flowering and pod formation stage

136. UPAS 120 is a variety of
   a) Wheat
   b) Pigeonpea
   c) Barley
   d) Urd

137. Intercropping of mustard with potato is recommended in
   a) Replacement series
   b) Additive series
   c) Replacement cum additive series
   d) None of these

138. Which clay mineral is rich in potash

a) Montmorillonite
b) Kaolinite
c) Illite
d) Chlorite

139. The carbohydrates produced in leaves can not be translocated to different growing parts of plant in the absence of

a) Manganese
b) Boron
c) Zinc
d) Iron

140. 'Aruna' is a mutant variety of

a) Pea
b) Cotton
c) Castor
d) Soybean

141. Which soil has the highest cation exchange capacity

a) Loam
b) Loamy sand
c) Sandy loam
d) Clay loam

142. During germination radical and plumule develop from

a) Embryo
b) Endosperm
c) Hilum
d) Seed coat

143. In which of the following, composite and synthetic cultivars are used

a) Rice
b) Wheat
c) Maize
d) Cotton

144. The recently developed terminator technology has been used in

a) Rice
b) Cotton
c) Tobacco
d) Wheat

145. Which cation has higher aggregation capacity in soil

a) Al^{3+}
b) Ca^{2+}
c) Mg^{2+}
d) K^{+}

146. Which cation has low adsorption capacity on clay,

a) H
b) Ca
c) Fe
d) Na

147. Black cotton soil is rich in

a) Montmorillonite
b) Kaolinite
c) Illite
d) Chlorite

148. Which form of nitrogen is a,bsorbed by paddy under waterlogged conditions

a) N_2
b) NH_4 ion
c) NO_2 ion
d) Nitrate ion

149. Phosphate containing mineral is
a) Dolomite
b) Aptatite
c) Marble
d) Pyrites

150. PDM 11 is a variety of
a) Urd
b) Arhar
c) Moong
d) Lobia

151. Excess uptake of which element is known as luxury consumption
a) N
b) P
c) K
d) Zn

152. The soil carried in saltation consists of particle size
a) 0.1 to 0.5 mm
b) 0.1 to 0.2 mm
c) 0.5 to 1.0 mm
d) None of these

153. Which one of the following causes pungency of mustard oil
a) Phenols
b) Amino acids
c) Glucosinolates
d) Erucic acid

154. Texture of the soil can be changed by
a) Use of fertilizer
b) Use of manures
c) Use of tillage practices
d) None of these

155. The slow growing species of legume root nodule bacteria are included in bacterial genus
a) Rhizobium
b) Brady rhizobium
c) Sino-rhizobium
d) Azo-rhizobium

156. Substances responsible for bread making quality of wheat is
a) Gluten
b) Globulin
c) Glycine
d) Lycine

157. Which one of the following increases phosphate solubility
a) Clostridium
b) Pseudomonas
c) Azotobacter
d) Nitrosomonas

158. Seed work board is required for
a) Viability test of seed
b) Germination test of seed
c) Purity test of seed
d) Blending of seed

159. The formula LER-l/LER x value of combined intercrop represents
a) MAI
b) LER
c) LAI
d) IER

160. LEISA is related to
a) Organic farming
b) Inorganic farming
c) Natural farming
d) All of these

161. Under drought the sorghum plants synthesize Dhurin in
a) Roots
b) Shoots
c) Leaves
d) All the plant parts

162. Which one of the following varieties of barley is nematode resistant
a) Jyoti
b) Ratna
c) Karan-19
d) Raj Kiran (RD-387)

163. All the grain legumes have a high photo respiration because of
a) C3 mechanism
b) More vegetative growth
c) Indeterminate
d) Pod position

164. Pollen viability of wheat is related to supply of
a) Zinc
b) Molybdenum
c) Boron
d) Magnesium

165. Which one of the following is a green house gas
a) Oxygen
b) Ammonia
c) Methane
d) Chlorine

166. Khaira disease in rice is caused due to
a) Fungal infection
b) Zinc deficiency
c) Excessive application of potassium
d) Bacterial infection

167. 'Tarai' soils are deficient in
a) N
b) P
c) S
d) Zn

168. Process of loosening for separating fibres from plant is known as
a) Curing
b) Retting
c) Wetting
d) Netting

169. Flow of water in saturated soil is described by
a) Poiseuille's law
b) Darcy's law
c) Fick's'law
d) Both a& b

170. The use of tensiometer is confined up to metric potential of
a) -0.8 bar
b) -0.6 bar
c) -0.4 bar
d) -0.2 bar

171. Each unit change in pH represents a change in activity of H+ or OH ions by

a) One fold
b) Ten fold
c) Hundred fold
d) Thousand fold

172. Relative weeds are

a) Plants of same crops but other variety in the field
b) Plants of other crop in the crop field
c) Seasonal weeds in the crop field
d) All of these

173. How many agroclimatic zones are categorized in India

a) 20
b) 18
c) 14
d) 15

174. Pusa Phalguni, Pusa Barsati, Pusa Rituraj and Pusa Dofasli are the improved varieties of

a) Moong
b) Urd
c) Pea
d) Cowpea

175. Linseed belongs to family

a) Liliaceae
b) Linaceae
c) Tiliaceae
d) Pedaliaceae

176. The optimum seed rate for berseem should be (kg/ha)

a) 15-20
b) 20-25
c) 25-30
d) 30-40

177. Lucerne *(Medicago sativa)* is a native of

a) Persia
b) India
c) Iraq
d) Afghanistan

178. Rice plant belongs to

a) C_3
b) C_4
c) CAM
d) None of the above

179. Stem nodulating green manure plants is

a) Sum hemp
b) Susbania aculiata
c) Sesbania rostrate
d) Tephrosea sp.

180. Schoefield is associated with

a) Neutron moisture metre
b) Electrical moisture box
c) Pressure plate apparatus
d) Longarithm of soil moisture

181. The first dry farming research station was started in the year 1936 at

a) Bijapur
b) Sholapur
c) Manjri
d) Rohtak

182. Law of minimum was proposed by
    a) Theodore de Saussure
    b) Justice Van Liebig
    c) Laes and Gilbert
    d) Jethro Tull

183. Orobanche is a parasite on
    a) Sorghum
    b) Sugarcane
    c) Safflower
    d) Tobacco

184. Chemical for inhibiting the process of nitrification
    a) 24-D
    b) N-serve
    c) CON_2H_4
    d) Paraquat

185. Dominant from of nitrogen absorbed by plants
    a) NH_4
    b) NO_3
    c) NO_2
    d) NH_3

186. India, first Dry Farming Research Station, Manjri was started in
    a) 1923
    b) 1933
    c) 1934
    d) 1935

187. Under stress condition, which amino acid accumulated in crop plants
    a) Methionine
    b) Tryptophan
    c) Proline
    d) Phenyl alanine

188. The most concentrated fertilizer used for nutrient supply is
    a) Urea
    b) DAP
    c) Anhydrous ammonia
    d) SSP

189. Diallate and Triallate herbicides come under:
    a) Nitrates group
    b) Phenols group
    c) Triazoles group
    d) Thiocarbamates group

190. ....is an annual stem parasite
    a) Dodder
    b) Broomrape
    c) Witchweed
    d) Lantana camera

191. Weeds which will grow well on alkali soil:
    a) Barnyard grass
    b) Nut grass
    c) Quack grass
    d) *Chenopodium album*

192. Quartiles are useful to separate:
    a) Extreme values
    b) Mean of the values
    c) Normalized values
    d) None of the above

193. The growth pattern of a cereal plant falls in:
a) Linear curve
b) Sigmoidal curve
c) Cubic curve
d) Binomial curve

194. Dividing experimental units into compact blocks satisfies:
a) Randomization
b) Uniformity
c) Local control
d) Replication

195. Bolting refers to:
a) Stem elongation
b) Flowering
c) Fruiting
d) Flowering dropping

196. Lithosphytes are plants which are usually found on:
a) Soil surface
b) Rock surface
c) Water bodies
d) None of the above

197. The origin of Toria is:
a) China
b) Afghanistan
c) India
d) Middle East

198. The equivalent acidity (per 100 kg of fertilizer) of urea is:
a) 80 kg
b) 95 kg
c) 110 kg
d) 128 kg

199. Luxury consumption is usually referred to nutrient:
a) Sulphur
b) Nitrogen
c) Phosphorus
d) Potassium

200. Mar is referred to:
a) Pig manure
b) Poultry manure
c) Raw humus
d) Iron

❑❑❑

Zeitfracht Medien GmbH
Ferdinand-Jühlke-Straße 7
99095 Erfurt, Deutschland
produktsicherheit@kolibri360.de